MANUAL OF
MATHEMATICS AND MECHANICS

MANUAL OF
MATHEMATICS
AND
MECHANICS

BY

GUY ROGER CLEMENTS, Ph. D.
Professor in the Department of Mathematics,
United States Naval Academy

AND

LEVI THOMAS WILSON, Ph. D.
Professor in the Department of Mathematics,
United States Naval Academy

WILDSIDE PRESS

www.wildsidepress.com

THE MAPLE PRESS COMPANY, YORK, PA.

PREFACE

This manual contains facts and formulas that are useful in courses in mathematics and mechanics in colleges and engineering schools, arranged and printed in a form that makes them readily available for *rapid work with minimum eye strain*. It will serve both as an outline of the things which it is essential that the student of mathematics should understand, and as a source of ready reference when he has need of these facts in his scientific and engineering practice.

Such topics as the solution of plane and spherical triangles, the convergence of series, methods of integration, and the solution of the ordinary differential equations of mechanics and physics have been outlined more completely than is usually done in such manuals.

There is ample provision for the simplifications that can be made in many theoretical discussions and computations by the use of hyperbolic functions.

Numerical tables should be easy to use and should give results with a degree of accuracy commensurate with that required in practical work. For this purpose four-figure accuracy is usually sufficient, and the use of more extended tables is unnecessarily cumbersome and time consuming.

We have tried to insure accuracy of statement in both formulas and tables. Our experience in checking our material, so far as possible, from many manuals and textbooks, indicates that errors may still exist in a book containing so many numbers and formulas. We shall be grateful to any one who will notify us of such errors as he may discover.

We wish to acknowledge the helpful suggestions of Captain W. W. Smith, U.S.N., and of our colleagues Professors C. L. Leiper, Paul Capron, J. B. Scarborough, and Dr. A. E. Currier, while the manuscript was in preparation.

Guy Roger Clements.
Levi Thomas Wilson.

Annapolis, Maryland,
August, 1937.

v

CONTENTS

MANUAL OF MATHEMATICS AND MECHANICS

NUMERICAL TABLES

Throughout this book

Logarithms to the base 10 are denoted by log.

Logarithms to the base e = 2.71828 are denoted by ln.

Table 1. Mantissas of Common Logarithms. Any positive number N can be written $N = 10^c M$, where c is an integer and M is between 1 and 10 (or $M = 1$). Therefore $\log N = c + m$, where $m = \log M$ is a positive proper fraction (or zero). The values of m (each to be preceded by a decimal point) are given in Table 1, to four significant figures. The integer c is called the *characteristic* of the logarithm, and m the mantissa.

A negative characteristic c is usually written as

(a positive integer between 1 and 10) − (an integral multiple of 10). Thus

$$\log 0.0356 = \log (10^{-2} \times 3.56) = -\overline{2} + .5514 = 8.5514 - 10.$$

This is sometimes written $\log 0.0356 = \overline{2}.5514$.

The *cologarithm* of a number is the logarithm of its reciprocal

$$\operatorname{colog} N = \log \frac{1}{N} = (10 - \log N) - 10.$$

Table 2. Mantissas of Common Logarithms (Special Table). This table is to be used to avoid interpolation in the first lines of Table 1, where the tabular differences are large.

Table 3. Natural Logarithms of Numbers, Base e. Each entry in Table 3 is to be preceded by 0., 1., or 2., as indicated in the column headed 0. The asterisk (*) signifies that the integral part from the next lower line is to be used. Thus $\ln 2.75 = 1.0116$.

1

Since $\ln (10^n N) = n \ln 10 + \ln N$, moving the decimal point n places to the right in the number adds $n \times (\ln 10 = 2.3026)$ to the logarithm; moving the decimal point n places to the left adds $n \times (\ln\ 0.1 = 0.6974 - 3 = \bar{3}.6974)$ to the logarithm. See auxiliary table of multiples of ln 10 at the foot of Table 3. The natural logarithms of the integers 1–1000 are given in Table 11.

Table 4. Values and Logarithms of Trigonometric Functions. To find log csc x and log sec x, use the relations

$$\log \csc x = \operatorname{colog} \sin x, \qquad \log \sec x = \operatorname{colog} \cos x.$$

Table 5. Logarithms of (a) sin x and (b) tan x, at Intervals of One Minute (0°–8°). The characteristic and, in general, the first figure of the mantissa are found in the column headed 0′; the remaining figures of the logarithm are found in the column headed by the unit-digit of the number of minutes. Thus

$$\log \sin 3°25' = 8.7752 - 10, \qquad \log \tan 5°48' = 9.0068 - 10.$$

Table 6. Values of the Trigonometric Functions for Angles in Radians. This table is designed for use in plotting the graphs of the trigonometric functions and in the solution by trial of numerical equations involving such functions. For more accurate values see Tables 4 and 8. The equivalent in degrees, of an angle given in radians, is accurate to the nearest minute.

Table 7. Conversion Tables (a) Degrees to Radians, (b) Radians to Degrees. See also Tables 4, 6, and 8.

Table 8. Exponential and Hyperbolic Functions. In addition to giving the exponential and hyperbolic functions of x, this table also shows the equivalence of the hyperbolic functions with the trigonometric functions of the gudermannian (see Art. 70). Furthermore, when more accurate values of the hyperbolic and exponential functions are needed, they may be obtained from the gudermannian by the use of a five- or six-place table of the trigonometric functions.

To find log e^{-x}, log csch x, log sech x, use

$$\log e^{-x} = \operatorname{colog} e^x,$$
$$\log \operatorname{csch} x = \operatorname{colog} \sinh x,$$
$$\log \operatorname{sech} x = \operatorname{colog} \cosh x.$$

The table also gives the number whose natural logarithm is x. For, if $N = e^x$, $x = \ln N$; if $M = e^{-x}$, $-x = \ln M$.

Table 9. Common Logarithms of Powers of e. Example:

$$\log e^{2.3456} = \log e^2 + \log e^{.3} + \log e^{.04} + \log e^{.005} + \log e^{.0006}$$
$$= 0.86859 + 0.13029 + 0.01737 + 0.00217 + 0.00026$$
$$= 1.01868.$$

Therefore $\log e^{2.3456} = 1.0187$.

Table 10. Values of Inverse Trigonometric and Hyperbolic Functions. This table is designed to save the labor of taking the inverse functions from Tables 4 and 8, for the values of x that are listed. When accurate interpolation is needed, Tables 4 and 8 should be used. The table is complete, at intervals of 0.01 for x, for $\sin^{-1} x$, $\cos^{-1} x$, $\tanh^{-1} x$, and $\operatorname{sech}^{-1} x$.

Table 11. Squares, Cubes, Square Roots, Cube Roots, Reciprocals, and Natural Logarithms of the Integers 1–1000.

Table 12. Some Functions of the Integers 1–100. The columns headed πn and $\pi n^2/4$, give the circumference and the area, respectively, of a circle whose diameter is n. The functions $1/\sqrt{n}$ and $\log n!$ are tabulated.

The columns headed $0.6745/\sqrt{n-1}$ and $0.6745/\sqrt{n(n-1)}$ are for use in computing the probable error of a single measurement and the probable error of the arithmetic mean. By use of the last column the value of $\Gamma(n)$ for any value of n may be obtained from the relation $\Gamma(n) = (n-1)\,\Gamma(n-1)$. For example $\Gamma(3.5) = 2.5 \times 1.5\,\Gamma(1.5)$, $\Gamma(0.3) = (1\%_3)\,\Gamma(1.3)$. If n is a positive integer, $\Gamma(n) = (n-1)!$; if n is positive, $\Gamma(n) = \int_0^\infty x^{n-1}e^{-x}\,dx$; if $n = 0, -1, -2, \ldots$, $\Gamma(n)$ is infinite.

Table 13. Amount of 1 at Compound Interest (see Art. 47).

Table 14. Present Value of 1 at Compound Interest (see Art. 48).

Table 15. Amount of an Annuity of 1 (see Art. 49).

Table 16. Present Value of an Annuity of 1 (see Art. 49).

Table 17. The Annuity That 1 Will Purchase (see Art. 49). This table is also to be used in finding the annuity that will amount to 1.

Table 18. (a) Compound Amount for Times Less Than a Year, (b) The Value of $j_{(p)} = p\left[(1 + i)^{\frac{1}{p}} - 1\right]$, (c) The value of $i/j_{(p)}$.

Table 19. American Experience Table of Mortality.

Table 20. Values of the Complete Elliptic Integrals, K and E.

Table 21. Values of the Probability Integral.

Table 22. Conversion Tables (a) Length, (b) Area, (c) Volume and Capacity, (d) Mass, (e) Velocity, (f) Energy, (g) Pressure.

Table 23. Binomial Coefficients.

SOME NOTES ON COMPUTATION

Accuracy. In computations such as are made in engineering, navigation, etc., the data are mostly the results of measurements. In all measurements errors exist due to physical limitations. The *absolute error* is the difference between the measured value and the exact value. Since the exact value is not known, the absolute error cannot be found; however, limits for this error can be determined. *Absolute accuracy* (accuracy to a certain number of decimal places) is measured by the absolute error; it is appropriate in calculations involving addition and subtraction.

The *relative error* is the ratio of the absolute error to the magnitude measured. *Relative accuracy* (accuracy to a certain number of significant figures) is measured by the relative error; it is appropriate in calculations involving multiplication and division.

Significant Figures. The digits 1, 2, . . . , 9 are always significant, but 0 (zero) may or may not be significant. Zeros to the left of all other digits are never significant (as in 0.0345 which has 3 significant figures). Zeros between other digits are always significant (as in 3005 which has 4 significant figures). Zeros to the right of all other digits are significant if any of them are also to the right of the decimal point (as in 40.0 and 0.0230, each having 3 significant figures). In a number like 45000 the best way in which to show the number of significant figures is to write the number in "standard form," 4.50×10^4, which shows that it has 3 significant figures.

Rounding Off. To round off a number to n significant figures, discard all digits to the right of the nth place. If the discarded number is less than half a unit in the nth place, leave the nth digit unchanged; if the discarded number is greater than half a unit in the nth place, add 1 to the nth digit. If the discarded number is *exactly* half a unit in the nth place, leave the nth digit unaltered if it is an even number, but increase it by 1 if it is an odd number.

Examples: (All rounded off to 4 significant figures)

29.6341   rounds off to 29.63
0.03789507   rounds off to 0.03790
1.286500   rounds off to 1.286
1.287500   rounds off to 1.288
268174.   rounds off to 268200 = 2.682×10^5

In interpolations by use of tables when there is a choice between two successive digits, the even one is to be chosen.

Multiplication and Division. In general, no increase in the accuracy of a result computed by multiplication and division is attained by using more significant figures for any number occurring in the computation than the number of significant figures in the least accurate datum.

Addition and Subtraction. In addition and subtraction accuracy to a certain number of decimal places is appropriate. Before adding, round off the given numbers so that not more than one column at the right is broken. Add, and then round off the sum so that its last digit comes in the last unbroken column. Thus the sum of 0.2056, 2.572, and 28.3 is properly found as shown. One is not justified in keeping more decimal places in the result than there are in the least accurate datum.

0.21
2.57
28.3
31.1

Logarithms. In general, when using logarithms, there is no gain in accuracy due to using more *decimal places* in the logarithms than the number of *significant figures* in the least accurate datum. When using four-place logarithms, round off all numbers to 4 significant figures and all logarithms to 4 decimal places.

TABLE 1.—MANTISSAS OF COMMON LOGARITHMS (BASE 10)

N	0	1	2	3	4	5	6	7	8	9
10	0000	0043	0086	0128	0170	0212	0253	0294	0334	0374
11	0414	0453	0492	0531	0569	0607	0645	0682	0719	0755
12	0792	0828	0864	0899	0934	0969	1004	1038	1072	1106
13	1139	1173	1206	1239	1271	1303	1335	1367	1399	1430
14	1461	1492	1523	1553	1584	1614	1644	1673	1703	1732
15	1761	1790	1818	1847	1875	1903	1931	1959	1987	2014
16	2041	2068	2095	2122	2148	2175	2201	2227	2253	2279
17	2304	2330	2355	2380	2405	2430	2455	2480	2504	2529
18	2553	2577	2601	2625	2648	2672	2695	2718	2742	2765
19	2788	2810	2833	2856	2878	2900	2923	2945	2967	2989
20	3010	3032	3054	3075	3096	3118	3139	3160	3181	3201
21	3222	3243	3263	3284	3304	3324	3345	3365	3385	3404
22	3424	3444	3464	3483	3502	3522	3541	3560	3579	3598
23	3617	3636	3655	3674	3692	3711	3729	3747	3766	3784
24	3802	3820	3838	3856	3874	3892	3909	3927	3945	3962
25	3979	3997	4014	4031	4048	4065	4082	4099	4116	4133
26	4150	4166	4183	4200	4216	4232	4249	4265	4281	4298
27	4314	4330	4346	4362	4378	4393	4409	4425	4440	4456
28	4472	4487	4502	4518	4533	4548	4564	4579	4594	4609
29	4624	4639	4654	4669	4683	4698	4713	4728	4742	4757
30	4771	4786	4800	4814	4829	4843	4857	4871	4886	4900
31	4914	4928	4942	4955	4969	4983	4997	5011	5024	5038
32	5051	5065	5079	5092	5105	5119	5132	5145	5159	5172
33	5185	5198	5211	5224	5237	5250	5263	5276	5289	5302
34	5315	5328	5340	5353	5366	5378	5391	5403	5416	5428
35	5441	5453	5465	5478	5490	5502	5514	5527	5539	5551
36	5563	5575	5587	5599	5611	5623	5635	5647	5658	5670
37	5682	5694	5705	5717	5729	5740	5752	5763	5775	5786
38	5798	5809	5821	5832	5843	5855	5866	5877	5888	5899
39	5911	5922	5933	5944	5955	5966	5977	5988	5999	6010
40	6021	6031	6042	6053	6064	6075	6085	6096	6107	6117
41	6128	6138	6149	6160	6170	6180	6191	6201	6212	6222
42	6232	6243	6253	6263	6274	6284	6294	6304	6314	6325
43	6335	6345	6355	6365	6375	6385	6395	6405	6415	6425
44	6435	6444	6454	6464	6474	6484	6493	6503	6513	6522
45	6532	6542	6551	6561	6571	6580	6590	6599	6609	6618
46	6628	6637	6646	6656	6665	6675	6684	6693	6702	6712
47	6721	6730	6739	6749	6758	6767	6776	6785	6794	6803
48	6812	6821	6830	6839	6848	6857	6866	6875	6884	6893
49	6902	6911	6920	6928	6937	6946	6955	6964	6972	6981
50	6990	6998	7007	7016	7024	7033	7042	7050	7059	7067
51	7076	7084	7093	7101	7110	7118	7126	7135	7143	7152
52	7160	7168	7177	7185	7193	7202	7210	7218	7226	7235
53	7243	7251	7259	7267	7275	7284	7292	7300	7308	7316
54	7324	7332	7340	7348	7356	7364	7372	7380	7388	7396
N	0	1	2	3	4	5	6	7	8	9

TABLE 1.—MANTISSAS OF COMMON LOGARITHMS.—(*Continued*)

N	0	1	2	3	4	5	6	7	8	9
55	7404	7412	7419	7427	7435	7443	7451	7459	7466	7474
56	7482	7490	7497	7505	7513	7520	7528	7536	7543	7551
57	7559	7566	7574	7582	7589	7597	7604	7612	7619	7627
58	7634	7642	7649	7657	7664	7672	7679	7686	7694	7701
59	7709	7716	7723	7731	7738	7745	7752	7760	7767	7774
60	7782	7789	7796	7803	7810	7818	7825	7832	7839	7846
61	7853	7860	7868	7875	7882	7889	7896	7903	7910	7917
62	7924	7931	7938	7945	7952	7959	7966	7973	7980	7987
63	7993	8000	8007	8014	8021	8028	8035	8041	8048	8055
64	8062	8069	8075	8082	8089	8096	8102	8109	8116	8122
65	8129	8136	8142	8149	8156	8162	8169	8176	8182	8189
66	8195	8202	8209	8215	8222	8228	8235	8241	8248	8254
67	8261	8267	8274	8280	8287	8293	8299	8306	8312	8319
68	8325	8331	8338	8344	8351	8357	8363	8370	8376	8382
69	8388	8395	8401	8407	8414	8420	8426	8432	8439	8445
70	8451	8457	8463	8470	8476	8482	8488	8494	8500	8506
71	8513	8519	8525	8531	8537	8543	8549	8555	8561	8567
72	8573	8579	8585	8591	8597	8603	8609	8615	8621	8627
73	8633	8639	8645	8651	8657	8663	8669	8675	8681	8686
74	8692	8698	8704	8710	8716	8722	8727	8733	8739	8745
75	8751	8756	8762	8768	8774	8779	8785	8791	8797	8802
76	8808	8814	8820	8825	8831	8837	8842	8848	8854	8859
77	8865	8871	8876	8882	8887	8893	8899	8904	8910	8915
78	8921	8927	8932	8938	8943	8949	8954	8960	8965	8971
79	8976	8982	8987	8993	8998	9004	9009	9015	9020	9025
80	9031	9036	9042	9047	9053	9058	9063	9069	9074	9079
81	9085	9090	9096	9101	9106	9112	9117	9122	9128	9133
82	9138	9143	9149	9154	9159	9165	9170	9175	9180	9186
83	9191	9196	9201	9206	9212	9217	9222	9227	9232	9238
84	9243	9248	9253	9258	9263	9269	9274	9279	9284	9289
85	9294	9299	9304	9309	9315	9320	9325	9330	9335	9340
86	9345	9350	9355	9360	9365	9370	9375	9380	9385	9390
87	9395	9400	9405	9410	9415	9420	9425	9430	9435	9440
88	9445	9450	9455	9460	9465	9469	9474	9479	9484	9489
89	9494	9499	9504	9509	9513	9518	9523	9528	9533	9538
90	9542	9547	9552	9557	9562	9566	9571	9576	9581	9586
91	9590	9595	9600	9605	9609	9614	9619	9624	9628	9633
92	9638	9643	9647	9652	9657	9661	9666	9671	9675	9680
93	9685	9689	9694	9699	9703	9708	9713	9717	9722	9727
94	9731	9736	9741	9745	9750	9754	9759	9763	9768	9773
95	9777	9782	9786	9791	9795	9800	9805	9809	9814	9818
96	9823	9827	9832	9836	9841	9845	9850	9854	9859	9863
97	9868	9872	9877	9881	9886	9890	9894	9899	9903	9908
98	9912	9917	9921	9926	9930	9934	9939	9943	9948	9952
99	9956	9961	9965	9969	9974	9978	9983	9987	9991	9996
N	0	1	2	3	4	5	6	7	8	9

TABLE 2.—MANTISSAS OF COMMON LOGARITHMS (SPECIAL TABLE)

N	0	1	2	3	4	5	6	7	8	9
100	0000	0004	0009	0013	0017	0022	0026	0030	0035	0039
101	0043	0048	0052	0056	0060	0065	0069	0073	0077	0082
102	0086	0090	0095	0099	0103	0107	0111	0116	0120	0124
103	0128	0133	0137	0141	0145	0149	0154	0158	0162	0166
104	0170	0175	0179	0183	0187	0191	0195	0199	0204	0208
105	0212	0216	0220	0224	0228	0233	0237	0241	0245	0249
106	0253	0257	0261	0265	0269	0273	0278	0282	0286	0290
107	0294	0298	0302	0306	0310	0314	0318	0322	0326	0330
108	0334	0338	0342	0346	0350	0354	0358	0362	0366	0370
109	0374	0378	0382	0386	0390	0394	0398	0402	0406	0410
110	0414	0418	0422	0426	0430	0434	0438	0441	0445	0449
111	0453	0457	0461	0465	0469	0473	0477	0481	0484	0488
112	0492	0496	0500	0504	0508	0512	0515	0519	0523	0527
113	0531	0535	0538	0542	0546	0550	0554	0558	0561	0565
114	0569	0573	0577	0580	0584	0588	0592	0596	0599	0603
115	0607	0611	0615	0618	0622	0626	0630	0633	0637	0641
116	0645	0648	0652	0656	0660	0663	0667	0671	0674	0678
117	0682	0686	0689	0693	0697	0700	0704	0708	0711	0715
118	0719	0722	0726	0730	0734	0737	0741	0745	0748	0752
119	0755	0759	0763	0766	0770	0774	0777	0781	0785	0788
120	0792	0795	0799	0803	0806	0810	0813	0817	0821	0824
121	0828	0831	0835	0839	0842	0846	0849	0853	0856	0860
122	0864	0867	0871	0874	0878	0881	0885	0888	0892	0896
123	0899	0903	0906	0910	0913	0917	0920	0924	0927	0931
124	0934	0938	0941	0945	0948	0952	0955	0959	0962	0966
125	0969	0973	0976	0980	0983	0986	0990	0993	0997	1000
126	1004	1007	1011	1014	1017	1021	1024	1028	1031	1035
127	1038	1041	1045	1048	1052	1055	1059	1062	1065	1069
128	1072	1075	1079	1082	1086	1089	1092	1096	1099	1103
129	1106	1109	1113	1116	1119	1123	1126	1129	1133	1136
130	1139	1143	1146	1149	1153	1156	1159	1163	1166	1169
131	1173	1176	1179	1183	1186	1189	1193	1196	1199	1202
132	1206	1209	1212	1216	1219	1222	1225	1229	1232	1235
133	1239	1242	1245	1248	1252	1255	1258	1261	1265	1268
134	1271	1274	1278	1281	1284	1287	1290	1294	1297	1300
135	1303	1307	1310	1313	1316	1319	1323	1326	1329	1332
136	1335	1339	1342	1345	1348	1351	1355	1358	1361	1364
137	1367	1370	1374	1377	1380	1383	1386	1389	1392	1396
138	1399	1402	1405	1408	1411	1414	1418	1421	1424	1427
139	1430	1433	1436	1440	1443	1446	1449	1452	1455	1458
140	1461	1464	1467	1471	1474	1477	1480	1483	1486	1489
141	1492	1495	1498	1501	1504	1508	1511	1514	1517	1520
142	1523	1526	1529	1532	1535	1538	1541	1544	1547	1550
143	1553	1556	1559	1562	1565	1569	1572	1575	1578	1581
144	1584	1587	1590	1593	1596	1599	1602	1605	1608	1611
N	0	1	2	3	4	5	6	7	8	9

TABLE 2.—MANTISSAS OF COMMON LOGARITHMS (SPECIAL TABLE).—
(Continued)

N	0	1	2	3	4	5	6	7	8	9
145	1614	1617	1620	1623	1626	1629	1632	1635	1638	1641
146	1644	1647	1649	1652	1655	1658	1661	1664	1667	1670
147	1673	1676	1679	1682	1685	1688	1691	1694	1697	1700
148	1703	1706	1708	1711	1714	1717	1720	1723	1726	1729
149	1732	1735	1738	1741	1744	1746	1749	1752	1755	1758
150	1761	1764	1767	1770	1772	1775	1778	1781	1784	1787
151	1790	1793	1796	1798	1801	1804	1807	1810	1813	1816
152	1818	1821	1824	1827	1830	1833	1836	1838	1841	1844
153	1847	1850	1853	1855	1858	1861	1864	1867	1870	1872
154	1875	1878	1881	1884	1886	1889	1892	1895	1898	1901
155	1903	1906	1909	1912	1915	1917	1920	1923	1926	1928
156	1931	1934	1937	1940	1942	1945	1948	1951	1953	1956
157	1959	1962	1965	1967	1970	1973	1976	1978	1981	1984
158	1987	1989	1992	1995	1998	2000	2003	2006	2009	2011
159	2014	2017	2019	2022	2025	2028	2030	2033	2036	2038
160	2041	2044	2047	2049	2052	2055	2057	2060	2063	2066
161	2068	2071	2074	2076	2079	2082	2084	2087	2090	2092
162	2095	2098	2101	2103	2106	2109	2111	2114	2117	2119
163	2122	2125	2127	2130	2133	2135	2138	2140	2143	2146
164	2148	2151	2154	2156	2159	2162	2164	2167	2170	2172
165	2175	2177	2180	2183	2185	2188	2191	2193	2196	2198
166	2201	2204	2206	2209	2212	2214	2217	2219	2222	2225
167	2227	2230	2232	2235	2238	2240	2243	2245	2248	2251
168	2253	2256	2258	2261	2263	2266	2269	2271	2274	2276
169	2279	2281	2284	2287	2289	2292	2294	2297	2299	2302
170	2304	2307	2310	2312	2315	2317	2320	2322	2325	2327
171	2330	2333	2335	2338	2340	2343	2345	2348	2350	2353
172	2355	2358	2360	2363	2365	2368	2370	2373	2375	2378
173	2380	2383	2385	2388	2390	2393	2395	2398	2400	2403
174	2405	2408	2410	2413	2415	2418	2420	2423	2425	2428
175	2430	2433	2435	2438	2440	2443	2445	2448	2450	2453
176	2455	2458	2460	2463	2465	2467	2470	2472	2475	2477
177	2480	2482	2485	2487	2490	2492	2494	2497	2499	2502
178	2504	2507	2509	2512	2514	2516	2519	2521	2524	2526
179	2529	2531	2533	2536	2538	2541	2543	2545	2548	2550
180	2553	2555	2558	2560	2562	2565	2567	2570	2572	2574
181	2577	2579	2582	2584	2586	2589	2591	2594	2596	2598
182	2601	2603	2605	2608	2610	2613	2615	2617	2620	2622
183	2625	2627	2629	2632	2634	2636	2639	2641	2643	2646
184	2648	2651	2653	2655	2658	2660	2662	2665	2667	2669
185	2672	2674	2676	2679	2681	2683	2686	2688	2690	2693
186	2695	2697	2700	2702	2704	2707	2709	2711	2714	2716
187	2718	2721	2723	2725	2728	2730	2732	2735	2737	2739
188	2742	2744	2746	2749	2751	2753	2755	2758	2760	2762
189	2765	2767	2769	2772	2774	2776	2778	2781	2783	2785
N	0	1	2	3	4	5	6	7	8	9

TABLE 3.—NATURAL LOGARITHS (LN N) OF NUMBERS
(BASE $e = 2.71828 \ldots$)

N	0	1	2	3	4	5	6	7	8	9
1.0	0.0000	0100	0198	0296	0392	0488	0583	0677	0770	0862
1.1	0953	1044	1133	1222	1310	1398	1484	1570	1655	1740
1.2	1823	1906	1989	2070	2151	2231	2311	2390	2469	2546
1.3	2624	2700	2776	2852	2927	3001	3075	3148	3221	3293
1.4	3365	3436	3507	3577	3646	3716	3784	3853	3920	3988
1.5	0.4055	4121	4187	4253	4318	4383	4447	4511	4574	4637
1.6	4700	4762	4824	4886	4947	5008	5068	5128	5188	5247
1.7	5306	5365	5423	5481	5539	5596	5653	5710	5766	5822
1.8	5878	5933	5988	6043	6098	6152	6206	6259	6313	6366
1.9	6419	6471	6523	6575	6627	6678	6729	6780	6831	6881
2.0	0.6931	6981	7031	7080	7129	7178	7227	7275	7324	7372
2.1	7419	7467	7514	7561	7608	7655	7701	7747	7793	7839
2.2	7885	7930	7975	8020	8065	8109	8154	8198	8242	8286
2.3	8329	8372	8416	8459	8502	8544	8587	8629	8671	8713
2.4	8755	8796	8838	8879	8920	8961	9002	9042	9083	9123
2.5	0.9163	9203	9243	9282	9322	9361	9400	9439	9478	9517
2.6	9555	9594	9632	9670	9708	9746	9783	9821	9858	9895
2.7	0.9933	9969	*0006	*0043	*0080	*0116	*0152	*0188	*0225	*0260
2.8	1.0296	0332	0367	0403	0438	0473	0508	0543	0578	0613
2.9	0647	0682	0716	0750	0784	0818	0852	0886	0919	0953
3.0	1.0986	1019	1053	1086	1119	1151	1184	1217	1249	1282
3.1	1314	1346	1378	1410	1442	1474	1506	1537	1569	1600
3.2	1632	1663	1694	1725	1756	1787	1817	1848	1878	1909
3.3	1939	1969	2000	2030	2060	2090	2119	2149	2179	2208
3.4	2238	2267	2296	2326	2355	2384	2413	2442	2470	2499
3.5	1.2528	2556	2585	2613	2641	2669	2698	2726	2754	2782
3.6	2809	2837	2865	2892	2920	2947	2975	3002	3029	3056
3.7	3083	3110	3137	3164	3191	3218	3244	3271	3297	3324
3.8	3350	3376	3403	3429	3455	3481	3507	3533	3558	3584
3.9	3610	3635	3661	3686	3712	3737	3762	3788	3813	3838
4.0	1.3863	3888	3913	3938	3962	3987	4012	4036	4061	4085
4.1	4110	4134	4159	4183	4207	4231	4255	4279	4303	4327
4.2	4351	4375	4398	4422	4446	4469	4493	4516	4540	4563
4.3	4586	4609	4633	4656	4679	4702	4725	4748	4770	4793
4.4	4816	4839	4861	4884	4907	4929	4951	4974	4996	5019
4.5	1.5041	5063	5085	5107	5129	5151	5173	5195	5217	5239
4.6	5261	5282	5304	5326	5347	5369	5390	5412	5433	5454
4.7	5476	5497	5518	5539	5560	5581	5602	5623	5644	5665
4.8	5686	5707	5728	5748	5769	5790	5810	5831	5851	5872
4.9	5892	5913	5933	5953	5974	5994	6014	6034	6054	6074
5.0	1.6094	6114	6134	6154	6174	6194	6214	6233	6253	6273
5.1	6292	6312	6332	6351	6371	6390	6409	6429	6448	6467
5.2	6487	6506	6525	6544	6563	6582	6601	6620	6639	6658
5.3	6677	6696	6715	6734	6752	6771	6790	6808	6827	6845
5.4	6864	6882	6901	6919	6938	6956	6974	6993	7011	7029

k	1	2	3	4	5	6
ln 10^{-k}	$\bar{3}.6974$	$\bar{5}.3948$	$\bar{7}.0922$	$\overline{10}.7897$	$\overline{12}.4871$	$\overline{14}.1845$

Table 3.—Natural Logarithms (ln *N*) of Numbers.—(*Continued*)

N	0	1	2	3	4	5	6	7	8	9
5.5	1.7047	7066	7084	7102	7120	7138	7156	7174	7192	7210
5.6	7228	7246	7263	7281	7299	7317	7334	7352	7370	7387
5.7	7405	7422	7440	7457	7475	7492	7509	7527	7544	7561
5.8	7579	7596	7613	7630	7647	7664	7681	7699	7716	7733
5.9	7750	7766	7783	7800	7817	7834	7851	7867	7884	7901
6.0	1.7918	7934	7951	7967	7984	8001	8017	8034	8050	8066
6.1	8083	8099	8116	8132	8148	8165	8181	8197	8213	8229
6.2	8245	8262	8278	8294	8310	8326	8342	8358	8374	8390
6.3	8405	8421	8437	8453	8469	8485	8500	8516	8532	8547
6.4	8563	8579	8594	8610	8625	8641	8656	8672	8687	8703
6.5	1.8718	8733	8749	8764	8779	8795	8810	8825	8840	8856
6.6	8871	8886	8901	8916	8931	8946	8961	8976	8991	9006
6.7	9021	9036	9051	9066	9081	9095	9110	9125	9140	9155
6.8	9169	9184	9199	9213	9228	9242	9257	9272	9286	9301
6.9	9315	9330	9344	9359	9373	9387	9402	9416	9430	9445
7.0	1.9459	9473	9488	9502	9516	9530	9544	9559	9573	9587
7.1	9601	9615	9629	9643	9657	9671	9685	9699	9713	9727
7.2	9741	9755	9769	9782	9796	9810	9824	9838	9851	9865
7.3	1.9879	9892	9906	9920	9933	9947	9961	9974	9988	*0001
7.4	2.0015	0028	0042	0055	0069	0082	0096	0109	0122	0136
7.5	2.0149	0162	0176	0189	0202	0215	0229	0242	0255	0268
7.6	0281	0295	0308	0321	0334	0347	0360	0373	0386	0399
7.7	0412	0425	0438	0451	0464	0477	0490	0503	0516	0528
7.8	0541	0554	0567	0580	0592	0605	0618	0631	0643	0656
7.9	0669	0681	0694	0707	0719	0732	0744	0757	0769	0782
8.0	2.0794	0807	0819	0832	0844	0857	0869	0882	0894	0906
8.1	0919	0931	0943	0956	0968	0980	0992	1005	1017	1029
8.2	1041	1054	1066	1078	1090	1102	1114	1126	1138	1150
8.3	1163	1175	1187	1199	1211	1223	1235	1247	1258	1270
8.4	1282	1294	1306	1318	1330	1342	1353	1365	1377	1389
8.5	2.1401	1412	1424	1436	1448	1459	1471	1483	1494	1506
8.6	1518	1529	1541	1552	1564	1576	1587	1599	1610	1622
8.7	1633	1645	1656	1668	1679	1691	1702	1713	1725	1736
8.8	1748	1759	1770	1782	1793	1804	1815	1827	1838	1849
8.9	1861	1872	1883	1894	1905	1917	1928	1939	1950	1961
9.0	2.1972	1983	1994	2006	2017	2028	2039	2050	2061	2072
9.1	2083	2094	2105	2116	2127	2138	2148	2159	2170	2181
9.2	2192	2203	2214	2225	2235	2246	2257	2268	2279	2289
9.3	2300	2311	2322	2332	2343	2354	2364	2375	2386	2396
9.4	2407	2418	2428	2439	2450	2460	2471	2481	2492	2502
9.5	2.2513	2523	2534	2544	2555	2565	2576	2586	2597	2607
9.6	2618	2628	2638	2649	2659	2670	2680	2690	2701	2711
9.7	2721	2732	2742	2752	2762	2773	2783	2793	2803	2814
9.8	2824	2834	2844	2854	2865	2875	2885	2895	2905	2915
9.9	2.2925	2935	2946	2956	2966	2976	2986	2996	3006	3016

k	1	2	3	4	5	6
ln 10^k	2.3026	4.6052	6.9078	9.2103	11.5129	13.8155

TABLE 4.—TRIGONOMETRIC FUNCTIONS

Angle		Sine		Cose-cant	Tangent		
Degrees	Radians	Value	Log*	Value	Value	Log*	
0° 00′	0.0000	0.0000	∞	∞	0.0000	∞	**90° 00′**
10′	0029	0029	7.4637	343.78	0029	7.4637	89 50
20	0058	0058	7648	171.89	0058	7648	40
30	0087	0087	7.9408	114.59	0087	7.9409	30
40	0116	0116	8.0658	85.946	0116	8.0658	20
0 50	0145	0145	1627	68.757	0145	1627	10
1° 00′	0.0175	0.0175	8.2419	57.299	0.0175	8.2419	**89° 00′**
10	0204	0204	3088	49.114	0204	3089	88 50
20	0233	0233	3668	42.976	0233	3669	40
30	0262	0262	4179	38.202	0262	4181	30
40	0291	0291	4637	34.382	0291	4638	20
1 50	0320	0320	5050	31.258	0320	5053	10
2° 00′	0.0349	0.0349	8.5428	28.654	0.0349	8.5431	**88° 00′**
10	0378	0378	5776	26.451	0378	5779	87 50
20	0407	0407	6097	24.562	0407	6101	40
30	0436	0436	6397	22.926	0437	6401	30
40	0465	0465	6677	21.494	0466	6682	20
2 50	0495	0494	6940	20.230	0495	6945	10
3° 00′	0.0524	0.0523	8.7188	19.107	0.0524	8.7194	**87° 00′**
10	0553	0552	7423	18.103	0553	7429	86 50
20	0582	0581	7645	17.198	0582	7652	40
30	0611	0610	7857	16.380	0612	7865	30
40	0640	0640	8059	15.637	0641	8067	20
3 50	0669	0669	8251	14.958	0670	8261	10
4° 00′	0.0698	0.0698	8.8436	14.336	0.0699	8.8446	**86° 00′**
10	0727	0727	8613	13.763	0729	8624	85 50
20	0756	0756	8783	13.235	0758	8795	40
30	0785	0785	8946	12.745	0787	8960	30
40	0814	0814	9104	12.291	0816	9118	20
4 50	0844	0843	9256	11.868	0846	9272	10
5° 00′	0.0873	0.0872	8.9403	11.474	0.0875	8.9420	**85° 00′**
10	0902	0901	9545	11.105	0904	9563	84 50
20	0931	0929	9682	10.758	0934	9701	40
30	0960	0958	·9816	10.433	0963	9836	30
40	0989	0987	8.9945	10.128	0992	8.9966	20
5 50	1018	1016	9.0070	9.8391	1022	9.0093	10
6° 00′	0.1047	0.1045	9.0192	9.5668	0.1051	9.0216	**84° 00′**
10	1076	1074	0311	3092	1080	0336	83 50
20	1105	1103	0426	9.0652	1110	0453	40
30	1134	1132	0539	8.8337	1139	0567	30
40	1164	1161	0648	6138	1169	0678	20
6 50	1193	1190	0755	4047	1198	0786	10
7° 00′	0.1222	0.1219	9.0859	8.2055	0.1228	9.0891	**83° 00′**
10	1251	1248	0961	8.0156	1257	0995	82 50
20	1280	1276	1060	7.8344	1287	1096	40
30	1309	1305	1157	6613	1317	1194	30
40	1338	1334	1252	4957	1346	1291	20
7 50	1367	1363	1345	3372	1376	1385	10
8° 00′	0.1396	0.1392	9.1436	7.1853	0.1405	9.1478	**82° 00′**
		Value	Log*	Value	Value	Log*	Degrees Angle
		Cosine		Secant	Cotangent		

* Attach −10 to entries in this column.

TABLE 4.—TRIGONOMETRIC FUNCTIONS.—(Continued)

Angle Degrees	Cotangent Value	Cotangent Log	Cosine Value	Cosine Log*	Secant Value	Radians	Degrees
0° 00′	∞	∞	1.0000	10.0000	1.0000	1.5708	90° 00′
10	343.77	2.5363	0000	0000	0000	5679	89 50
20	171.89	2352	0000	0000	0000	5650	40
30	114.59	2.0591	1.0000	0000	0000	5621	30
40	85.940	1.9342	0.9999	0000	0001	5592	20
0 50	68.750	8373	9999	10.0000	0001	5563	10
1° 00′	57.290	1.7581	0.9998	9.9999	1.0002	1.5533	89° 00′
10	49.104	6911	9998	9999	0002	5504	88 50
20	42.964	6331	9997	9999	0003	5475	40
30	38.188	5819	9997	9999	0003	5446	30
40	34.368	5362	9996	9998	0004	5417	20
1 50	31.242	4947	9995	9998	0005	5388	10
2° 00′	28.636	1.4569	0.9994	9.9997	1.0006	1.5359	88° 00′
10	26.432	4221	9993	9997	0007	5330	87 50
20	24.542	3899	9992	9996	0008	5301	40
30	22.904	3599	9990	9996	0010	5272	30
40	21.470	3318	9989	9995	0011	5243	20
2 50	20.206	3055	9988	9995	0012	5213	10
3° 00′	19.081	1.2806	0.9986	9.9994	1.0014	1.5184	87° 00′
10	18.075	2571	9985	9993	0015	5155	86 50
20	17.169	2348	9983	9993	0017	5126	40
30	16.350	2135	9981	9992	0019	5097	30
40	15.605	1933	9980	9991	0021	5068	20
3 50	14.924	1739	9978	9990	0022	5039	10
4° 00′	14.301	1.1554	0.9976	9.9989	1.0024	1.5010	86° 00′
10	13.727	1376	9974	9989	0027	4981	85 50
20	13.197	1205	9971	9988	0029	4952	40
30	12.706	1040	9969	9987	0031	4923	30
40	12.251	0882	9967	9986	0033	4893	20
4 50	11.826	0728	9964	9985	0036	4864	10
5° 00′	11.430	1.0580	0.9962	9.9983	1.0038	1.4835	85° 00′
10	11.059	0437	9959	9982	0041	4806	84 50
20	10.712	0299	9957	9981	0043	4777	40
30	10.385	0164	9954	9980	0046	4748	30
40	10.078	1.0034	9951	9979	0049	4719	20
5 50	9.7882	0.9907	9948	9977	0052	4690	10
6° 00′	9.5144	0.9784	0.9945	9.9976	1.0055	1.4661	84° 00′
10	9.2553	9664	9942	9975	0058	4632	83 50
20	9.0098	9547	9939	9973	0061	4603	40
30	8.7769	9433	9936	9972	0065	4573	30
40	8.5555	9322	9932	9971	0068	4544	20
6 50	8.3450	9214	9929	9969	0072	4515	10
7° 00′	8.1443	0.9109	0.9925	9.9968	1.0075	1.4486	83° 00′
10	7.9530	9005	9922	9966	0079	4457	82 50
20	7.7704	8904	9918	9964	0082	4428	40
30	7.5958	8806	9914	9963	0086	4399	30
40	7.4287	8709	9911	9961	0090	4370	20
7 50	7.2687	8615	9907	9959	0094	4341	10
8° 00′	7.1154	0.8522	0.9903	9.9958	1.0098	1.4312	82° 00′
	Value — Log		Value — Log*		Value	Radians — Degrees	
	Tangent		Sine		Cosecant	Angle	

* Attach −10 to entries in this column.

TABLE 4.—TRIGONOMETRIC FUNCTIONS.—(Continued)

Angle		Sine		Cosecant	Tangent		
Degrees	Radians	Value	Log*	Value	Value	Log*	
8° 00′	0.1396	0.1392	9.1436	7.1853	0.1405	9.1478	82° 00′
10	1425	1421	1525	7.0396	1435	1569	81 50
20	1454	1449	1612	6.8998	1465	1658	40
30	1484	1478	1697	7655	1495	1745	30
40	1513	1507	1781	6363	1524	1831	20
8 50	1542	1536	1863	5121	1554	1915	10
9° 00′	0.1571	0.1564	9.1943	6.3925	0.1584	9.1997	81° 00′
10	1600	1593	2022	2772	1614	2078	80 50
20	1629	1622	2100	1661	1644	2158	40
30	1658	1650	2176	6.0589	1673	2236	30
40	1687	1679	2251	5.9554	1703	2313	20
9 50	1716	1708	2324	8554	1733	2389	10
10° 00′	0.1745	0.1736	9.2397	5.7588	0.1763	9.2463	80° 00′
10	1774	1765	2468	6653	1793	2536	79 50
20	1804	1794	2538	5749	1823	2609	40
30	1833	1822	2606	4874	1853	2680	30
40	1862	1851	2674	4026	1883	2750	20
10 50	1891	1880	2740	3205	1914	2819	10
11° 00′	0.1920	0.1908	9.2806	5.2408	0.1944	9.2887	79° 00′
10	1949	1937	2870	1636	1974	2953	78 50
20	1978	1965	2934	0886	2004	3020	40
30	2007	1994	2997	5.0159	2035	3085	30
40	2036	2022	3058	4.9452	2065	3149	20
11 50	2065	2051	3119	8765	2095	3212	10
12° 00′	0.2094	0.2079	9.3179	4.8097	0.2126	9.3275	78° 00′
10	2123	2108	3238	7448	2156	3336	77 50
20	2153	2136	3296	6817	2186	3397	40
30	2182	2164	3353	6202	2217	3458	30
40	2211	2193	3410	5604	2247	3517	20
12 50	2240	2221	3466	5022	2278	3576	10
13° 00′	0.2269	0.2250	9.3521	4.4454	0.2309	9.3634	77° 00′
10	2298	2278	3575	3901	2339	3691	76 50
20	2327	2306	3629	3362	2370	3748	40
30	2356	2334	3682	2837	2401	3804	30
40	2385	2363	3734	2324	2432	3859	20
13 50	2414	2391	3786	1824	2462	3914	10
14° 00′	0.2443	0.2419	9.3837	4.1336	0.2493	9.3968	76° 00′
10	2473	2447	3887	0859	2524	4021	75 50
20	2502	2476	3937	4.0394	2555	4074	40
30	2531	2504	3986	3.9939	2586	4127	30
40	2560	2532	4035	9495	2617	4178	20
14 50	2589	2560	4083	9061	2648	4230	10
15° 00′	0.2618	0.2588	9.4130	3.8637	0.2679	9.4281	75° 00′
10	2647	2616	4177	8222	2711	4331	74 50
20	2676	2644	4223	7817	2742	4381	40
30	2705	2672	4269	7420	2773	4430	30
40	2734	2700	4314	7032	2805	4479	20
15 50	2763	2728	4359	6652	2836	4527	10
16° 00′	0.2793	0.2756	9.4403	3.6280	0.2867	9.4575	74° 00′
		Value	Log*	Value	Value	Log*	Degrees
		Cosine		Secant	Cotangent		Angle

* Attach −10 to entries in this column.

TABLE 4.—TRIGONOMETRIC FUNCTIONS.—(*Continued*)

Angle Degrees	Cotangent		Cosine		Secant		
	Value	Log	Value	Log*	Value		
8° 00'	7.1154	0.8522	0.9903	9.9958	1.0098	1.4312	82° 00'
10	6.9682	8431	9899	9956	0102	4283	81 50
20	8269	8342	9894	9954	0107	4254	40
30	6912	8255	9890	9952	0111	4224	30
40	5606	8169	9886	9950	0116	4195	20
8 50	4348	8085	9881	9948	0120	4166	10
9° 00'	6.3138	0.8003	0.9877	9.9946	1.0125	1.4137	81° 00'
10	6.1970	7922	9872	9944	0129	4108	80 50
20	6.0844	7842	9868	9942	0134	4079	40
30	5.9758	7764	9863	9940	0139	4050	30
40	8708	7687	9858	9938	0144	4021	20
9 50	7694	7611	9853	9936	0149	3992	10
10° 00'	5.6713	0.7537	0.9848	9.9934	1.0154	1.3963	80° 00'
10	5764	7464	9843	9931	0160	3934	79 50
20	4845	7391	9838	9929	0165	3904	40
30	3955	7320	9833	9927	0170	3875	30
40	3093	7250	9827	9924	0176	3846	20
10 50	2257	7181	9822	9922	0181	3817	10
11° 00'	5.1446	0.7113	0.9816	9.9919	1.0187	1.3788	79° 00'
10	5.0658	7047	9811	9917	0193	3759	78 50
20	4.9894	6980	9805	9914	0199	3730	40
30	9152	6915	9799	9912	0205	3701	30
40	8430	6851	9793	9909	0211	3672	20
11 50	7729	6788	9787	9907	0217	3643	10
12° 00'	4.7046	0.6725	0.9781	9.9904	1.0223	1.3614	78° 00'
10	6382	6664	9775	9901	0230	3584	77 50
20	5736	6603	9769	9899	0236	3555	40
30	5107	6542	9763	9896	0243	3526	30
40	4494	6483	9757	9893	0249	3497	20
12 50	3897	6424	9750	9890	0256	3468	10
13° 00'	4.3315	0.6366	0.9744	9.9887	1.0263	1.3439	77° 00'
10	2747	6309	9737	9884	0270	3410	76 50
20	2193	6252	9730	9881	0277	3381	40
30	1653	6196	9724	9878	0284	3352	30
40	1126	6141	9717	9875	0291	3323	20
13 50	0611	6086	9710	9872	0299	3294	10
14° 00'	4.0108	0.6032	0.9703	9.9869	1.0306	1.3265	76° 00'
10	3.9617	5979	9696	9866	0314	3235	75 50
20	9136	5926	9689	9863	0321	3206	40
30	8667	5873	9681	9859	0329	3177	30
40	8208	5822	9674	9856	0337	3148	20
14 50	7760	5770	9667	9853	0345	3119	10
15° 00'	3.7321	0.5719	0.9659	9.9849	1.0353	1.3090	75° 00'
10	6891	5669	9652	9846	0361	3061	74 50
20	6470	5619	9644	9843	0369	3032	40
30	6059	5570	9636	9839	0377	3003	30
40	5656	5521	9628	9836	0386	2974	20
15 50	5261	5473	9621	9832	0394	2945	10
16° 00'	3.4874	0.5425	0.9613	9.9828	1.0403	1.2915	74° 00'
	Value	Log	Value	Log*	Value	Radians	Degrees
	Tangent		Sine		Cose-cant	Angle	

* Attach −10 to entries in this column.

TABLE 4.—TRIGONOMETRIC FUNCTIONS.—(*Continued*)

Angle		Sine		Cosecant	Tangent		
Degrees	Radians	Value	Log*	Value	Value	Log*	
16° 00′	0.2793	0.2756	9.4403	3.6280	0.2867	9.4575	74° 00′
10	2822	2784	4447	5915	2899	4622	73 50
20	2851	2812	4491	5559	2931	4669	40
30	2880	2840	4533	5209	2962	4716	30
40	2909	2868	4576	4867	2994	4762	20
16 50	2938	2896	4618	4532	3026	4808	10
17° 00′	0.2967	0.2924	9.4659	3.4203	0.3057	9.4853	73° 00′
10	2996	2952	4700	3881	3089	4898	72 50
20	3025	2979	4741	3565	3121	4943	40
30	3054	3007	4781	3255	3153	4987	30
40	3083	3035	4821	2951	3185	5031	20
17 50	3113	3062	4861	2653	3217	5075	10
18° 00′	0.3142	0.3090	9.4900	3.2361	0.3249	9.5118	72° 00′
10	3171	3118	4939	2074	3281	5161	71 50
20	3200	3145	4977	1792	3314	5203	40
30	3229	3173	5015	1515	3346	5245	30
40	3258	3201	5052	1244	3378	5287	20
18 50	3287	3228	5090	0977	3411	5329	10
19° 00′	0.3316	0.3256	9.5126	3.0716	0.3443	9.5370	71° 00′
10	3345	3283	5163	0458	3476	5411	70 50
20	3374	3311	5199	3.0206	3508	5451	40
30	3403	3338	5235	2.9957	3541	5491	30
40	3432	3365	5270	9713	3574	5531	20
19 50	3462	3393	5306	9474	3607	5571	10
20° 00′	0.3491	0.3420	9.5341	2.9238	0.3640	9.5611	70° 00′
10	3520	3448	5375	9006	3673	5650	69 50
20	3549	3475	5409	8779	3706	5689	40
30	3578	3502	5443	8555	3739	5727	30
40	3607	3529	5477	8334	3772	5766	20
20 50	3636	3557	5510	8117	3805	5804	10
21° 00′	0.3665	0.3584	9.5543	2.7904	0.3839	9.5842	69° 00′
10	3694	3611	5576	7695	3872	5879	68 50
20	3723	3638	5609	7488	3906	5917	40
30	3752	3665	5641	7285	3939	5954	30
40	3782	3692	5673	7085	3973	5991	20
21 50	3811	3719	5704	6888	4006	6028	10
22° 00′	0.3840	0.3746	9.5736	2.6695	0.4040	9.6064	68° 00′
10	3869	3773	5767	6504	4074	6100	67 50
20	3898	3800	5798	6316	4108	6136	40
30	3927	3827	5828	6131	4142	6172	30
40	3956	3854	5859	5949	4176	6208	20
22 50	3985	3881	5889	5770	4210	6243	10
23° 00′	0.4014	0.3907	9.5919	2.5593	0.4245	9.6279	67° 00′
10	4043	3934	5948	5419	4279	6314	66 50
20	4072	3961	5978	5247	4314	6348	40
30	4102	3987	6007	5078	4348	6383	30
40	4131	4014	6036	4912	4383	6417	20
23 50	4160	4041	6065	4748	4417	6452	10
24° 00′	0.4189	0.4067	9.6093	2.4586	0.4452	9.6486	66° 00′
		Value	Log*	Value	Value	Log*	Degrees
		Cosine		Secant	Cotangent		Angle

* Attach −10 to entries in this column.

TABLE 4.—TRIGONOMETRIC FUNCTIONS.—(*Continued*)

Angle Degrees	Cotangent		Cosine		Secant		
	Value	Log	Value	Log*	Value		
16° 00′	3.4874	0.5425	0.9613	9.9828	1.0403	1.2915	**74° 00′**
10	4495	5378	9605	9825	0412	2886	73 50
20	4124	5331	9596	9821	0421	2857	40
30	3759	5284	9588	9817	0429	2828	30
40	3402	5238	9580	9814	0439	2799	20
16 50	3052	5192	9572	9810	0448	2770	10
17° 00′	3.2709	0.5147	0.9563	9.9806	1.0457	1.2741	**73° 00′**
10	2371	5102	9555	9802	0466	2712	72 50
20	2041	5057	9546	9798	0476	2683	40
30	1716	5013	9537	9794	0485	2654	30
40	1397	4969	9528	9790	0495	2625	20
17 50	1084	4925	9520	9786	0505	2595	10
18° 00′	3.0777	0.4882	0.9511	9.9782	1.0515	1.2566	**72° 00′**
10	0475	4839	9502	9778	0525	2537	71 50
20	3.0178	4797	9492	9774	0535	2508	40
30	2.9887	4755	9483	9770	0545	2479	30
40	9600	4713	9474	9765	0555	2450	20
18 50	9319	4671	9465	9761	0566	2421	10
19° 00′	2.9042	0.4630	0.9455	9.9757	1.0576	1.2392	**71° 00′**
10	8770	4589	9446	9752	0587	2363	70 50
20	8502	4549	9436	9748	0598	2334	40
30	8239	4509	9426	9743	0608	2305	30
40	7980	4469	9417	9739	0619	2275	20
19 50	7725	4429	9407	9734	0631	2246	10
20° 00′	2.7475	0.4389	0.9397	9.9730	1.0642	1.2217	**70° 00′**
10	7228	4350	9387	9725	0653	2188	69 50
20	6985	4311	9377	9721	0665	2159	40
30	6746	4273	9367	9716	0676	2130	30
40	6511	4234	9356	9711	0688	2101	20
20 50	6279	4196	9346	9706	0700	2072	10
21° 00′	2.6051	0.4158	0.9336	9.9702	1.0711	1.2043	**69° 00′**
10	5826	4121	9325	9697	0723	2014	68 50
20	5605	4083	9315	9692	0736	1985	40
30	5386	4046	9304	9687	0748	1956	30
40	5172	4009	9293	9682	0760	1926	20
21 50	4960	3972	9283	9677	0773	1897	10
22° 00′	2.4751	0.3936	0.9272	9.9672	1.0785	1.1868	**68° 00′**
10	4545	3900	9261	9667	0798	1839	67 50
20	4342	3864	9250	9661	0811	1810	40
30	4142	3828	9239	9656	0824	1781	30
40	3945	3792	9228	9651	0837	1752	20
22 50	3750	3757	9216	9646	0850	1723	10
23° 00′	2.3559	0.3721	0.9205	9.9640	1.0864	1.1694	**67° 00′**
10	3369	3686	9194	9635	0877	1665	66 50
20	3183	3652	9182	9629	0891	1636	40
30	2998	3617	9171	9624	0904	1606	30
40	2817	3583	9159	9618	0918	1577	20
23 50	2637	3548	9147	9613	0932	1548	10
24° 00′	2.2460	0.3514	0.9135	9.9607	1.0946	1.1519	**66° 00′**
	Value	Log	Value	Log*	Value	Radians	Degrees
	Tangent		Sine		Cose-cant	Angle	

* Attach −10 to entries in this column.

18 · MANUAL OF MATHEMATICS AND MECHANICS

TABLE 4.—TRIGONOMETRIC FUNCTIONS.—(Continued)

Angle		Sine		Cosecant	Tangent		
Degrees	Radians	Value	Log*	Value	Value	Log*	
24° 00′	0.4189	0.4067	9.6093	2.4586	0.4452	9.6486	66° 00′
10	4218	4094	6121	4426	4487	6520	65 50
20	4247	4120	6149	4269	4522	6553	40
30	4276	4147	6177	4114	4557	6587	30
40	4305	4173	6205	3961	4592	6620	20
24 50	4334	4200	6232	3811	4628	6654	10
25° 00′	0.4363	0.4226	9.6259	2.3662	0.4663	9.6687	65° 00′
10	4392	4253	6286	3515	4699	6720	64 50
20	4422	4279	6313	3371	4734	6752	40
30	4451	4305	6340	3228	4770	6785	30
40	4480	4331	6366	3088	4806	6817	20
25 50	4509	4358	6392	2949	4841	6850	10
26° 00′	0.4538	0.4384	9.6418	2.2812	0.4877	9.6882	64° 00′
10	4567	4410	6444	2677	4913	6914	63 50
20	4596	4436	6470	2543	4950	6946	40
30	4625	4462	6495	2412	4986	6977	30
40	4654	4488	6521	2282	5022	7009	20
26 50	4683	4514	6546	2153	5059	7040	10
27° 00′	0.4712	0.4540	9.6570	2.2027	0.5095	9.7072	63° 00′
10	4741	4566	6595	1902	5132	7103	62 50
20	4771	4592	6620	1779	5169	7134	40
30	4800	4617	6644	1657	5206	7165	30
40	4829	4643	6668	1537	5243	7196	20
27 50	4858	4669	6692	1418	5280	7226	10
28° 00′	0.4887	0.4695	9.6716	2.1301	0.5317	9.7257	62° 00′
10	4916	4720	6740	1185	5354	7287	61 50
20	4945	4746	6763	1070	5392	7317	40
30	4974	4772	6787	0957	5430	7348	30
40	5003	4797	6810	0846	5467	7378	20
28 50	5032	4823	6833	0736	5505	7408	10
29° 00′	0.5061	0.4848	9.6856	2.0627	0.5543	9.7438	61° 00′
10	5091	4874	6878	0519	5581	7467	60 50
20	5120	4899	6901	0413	5619	7497	40
30	5149	4924	6923	0308	5658	7526	30
40	5178	4950	6946	0204	5696	7556	20
29 50	5207	4975	6968	0101	5735	7585	10
30° 00′	0.5236	0.5000	9.6990	2.0000	0.5774	9.7614	60° 00′
10	5265	5025	7012	1.9900	5812	7644	59 50
20	5294	5050	7033	9801	5851	7673	40
30	5323	5075	7055	9703	5890	7701	30
40	5352	5100	7076	9606	5930	7730	20
30 50	5381	5125	7097	9511	5969	7759	10
31° 00′	0.5411	0.5150	9.7118	1.9416	0.6009	9.7788	59° 00′
10	5440	5175	7139	9323	6048	7816	58 50
20	5469	5200	7160	9230	6088	7845	40
30	5498	5225	7181	9139	6128	7873	30
40	5527	5250	7201	9048	6168	7902	20
31 50	5556	5275	7222	8959	6208	7930	10
32° 00′	0.5585	0.5299	9.7242	1.8871	0.6249	9.7958	58° 00′
		Value	Log*	Value	Value	Log*	Degrees
		Cosine		Secant	Cotangent		Angle

* Attach −10 to entries in this column.

TABLE 4.—TRIGONOMETRIC FUNCTIONS.—(*Continued*)

Angle Degrees	Cotangent Value	Cotangent Log	Cosine Value	Cosine Log*	Secant Value		
24° 00'	2.2460	0.3514	0.9135	9.9607	1.0946	1.1519	66° 00'
10	2286	3480	9124	9602	0961	1490	65 50
20	2113	3447	9112	9596	0975	1461	40
30	1943	3413	9100	9590	0989	1432	30
40	1775	3380	9088	9584	1004	1403	20
24 50	1609	3346	9075	9579	1019	1374	10
25° 00'	2.1445	0.3313	0.9063	9.9573	1.1034	1.1345	65° 00'
10	1283	3280	9051	9567	1049	1316	64 50
20	1123	3248	9038	9561	1064	1286	40
30	0965	3215	9026	9555	1079	1257	30
40	0809	3183	9013	9549	1095	1228	20
25 50	0655	3150	9001	9543	1110	1199	10
26° 00'	2.0503	0.3118	0.8988	9.9537	1.1126	1.1170	64° 00'
10	0353	3086	8975	9530	1142	1141	63 50
20	0204	3054	8962	9524	1158	1112	40
30	2.0057	3023	8949	9518	1174	1083	30
40	1.9912	2991	8936	9512	1190	1054	20
26 50	9768	2960	8923	9505	1207	1025	10
27° 00'	1.9626	0.2928	0.8910	9.9499	1.1223	1.0996	63° 00'
10	9486	2897	8897	9492	1240	0966	62 50
20	9347	2866	8884	9486	1257	0937	40
30	9210	2835	8870	9479	1274	0908	30
40	9074	2804	8857	9473	1291	0879	20
27 50	8940	2774	8843	9466	1308	0850	10
28° 00'	1.8807	0.2743	0.8829	9.9459	1.1326	1.0821	62° 00'
10	8676	2713	8816	9453	1343	0792	61 50
20	8546	2683	8802	9446	1361	0763	40
30	8418	2652	8788	9439	1379	0734	30
40	8291	2622	8774	9432	1397	0705	20
28 50	8165	2592	8760	9425	1415	0676	10
29° 00'	1.8040	0.2562	0.8746	9.9418	1.1434	1.0647	61° 00'
10	7917	2533	8732	9411	1452	0617	60 50
20	7796	2503	8718	9404	1471	0588	40
30	7675	2474	8704	9397	1490	0559	30
40	7556	2444	8689	9390	1509	0530	20
29 50	7437	2415	8675	9383	1528	0501	10
30° 00'	1.7321	0.2386	0.8660	9.9375	1.1547	1.0472	60° 00'
10	7205	2356	8646	9368	1566	0443	59 50
20	7090	2327	8631	9361	1586	0414	40
30	6977	2299	8616	9353	1606	0385	30
40	6864	2270	8601	9346	1626	0356	20
30 50	6753	2241	8587	9338	1646	0327	10
31° 00'	1.6643	0.2212	0.8572	9.9331	1.1666	1.0297	59° 00'
10	6534	2184	8557	9323	1687	0268	58 50
20	6426	2155	8542	9315	1707	0239	40
30	6319	2127	8526	9308	1728	0210	30
40	6212	2098	8511	9300	1749	0181	20
31 50	6107	2070	8496	9292	1770	0152	10
32° 00'	1.6003	0.2042	0.8480	9.9284	1.1792	1.0123	58° 00'
	Value Log		Value Log*		Value Cosecant	Radians Degrees	
	Tangent		Sine			Angle	

* Attach −10 to entries in this column.

TABLE 4.—TRIGONOMETRIC FUNCTIONS.—(Continued)

Angle		Sine		Cose-cant	Tangent		
Degrees	Radians	Value	Log*	Value	Value	Log*	
32° 00'	0.5585	0.5299	9.7242	1.8871	0.6249	9.7958	58° 00'
10	5614	5324	7262	8783	6289	7986	57 50
20	5643	5348	7282	8697	6330	8014	40
30	5672	5373	7302	8612	6371	8042	30
40	5701	5398	7322	8527	6412	8070	20
32 50	5730	5422	7342	8443	6453	8097	10
33° 00'	0.5760	0.5446	9.7361	1.8361	0.6494	9.8125	57° 00'
10	5789	5471	7380	8279	6536	8153	56 50
20	5818	5495	7400	8198	6577	8180	40
30	5847	5519	7419	8118	6619	8208	30
40	5876	5544	7438	8039	6661	8235	20
33 50	5905	5568	7457	7960	6703	8263	10
34° 00'	0.5934	0.5592	9.7476	1.7883	0.6745	9.8290	56° 00'
10	5963	5616	7494	7806	6787	8317	55 50
20	5992	5640	7513	7730	6830	8344	40
30	6021	5664	7531	7655	6873	8371	30
40	6050	5688	7550	7581	6916	8398	20
34 50	6080	5712	7568	7507	6959	8425	10
35° 00'	0.6109	0.5736	9.7586	1.7434	0.7002	9.8452	55° 00'
10	6138	5760	7604	7362	7046	8479	54 50
20	6167	5783	7622	7291	7089	8506	40
30	6196	5807	7640	7221	7133	8533	30
40	6225	5831	7657	7151	7177	8559	20
35 50	6254	5854	7675	7081	7221	8586	10
36° 00'	0.6283	0.5878	9.7692	1.7013	0.7265	9.8613	54° 00'
10	6312	5901	7710	6945	7310	8639	53 50
20	6341	5925	7727	6878	7355	8666	40
30	6370	5948	7744	6812	7400	8692	30
40	6400	5972	7761	6746	7445	8718	20
36 50	6429	5995	7778	6681	7490	8745	10
37° 00'	0.6458	0.6018	9.7795	1.6616	0.7536	9.8771	53° 00'
10	6487	6041	7811	6553	7581	8797	52 50
20	6516	6065	7828	6489	7627	8824	40
30	6545	6088	7844	6427	7673	8850	30
40	6574	6111	7861	6365	7720	8876	20
37 50	6603	6134	7877	6303	7766	8902	10
38° 00'	0.6632	0.6157	9.7893	1.6243	0.7813	9.8928	52° 00'
10	6661	6180	7910	6183	7860	8954	51 50
20	6690	6202	7926	6123	7907	8980	40
30	6720	6225	7941	6064	7954	9006	30
40	6749	6248	7957	6005	8002	9032	20
38 50	6778	6271	7973	5948	8050	9058	10
39° 00'	0.6807	0.6293	9.7989	1.5890	0.8098	9.9084	51° 00'
10	6836	6316	8004	5833	8146	9110	50 50
20	6865	6338	8020	5777	8195	9135	40
30	6894	6361	8035	5721	8243	9161	30
40	6923	6383	8050	5666	8292	9187	20
39 50	6952	6406	8066	5611	8342	9212	10
40° 00'	0.6981	0.6428	9.8081	1.5557	0.8391	9.9238	50° 00'
		Value	Log*	Value	Value	Log*	Degrees
		Cosine		Secant	Cotangent		Angle

* Attach −10 to entries in this column.

TABLE 4.—TRIGONOMETRIC FUNCTIONS.—(*Continued*)

Angle Degrees	Cotangent Value	Cotangent Log	Cosine Value	Cosine Log*	Secant Value		
32° 00′	1.6003	0.2042	0.8480	9.9284	1.1792	1.0123	58° 00′
10	5900	2014	8465	9276	1813	0094	57 50
20	5798	1986	8450	9268	1835	0065	40
30	5697	1958	8434	9260	1857	0036	30
40	5597	1930	8418	9252	1879	1.0007	20
32 50	5497	1903	8403	9244	1901	0.9977	10
33° 00′	1.5399	0.1875	0.8387	9.9236	1.1924	0.9948	57° 00′
10	5301	1847	8371	9228	1946	9919	56 50
20	5204	1820	8355	9219	1969	9890	40
30	5108	1792	8339	9211	1992	9861	30
40	5013	1765	8323	9203	2015	9832	20
33 50	4919	1737	8307	9194	2039	9803	10
34° 00′	1.4826	0.1710	0.8290	9.9186	1.2062	0.9774	56° 00′
10	4733	1683	8274	9177	2086	9745	55 50
20	4641	1656	8258	9169	2110	9716	40
30	4550	1629	8241	9160	2134	9687	30
40	4460	1602	8225	9151	2158	9657	20
34 50	4370	1575	8208	9142	2183	9628	10
35° 00′	1.4281	0.1548	0.8192	9.9134	1.2208	0.9599	55° 00′
10	4193	1521	8175	9125	2233	9570	54 50
20	4106	1494	8158	9116	2258	9541	40
30	4019	1467	8141	9107	2283	9512	30
40	3934	1441	8124	9098	2309	9483	20
35 50	3848	1414	8107	9089	2335	9454	10
36° 00′	1.3764	0.1387	0.8090	9.9080	1.2361	0.9425	54° 00′
10	3680	1361	8073	9070	2387	9396	53 50
20	3597	1334	8056	9061	2413	9367	40
30	3514	1308	8039	9052	2440	9338	30
40	3432	1282	8021	9042	2467	9308	20
36 50	3351	1255	8004	9033	2494	9279	10
37° 00′	1.3270	0.1229	0.7986	9.9023	1.2521	0.9250	53° 00′
10	3190	1203	7969	9014	2549	9221	52 50
20	3111	1176	7951	9004	2577	9192	40
30	3032	1150	7934	8995	2605	9163	30
40	2954	1124	7916	8985	2633	9134	30
37 50	2876	1098	7898	8975	2661	9105	10
38° 00′	1.2799	0.1072	0.7880	9.8965	1.2690	0.9076	52° 00′
10	2723	1046	7862	8955	2719	9047	51 50
20	2647	1020	7844	8945	2748	9018	40
30	2572	0994	7826	8935	2778	8988	30
40	2497	0968	7808	8925	2807	8959	20
38 50	2423	0942	7790	8915	2837	8930	10
39° 00′	1.2349	0.0916	0.7771	9.8905	1.2868	0.8901	51° 00′
10	2276	0890	7753	8895	2898	8872	50 50
20	2203	0865	7735	8884	2929	8843	40
30	2131	0839	7716	8874	2960	8814	30
40	2059	0813	7698	8864	2991	8785	20
39 50	1988	0788	7679	8853	3022	8756	10
40° 00′	1.1918	0.0762	0.7660	9.8843	1.3054	0.8727	50° 00′
	Value	Log	Value	Log*	Value	Radians	Degrees
	Tangent		Sine		Cosecant	Angle	

* Attach − 10 to entries in this column.

TABLE 4.—TRIGONOMETRIC FUNCTIONS.—(*Continued*)

Angle		Sine		Cosecant	Tangent		
Degrees	Radians	Value	Log*	Value	Value	Log*	
40° 00′	0.6981	0.6428	9.8081	1.5557	0.8391	9.9238	50° 00′
10	7010	6450	8096	5504	8441	9264	49 50
20	7039	6472	8111	5450	8491	9289	40
30	7069	6494	8125	5398	8541	9315	30
40	7098	6517	8140	5345	8591	9341	20
40 50	7127	6539	8155	5294	8642	9366	10
41° 00′	0.7156	0.6561	9.8169	1.5243	0.8693	9.9392	49° 00′
10	7185	6583	8184	5192	8744	9417	48 50
20	7214	6604	8198	5141	8796	9443	40
30	7243	6626	8213	5092	8847	9468	30
40	7272	6648	8227	5042	8899	9494	20
41 50	7301	6670	8241	4993	8952	9519	10
42° 00′	0.7330	0.6691	9.8255	1.4945	0.9004	9.9544	48° 00′
10	7359	6713	8269	4897	9057	9570	47 50
20	7389	6734	8283	4849	9110	9595	40
30	7418	6756	8297	4802	9163	9621	30
40	7447	6777	8311	4755	9217	9646	20
42 50	7476	6799	8324	4709	9271	9671	10
43° 00′	0.7505	0.6820	9.8338	1.4663	0.9325	9.9697	47° 00′
10	7534	6841	8351	4617	9380	9722	46 50
20	7563	6862	8365	4572	9435	9747	40
30	7592	6884	8378	4527	9490	9772	30
40	7621	6905	8391	4483	9545	9798	20
43 50	7650	6926	8405	4439	9601	9823	10
44° 00′	0.7679	0.6947	9.8418	1.4396	0.9657	9.9848	46° 00′
10	7709	6967	8431	4352	9713	9874	45 50
20	7738	6988	8444	4310	9770	9899	40
30	7767	7009	8457	4267	9827	9924	30
40	7796	7030	8469	4225	9884	9949	20
44 50	7825	7050	8482	4183	0.9942	9.9975	10
45° 00′	0.7854	0.7071	9.8495	1.4142	1.0000	10.0000	45° 00′
		Value	Log*	Value	Value	Log*	Degrees
		Cosine		Secant	Cotangent		Angle

* Attach −10 to entries in this column.

To find the trigonometric functions of a negative angle, or of an angle greater than 90°, use the following relations:

θ	sin θ	csc θ	tan θ
$-\varphi$	$-\sin\varphi$	$-\csc\varphi$	$-\tan\varphi$
$90° + \varphi$	$\cos\varphi$	$\sec\varphi$	$-\cot\varphi$
$180° - \varphi$	$\sin\varphi$	$\csc\varphi$	$-\tan\varphi$
$180° + \varphi$	$-\sin\varphi$	$-\csc\varphi$	$\tan\varphi$
$270° - \varphi$	$-\cos\varphi$	$-\sec\varphi$	$\cot\varphi$
$270° + \varphi$	$-\cos\varphi$	$-\sec\varphi$	$-\cot\varphi$
$360° - \varphi$	$-\sin\varphi$	$-\csc\varphi$	$-\tan\varphi$
$n \times 360° + \varphi$†	$\sin\varphi$	$\csc\varphi$	$\tan\varphi$

† Where n is any integer.

TABLE 4.—TRIGONOMETRIC FUNCTIONS.—(*Continued*)

Angle Degrees	Cotangent		Cosine		Secant			
	Value	Log	Value	Log*	Value			
40° 00'	1.1918	0.0762	0.7660	9.8843	1.3054	0.8727	**50° 00'**	
10	1847	0736	7642	8832	3086	8698	49 50	
20	1778	0711	7623	8821	3118	8668	40	
30	1708	0685	7604	8810	3151	8639	30	
40	1640	0659	7585	8800	3184	8610	20	
40 50	1571	0634	7566	8789	3217	8581	10	
41° 00'	1.1504	0.0608	0.7547	9.8778	1.3250	0.8552	**49° 00'**	
10	1436	0583	7528	8767	3284	8523	48 50	
20	1369	0557	7509	8756	3318	8494	40	
30	1303	0532	7490	8745	3352	8465	30	
40	1237	0506	7470	8733	3386	8436	20	
41 50	1171	0481	7451	8722	3421	8407	10	
42° 00'	1.1106	0.0456	0.7431	9.8711	1.3456	0.8378	**48° 00'**	
10	1041	0430	7412	8699	3492	8348	47 50	
20	0977	0405	7392	8688	3527	8319	40	
30	0913	0379	7373	8676	3563	8290	30	
40	0850	0354	7353	8665	3600	8261	20	
42 50	0786	0329	7333	8653	3636	8232	10	
43° 00'	1.0724	0.0303	0.7314	9.8641	1.3673	0.8203	**47° 00'**	
10	0661	0278	7294	8629	3711	8174	46 50	
20	0599	0253	7274	8618	3748	8145	40	
30	0538	0228	7254	8606	3786	8116	30	
40	0477	0202	7234	8594	3824	8087	20	
43 50	0416	0177	7214	8582	3863	8058	10	
44° 00'	1.0355	0.0152	0.7193	9.8569	1.3902	0.8029	**46° 00'**	
10	0295	0126	7173	8557	3941	7999	45 50	
20	0235	0101	7153	8545	3980	7970	40	
30	0176	0076	7133	8532	4020	7941	30	
40	0117	0051	7112	8520	4061	7912	20	
44 50	0058	0025	7092	8507	4101	7883	10	
45° 00'	1.0000	0.0000	0.7071	9.8495	1.4142	0.7854	**45° 00'**	
	Value	Log	Value	Log*	Value	Radians	Degrees	
	Tangent		Sine		Cose-cant	Angle		

* Attach − 10 to entries in this column.

To find the trigonometric functions of a negative angle, or of an angle greater than 90°, use the following relations:

θ	cot θ	cos θ	sec θ
− φ	−cot φ	cos φ	sec φ
90° + φ	−tan φ	−sin φ	−csc φ
180° − φ	−cot φ	−cos φ	−sec φ
180° + φ	cot φ	−cos φ	−sec φ
270° − φ	tan φ	−sin φ	−csc φ
270° + φ	−tan φ	sin φ	csc φ
360° − φ	−cot φ	cos φ	sec φ
n × 360° + φ†	cot φ	cos φ	sec φ

† Where n is any integer.

TABLE 5a.—LOGARITHMS† OF SIN x (0°–8°) AT INTERVALS OF ONE MINUTE

x	\multicolumn{10}{c}{$\log\ \sin x = \log \cos (90° - x)$}
° ′	0′	1′	2′	3′	4′	5′	6′	7′	8′	9′
0 00	∞	6.4637	7648	9408	*0658	*1627	*2419	*3088	*3668	*4180
10	7.4637	5051	5429	5777	6099	6398	6678	6942	7190	7425
20	7.7648	7859	8061	8255	8439	8617	8787	8951	*109	*261
30	7.9408	551	689	822	952	*078	*200	*319	*435	*548
40	8.0658	765	870	972	*072	*169	*265	*358	*450	*539
50	8.1627	713	797	880	961	*041	*119	*196	*271	*346
1 00	8.2419	490	561	630	699	766	832	898	962	*025
10	8.3088	150	210	270	329	388	445	502	558	613
20	8.3668	722	775	828	880	931	982	*032	*082	*131
30	8.4179	227	275	322	368	414	459	504	549	593
40	8.4637	680	723	765	807	848	890	930	971	*011
50	8.5050	090	129	167	206	243	281	318	355	392
2 00	8.5428	464	500	535	571	605	640	674	708	742
10	8.5776	809	842	875	907	939	972	*003	*035	*066
20	8.6097	128	159	189	220	250	279	309	339	368
30	397	426	454	483	511	539	567	595	622	650
40	677	704	731	758	784	810	837	863	889	914
50	8.6940	965	991	*016	*041	*066	*090	*115	*140	*164
3 00	8.7188	212	236	260	283	307	330	354	377	400
10	423	445	468	491	513	535	557	580	602	623
20	645	667	688	710	731	752	773	794	815	836
30	8.7857	877	898	918	939	959	979	999	*019	*039
40	8.8059	078	098	117	137	156	175	194	213	232
50	251	270	289	307	326	345	363	381	400	418
4 00	8.8436	454	472	490	508	525	543	560	578	595
10	613	630	647	665	682	699	716	733	749	766
20	783	799	816	833	849	865	882	898	914	930
30	8.8946	962	978	994	*010	*026	*042	*057	*073	*089
40	8.9104	119	135	150	166	181	196	211	226	241
50	256	271	286	301	315	330	345	359	374	388
5 00	8.9403	417	432	446	460	475	489	503	517	531
10	545	559	573	587	601	614	628	642	655	669
20	682	696	709	723	736	750	763	776	789	803
30	816	829	842	855	868	881	894	907	919	932
40	8.9945	958	970	983	996	*008	*021	*033	*046	*058
50	9.0070	083	095	107	120	132	144	156	168	180
6 00	9.0192	204	216	228	240	252	264	276	287	299
10	311	323	334	346	357	369	380	392	403	415
20	426	438	449	460	472	483	494	505	516	527
30	539	550	561	572	583	594	605	616	626	637
40	648	659	670	680	691	702	712	723	734	744
50	755	765	776	786	797	807	818	828	838	849
7 00	9.0859	869	879	890	900	910	920	930	940	951
10	9.0961	971	981	991	*001	*011	*020	*030	*040	*050
20	9.1060	070	080	089	099	109	118	128	138	147
30	157	167	176	186	195	205	214	224	233	242
40	252	261	271	280	289	299	308	317	326	336
50	345	354	363	372	381	390	399	409	418	427
8 00	9.1436									

† Attach −10 to entries in this Table.

TABLE 5b.—LOGARITHMS† OF TAN $x(0°-8°)$ AT INTERVALS OF ONE MINUTE

x	$\log \tan x = \log \cot (90° - x)$									
° ′	0′	1′	2′	3′	4′	5′	6′	7′	8′	9′
0 00	∞	6.4637	7648	9408	*0658	*1627	*2419	*3088	*3668	*4180
10	7.4637	5051	5429	5777	6099	6398	6678	6942	7190	7425
20	7.7648	7860	8062	8255	8439	8617	8787	8951	*109	*261
30	7.9409	551	689	823	952	*078	*200	*319	*435	*548
40	8.0658	765	870	972	*072	*170	*265	*359	*450	*540
50	8.1627	713	798	880	962	*041	*120	*196	*272	*346
1 00	8.2419	491	562	631	700	767	833	899	963	*026
10	8.3089	150	211	271	330	389	446	503	559	614
20	8.3669	723	776	829	881	932	983	*033	*083	132
30	8.4181	229	276	323	370	416	461	506	551	595
40	8.4638	682	725	767	809	851	892	933	973	*013
50	8.5053	092	131	170	208	246	283	321	358	394
2 00	8.5431	467	503	538	573	608	643	677	711	745
10	8.5779	812	845	878	911	943	975	*007	*038	*070
20	8.6101	132	163	193	223	254	283	313	343	372
30	401	430	459	487	515	544	571	599	627	654
40	682	709	736	762	789	815	842	868	894	920
50	8.6945	971	996	*021	*046	*071	*096	*121	*145	*170
3 00	8.7194	218	242	266	290	313	337	360	383	406
10	429	452	475	497	520	542	565	587	609	631
20	652	674	696	717	739	760	781	802	823	844
30	8.7865	886	906	927	947	967	988	*008	*028	*048
40	8.8067	087	107	126	146	165	185	204	223	242
50	261	280	299	317	336	355	373	392	410	428
4 00	8.8446	465	483	501	518	536	554	572	589	607
10	624	642	659	676	694	711	728	745	762	778
20	795	812	829	845	862	878	895	911	927	944
30	8.8960	976	992	*008	*024	*040	*056	*071	*087	*103
40	8.9118	134	150	165	180	196	211	226	241	256
50	272	287	302	316	331	346	361	376	390	405
5 00	8.9420	434	449	463	477	492	506	520	534	549
10	563	577	591	605	619	633	646	660	674	688
20	701	715	729	742	756	769	782	796	809	823
30	836	849	862	875	888	901	915	928	940	953
40	8.9966	979	992	*005	*017	*030	*043	*055	*068	*080
50	9.0093	105	118	130	143	155	167	180	192	204
6 00	9.0216	228	240	253	265	277	289	300	312	324
10	336	348	360	371	383	395	407	418	430	441
20	453	464	476	487	499	510	521	533	544	555
30	567	578	589	600	611	622	633	645	656	667
40	678	688	699	710	721	732	743	754	764	775
50	786	796	807	818	828	839	849	860	871	881
7 00	9.0891	902	912	923	933	943	954	964	974	984
10	9.0995	*005	*015	*025	*035	*045	*055	*066	*076	*086
20	9.1096	106	116	125	135	145	155	165	175	185
30	194	204	214	223	233	243	252	262	272	281
40	291	300	310	319	329	338	348	357	367	376
50	385	395	404	413	423	432	441	450	460	469
8 00	9.1478									

† Attach −10 to entries in this table.

TABLE 6.—TRIGONOMETRIC FUNCTIONS FOR ANGLES IN RADIANS
(0.00 TO 1.60) (SEE ALSO TABLE 4)

Rad.	Equivalent ° ′	sin	cos	tan	cot	sec	csc	Rad.
0.00	0 00	0.000	1.000	0.000	∞	1.000	∞	**0.00**
02	1 09	020	1.000	020	49.99	000	50.00	02
04	2 18	040	0.999	040	24.99	001	25.01	04
06	3 26	060	998	060	16.65	002	16.68	06
08	4 35	080	997	080	12.47	003	12.51	08
0.10	5 44	0.100	0.995	0.100	9.967	1.005	10.02	0.10
12	6 53	120	993	121	8.293	007	8.353	12
14	8 01	140	990	141	7.096	010	7.166	14
16	9 10	159	987	161	6.197	013	6.277	16
18	10 19	179	984	182	5.495	016	5.586	18
0.20	11 28	0.199	0.980	0.203	4.933	1.020	5.033	**0.20**
22	12 36	218	976	224	472	025	4.582	22
24	13 45	238	971	245	4.086	030	4.207	24
26	14 54	257	966	266	3.759	035	3.890	26
28	16 03	276	961	288	478	041	619	28
0.30	17 11	0.296	0.955	0.309	3.233	1.047	3.384	0.30
35	20 03	343	939	365	2.740	065	2.916	35
40	22 55	389	921	423	365	086	568	40
45	25 47	435	900	483	2.070	111	299	45
50	28 39	479	878	546	1.830	140	2.086	50
0.55	31 31	0.523	0.853	0.613	1.631	1.173	1.913	**0.55**
60	34 23	565	825	684	462	212	771	60
65	37 15	605	796	760	315	256	652	65
70	40 06	644	765	842	187	307	552	70
75	42 58	682	732	0.932	1.073	367	467	75
0.80	45 50	0.717	0.697	1.030	0.971	1.435	1.394	0.80
85	48 42	751	660	138	878	515	331	85
90	51 34	783	622	260	·794	609	277	90
0.95	54 26	813	582	398	715	719	229	0.95
1.00	57 18	841	540	557	642	1.851	188	1.00
1.05	60 10	0.867	0.498	1.743	0.574	2.010	1.153	**1.05**
10	63 02	891	454	1.965	509	205	122	10
15	65 53	913	408	2.234	448	448	096	15
20	68 45	932	362	2.572	389	2.760	073	20
25	71 37	949	315	3.010	332	3.171	054	25
1.30	74 29	0.964	0.267	3.602	0.278	3.738	1.038	1.30
32	75 38	969	248	3.903	256	4.029	032	32
34	76 47	973	229	4.256	235	372	027	34
36	77 55	978	209	4.673	214	4.779	023	36
38	79 04	982	190	5.177	193	5.273	018	38
1.40	80 13	0.985	0.170	5.798	0.172	5.883	1.015	**1.40**
42	81 22	989	150	6.581	152	6.657	011	42
44	82 30	991	130	7.602	132	7.667	009	44
46	83 39	994	111	8.989	111	9.044	006	46
48	84 48	996	091	10.98	091	11.03	004	48
1.50	85 57	0.997	0.071	14.10	0.071	14.14	1.002	1.50
52	87 05	0.999	051	19.67	051	19.69	001	52
54	88 14	1.000	031	32.46	031	32.48	000	54
56	89 23	1.000	0.011	92.62	0.011	92.63	000	56
58	90 32	1.000	−.009	−109	−.009	−109	000	58
1.60	91 40	1.000	−.029	−34.2	−.029	−34.2	1.000	**1.60**
Rad.	Equivalent	sin	cos	tan	cot	sec	csc	**Rad.**

TABLE 6.—TRIGONOMETRIC FUNCTIONS FOR ANGLES IN RADIANS.—
(*Continued*)
(1.60–3.20)

Rad.	Equiva-lent	sin	cos	tan	cot	sec	csc	Rad.
	° ′							
1.60	91 40	1.000	−.029	−34.2	−.029	−34.2	1.00	**1.60**
62	92 49	0.999	−.049	−20.3	−.049	−20.4	00	62
64	93 58	998	−.069	−14.4	−.069	−14.5	00	64
66	95 07	996	−.089	−11.2	−.089	−11.3	00	66
68	96 15	994	−.109	−9.12	−.110	−9.19	01	68
1.70	97 24	0.992	−.129	−7.70	−.130	−7.76	1.01	**1.70**
72	98 33	989	−.149	−6.65	−.150	−6.73	01	72
74	99 42	986	−.168	−5.85	−.171	−5.94	01	74
76	100 50	982	−.188	−5.22	−.191	−5.32	02	76
78	101 59	978	−.208	−4.71	−.212	−4.81	02	78
1.80	103 08	0.974	−.227	−4.29	−.233	−4.40	1.03	**1.80**
82	104 17	969	−.247	−3.93	−.254	−4.05	03	82
84	105 25	964	−.266	−3.62	−.276	−3.76	04	84
86	106 34	958	−.285	−3.36	−.298	−3.51	04	86
88	107 43	953	−.304	−3.13	−.319	−3.29	05	88
1.90	108 52	0.946	−.323	−2.93	−.342	−3.09	1.06	1.90
1.95	111 44	929	−.370	−2.51	−.398	−2.70	08	1.95
2.00	114 35	909	−.416	−2.19	−.458	−2.40	10	2.00
05	117 27	887	−.461	−1.92	−.520	−2.17	13	05
10	120 19	863	−.505	−1.71	−.585	−1.98	16	10
2.15	123 11	0.837	−.547	−1.53	−.654	−1.83	1.20	**2.15**
20	126 03	808	−.588	−1.37	−.728	−1.70	24	20
25	128 55	778	−.628	−1.24	−.807	−1.59	28	25
30	131 47	746	−.666	−1.12	−.894	−1.50	34	30
35	134 39	711	−.703	−1.01	−.988	−1.43	41	35
2.40	137 31	0.675	−.737	−.916	−1.09	−1.36	1.48	2.40
45	140 22	638	−.770	−.828	−1.21	−1.30	57	45
50	143 14	598	−.801	−.747	−1.34	−1.25	67	50
55	146 06	558	−.830	−.672	−1.49	−1.20	79	55
60	148 58	515	−.857	−.602	−1.66	−1.17	1.94	60
2.65	151 50	0.472	−.881	−.535	−1.87	−1.13	2.12	**2.65**
70	154 42	427	−.904	−.473	−2.12	−1.11	34	70
75	157 34	382	−.924	−.413	−2.42	−1.08	62	75
80	160 26	335	−.942	−.356	−2.81	−1.06	2.99	80
85	163 18	287	−.958	−.300	−3.33	−1.04	3.48	85
2.90	166 09	0.239	−.971	−.246	−4.06	−1.03	4.18	2.90
92	167 18	220	−.976	−.225	−4.44	−1.03	55	92
94	168 27	200	−.980	−.204	−4.89	−1.02	4.88	94
96	169 36	181	−.984	−.184	−5.45	−1.02	5.54	96
2.98	170 44	161	−.987	−.163	−6.13	−1.01	6.21	2.98
3.00	171 53	0.141	−.990	−.143	−7.01	−1.01	7.08	**3.00**
02	173 02	121	−.993	−.122	−8.18	−1.01	8.24	02
04	174 11	101	−.995	−.102	−9.82	−1.01	9.87	04
06	175 20	081	−.997	−.082	−12.3	−1.00	12.3	06
08	176 28	062	−.998	−.062	−16.2	−1.00	16.2	08
3.10	177 37	0.042	−.999	−.042	−24.0	−1.00	24.0	3.10
12	178 46	022	−1.000	−.022	−46.4	−1.00	46.5	12
14	179 55	0.001	−1.000	−.001	−688	−1.00	688	14
16	181 03	−.018	−.999	0.018	54.6	−1.00	−54.6	16
18	182 12	−.038	−.999	.038	26.0	−1.00	−26.0	18
3.20	183 21	−.058	−.998	0.059	17.1	−1.00	−17.1	**'3.20**
Rad.	Equiva-lent	sin	cos	tan	cot	sec	csc	Rad.

TABLE 6.—TRIGONOMETRIC FUNCTIONS FOR ANGLES IN RADIANS.—
(*Continued*)
(3.20–10)

Rad.	Equiva-lent	sin	cos	tan	cot	sec	csc	Rad.
	° ′							
3.20	183 21	−.058	−.998	0.058	17.1	−1.00	−17.1	**3.20**
25	186 13	−.108	−.994	109	9.19	−1.01	−9.24	25
30	189 05	−.158	−.987	160	6.26	−1.01	−6.34	30
35	191 56	−.207	−.978	211	4.73	−1.02	−4.84	35
40	194 48	−.256	−.967	264	3.78	−1.03	−3.91	40
3.50	200 32	−.351	−.936	0.375	2.67	−1.07	−2.85	3.50
60	206 16	−.443	−.897	493	2.03	−1.12	−2.26	60
70	212 00	−.530	−.848	625	1.60	−1.18	−1.89	70
80	217 43	−.612	−.791	774	1.29	−1.26	−1.63	80
3.90	223 27	−.688	−.726	0.947	1.06	−1.38	−1.45	3.90
4.00	229 11	−.757	−.654	1.16	0.864	−1.53	−1.32	**4.00**
10	234 55	−.818	−.575	1.42	702	−1.74	−1.22	10
20	240 39	−.872	−.490	1.78	562	−2.04	−1.15	20
30	246 22	−.916	−.401	2.29	437	−2.49	−1.09	30
40	252 06	−.952	−.307	3.10	323	−3.25	−1.05	40
4.45	254 58	−.966	−.259	3.72	0.269	−3.86	−1.04	4.45
50	257 50	−.978	−.211	4.64	216	−4.74	−1.02	50
55	260 42	−.987	−.162	6.10	164	−6.19	−1.01	55
60	263 34	−.994	−.112	8.86	113	−8.92	−1.01	60
65	266 26	−.998	−.062	16.0	062	−16.0	−1.00	65
4.70	269 17	−1.00	−.012	80.7	0.012	−80.7	−1.00	**4.70**
75	272 09	−.999	0.038	−26.6	−.038	26.6	−1.00	75
80	275 01	−.996	087	−11.4	−.088	11.4	−1.00	80
85	277 53	−.991	137	−7.22	−.138	7.29	−1.01	85
4.90	280 45	−.982	187	−5.27	−.190	5.36	−1.02	4.90
5.00	286 29	−.959	0.284	−3.38	−.296	3.53	−1.04	**5.00**
10	292 13	−.926	378	−2.45	−.408	2.65	−1.08	10
20	297 56	−.883	469	−1.89	−.530	2.13	−1.13	20
30	303 40	−.832	554	−1.50	−.666	1.80	−1.20	30
40	309 24	−.773	635	−1.22	−.821	1.58	−1.29	40
5.50	315 08	−.706	0.709	−.996	−1.00	1.41	−1.42	**5.50**
60	320 51	−.631	776	−.814	−1.23	1.29	−1.58	60
70	326 35	−.551	835	−.660	−1.52	1.20	−1.82	70
80	332 19	−.465	886	−.525	−1.91	1.13	−2.15	80
5.90	338 03	−.374	927	−.403	−2.48	1.08	−2.67	5.90
6.00	343 46	−.279	0.960	−.291	−3.44	1.04	−3.58	**6.00**
05	346 38	−.231	973	−.238	−4.21	1.03	−4.33	05
10	349 30	−.182	983	−.185	−5.40	1.02	−5.49	10
15	352 22	−.133	991	−.134	−7.46	1.01	−7.53	15
20	355 14	−.083	997	−.083	−12.0	1.00	−12.0	20
6.25	358 06	−.033	999	−.033	−30.1	1.00	−30.1	**6.25**
6.30	360 58	0.017	1.000	0.017	59.5	1.00	59.5	6.30
6.50	372 25	215	0.977	.220	4.54	1.02	4.65	6.50
7.00	401 04	657	754	0.871	1.15	1.33	1.52	7.00
7.50	429 43	938	0.347	2.71	0.370	2.88	1.07	7.50
8.00	458 22	0.989	−.146	−6.80	−.147	−6.87	1.01	8.00
8.50	487 01	798	−.602	−1.33	−.754	−1.66	1.25	8.50
9.00	515 40	0.412	−.911	−.452	−2.21	−1.10	2.43	9.00
9.50	544 19	−.075	−.997	0.075	13.3	−1.00	−13.4	9.50
10.0	572 57	−.544	−.839	0.648	1.54	−1.19	−1.84	10.0
Rad.	Equiva-lent	sin	cos	tan	cot	sec	csc	Rad.

Table 7a.—Conversion Table, Degrees, Minutes and Seconds to
Radians
(See also Table 4)

De-grees	Radians	De-grees	Radians	Min-utes	Radians	Sec-onds	Radians
1	0.01745	70	1.22173	1	0.00029	1	0.00000
2	03491	80	39626	2	00058	2	00001
3	05236	90	57080	3	00087	3	00001
4	06981	100	74533	4	00116	4	00002
5	0.08727	110	1.91986	5	0.00145	5	0.00002
6	10472	120	2.09440	6	00175	6	00003
7	12217	130	26893	7	00204	7	00003
8	13963	140	44346	8	00233	8	00004
9	15708	150	61799	9	00262	9	00004
10	0.17453	160	2.79253	10	0.00291	10	0.00005
20	34907	170	2.96706	20	00582	20	00010
30	52360	180	3.14159	30	00873	30	00015
40	69813	190	31613	40	01164	40	00019
50	0.87266	200	49066	50	01454	50	00024
60	1.04720	210	3.66519	60	0.01745	60	0.00029

1 degree = 0.01745 32925 19943 radians.
1 minute = 0.00029 08882 08666 radians.
1 second = 0.00000 48481 36811 radians.

Table 7b.—Conversion Table, Radians to Degrees, Minutes and
Seconds
(See also Table 6)

	Radians			Tenths			Hundredths			Thousandths			Ten-thousandths		
	°	′	″	°	′	″	°	′	″	°	′	″	°	′	″
1	57	17	44.8	5	43	46.5	0	34	22.6	0	3	26.3	0	0	20.6
2	114	35	29.6	11	27	33.0	1	8	45.3	0	6	52.5	0	0	41.3
3	171	53	14.4	17	11	19.4	1	43	07.9	0	10	18.8	0	1	01.9
4	229	10	59.2	22	55	05.9	2	17	30.6	0	13	45.1	0	1	22.5
5	286	28	44.0	28	38	52.4	2	51	53.2	0	17	11.3	0	1	43.1
6	343	46	28.8	34	22	38.9	3	26	15.9	0	20	37.6	0	2	03.8
7	401	4	13.6	40	6	25.4	4	0	38.5	0	24	03.9	0	2	24.4
8	458	21	58.4	45	50	11.8	4	35	01.2	0	27	30.1	0	2	45.0
9	515	39	43.3	51	33	58.3	5	9	23.8	0	30	56.4	0	3	05.6

1 radian = 57.29577 95131 degrees
= 3,437.74677 07849 minutes
= 206,264.80625 seconds.

TABLE 8.—EXPONENTIAL AND HYPERBOLIC FUNCTIONS

x	e^x	log e^x	e^{-x}	sinh x	log* sinh x	csch x	gd x (rad.)
0.00	1.0000	0.0000	1.0000	0.0000	∞	∞	0.0000
01	0101	0043	0.9900	0100	8.0000	100.0	0100
02	0202	0087	9802	0200	3011	50.00	0200
03	0305	0130	9704	0300	4772	33.33	0300
04	0408	0174	9608	0400	6022	24.99	0400
0.05	1.0513	0.0217	0.9512	0.0500	8.6992	19.99	0.0500
06	0618	0261	9418	0600	7784	16.66	0600
07	0725	0304	9324	0701	8455	14.27	0699
08	0833	0347	9231	0801	9036	12.49	0799
09	0942	0391	9139	0901	8.9548	11.10	0899
0.10	1.1052	0.0434	0.9048	0.1002	9.0007	9.983	0.0998
11	1163	0478	8958	1102	0423	9.073	1098
12	1275	0521	8869	1203	0802	8.313	1197
13	1388	0565	8781	1304	1152	7.671	1296
14	1503	0608	8694	1405	1475	7.120	1395
0.15	1.1618	0.0651	0.8607	0.1506	9.1777	6.642	0.1494
16	1735	0695	8521	1607	2060	6.223	1593
17	1853	0738	8437	1708	2325	5.854	1692
18	1972	0782	8353	1810	2576	5.526	1790
19	2092	0825	8270	1911	2814	5.232	1889
0.20	1.2214	0.0869	0.8187	0.2013	9.3039	4.967	0.1987
21	2337	0912	8106	2115	3254	727	2085
22	2461	0955	8025	2218	3459	509	2182
23	2586	0999	7945	2320	3656	310	2280
24	2712	1042	7866	2423	3844	4.127	2377
0.25	1.2840	0.1086	0.7788	0.2526	9.4025	3.959	0.2474
26	2969	1129	7711	2629	4199	803	2571
27	3100	1173	7634	2733	4366	659	2668
28	3231	1216	7558	2837	4528	525	2764
29	3364	1259	7483	2941	4685	400	2860
0.30	1.3499	0.1303	0.7408	0.3045	9.4836	3.284	0.2956
31	3634	1346	7334	3150	4983	175	3052
32	3771	1390	7261	3255	5125	3.072	3147
33	3910	1433	7189	3360	5264	2.976	3242
34	4049	1477	7118	3466	5398	885	3336
0.35	1.4191	0.1520	0.7047	0.3572	9.5529	2.800	0.3431
36	4333	1563	6977	3678	5656	719	3525
37	4477	1607	6907	3785	5781	642	3618
38	4623	1650	6839	3892	5902	569	3712
39	4770	1694	6771	4000	6020	500	3805
0.40	1.4918	0.1737	0.6703	0.4108	9.6136	2.435	0.3897
41	5068	1781	6637	4216	6249	372	3990
42	5220	1824	6570	4325	6359	312	4082
43	5373	1867	6505	4434	6468	255	4173
44	5527	1911	6440	4543	6574	201	4265
0.45	1.5683	0.1954	0.6376	0.4653	9.6678	2.149	0.4355
46	5841	1998	6313	4764	6780	099	4446
47	6000	2041	6250	4875	6880	051	4536
48	6161	2085	6188	4986	6978	2.005	4626
49	6323	2128	6126	5098	7074	1.961	4715
0.50	1.6487	0.2171	0.6065	0.5211	9.7169	1.919	0.4804
x	e^x	log e^x	e^{-x}	tan gd x	log* tan gd x	cot gd x	gd x (rad.)

* Attach −10 to entries in this column.

TABLE 8.—EXPONENTIAL AND HYPERBOLIC FUNCTIONS.—(*Continued*)

x	tanh x	log* tanh x	coth x	cosh x	log cosh x	sech x	gd x
							° ′ ″
0.00	0.0000	∞	∞	1.0000	0.0000	1.0000	0 00 00
01	0100	8.0000	100.00	0001	0000	0.9999	. 0 34 23
02	0200	3010	50.007	0002	0001	0.9998	1 08 45
03	0300	4770	33.343	0005	0002	9995	1 43 07
04	0400	6018	25.013	0008	0003	9992	2 17 28
0.05	0.0500	8.6986	20.017	1.0013	0.0005	0.9988	2 51 49
06	0599	7776	16.687	0018	0008	9982	3 26 08
07	0699	8444	14.309	0025	0011	9976	4 00 27
08	0798	9022	12.527	0032	0014	9968	4 34 44
09	0898	9531	11.141	0041	0018	9960	5 08 59
0.10	0.0997	8.9986	10.033	1.0050	0.0022	0.9950	5 43 12
11	1096	9.0396	9.1275	0061	0026	9940	6 17 24
12	1194	0771	8.3733	0072	0031	9928	6 51 33
13	1293	1115	7.7356	0085	0037	9916	7 25 39
14	1391	1433	7.1895	0098	0042	9903	7 59 43
0.15	0.1489	9.1729	6.7166	1.0113	0.0049	0.9889	8 33 44
16	1586	2004	6.3032	0128	0055	9873	9 07 42
17	1684	2263	5.9389	0145	0062	9857	9 41 37
18	1781	2506	5.6154	0162	0070	9840	10 15 29
19	1877	2736	5.3263	0181	0078	9822	10 49 17
0.20	0.1974	9.2953	5.0665	1.0201	0.0086	0.9803	11 23 01
21	2070	3159	4.8317	0221	0095	9783	11 56 41
22	2165	3355	6186	0243	0104	9763	12 30 17
23	2260	3542	4242	0266	0114	9741	13 03 48
24	2355	3720	2464	0289	0124	9719	13 37 15
0.25	0.2449	9.3890	4.0830	1.0314	0.0134	0.9695	14 10 37
26	2543	4053	3.9324	0340	0145	9671	14 43 55
27	2636	4210	7933	0367	0156	9646	15 17 07
28	2729	4360	6643	0395	0168	9620	15 50 14
29	2821	4505	5444	0423	0180	9594	16 23 16
0.30	0.2913	9.4644	3.4327	1.0453	0.0193	0.9566	16 56 12
31	3004	4778	3285	0484	0205	9538	17 29 02
32	3095	4907	2309	0516	0219	9509	18 01 46
33	3185	5031	1395	0549	0232	9479	18 34 25
34	3275	5152	3.0536	0584	0246	9449	19 06 57
0.35	0.3364	9.5268	2.9729	1.0619	0.0261	0.9417	19 39 22
36	3452	5381	8968	0655	0276	9385	20 11 42
37	3540	5490	8249	0692	0291	9353	20 43 54
38	3627	5596	7570	0731	0306	9319	21 16 00
39	3714	5698	6928	0770	0322	9285	21 47 58
0.40	0.3799	9.5797	2.6319	1.0811	0.0339	0.9250	22 19 50
41	3885	5894	5742	0852	0355	9215	22 51 34
42	3969	5987	5193	0895	0372	9178	23 23 11
43	4053	6078	4672	0939	0390	9142	23 54 41
44	4136	6166	4175	0984	0407	9104	24 26 02
0.45	0.4219	9.6252	2.3702	1.1030	0.0426	0.9066	24 57 16
46	4301	6336	3251	1077	0444	9028	25 28 22
47	4382	6417	2821	1125	0463	8989	25 59 21
48	4462	6496	2409	1174	0482	8949	26 30 11
49	4542	6573	2016	1225	0502	8909	27 00 52
0.50	0.4621	9.6648	2.1640	1.1276	0.0522	0.8868	27 31 26
x	sin gd x	log* sin gd x	csc gd x	sec gd x	log sec gd x	cos gd x	gd x

* Attach −10 to entries in this column.

TABLE 8.—EXPONENTIAL AND HYPERBOLIC FUNCTIONS.—(Continued)

x	e^x	log e^x	e^{-x}	sinh x	log* sinh x	csch x	gd x (rad.)
0.50	1.6487	0.2171	0.6065	0.5211	9.7169	1.9190	0.4804
51	6653	2215	6005	5324	7262	8783	4892
52	6820	2258	5945	5438	7354	8391	4980
53	6989	2302	5886	5552	7444	8013	5068
54	7160	2345	5827	5666	7533	7648	5155
0.55	1.7333	0.2389	0.5769	0.5782	9.7620	1.7297	0.5242
56	7507	2432	5712	5897	7707	6957	5328
57	7683	2475	5655	6014	7791	6629	5414
58	7860	2519	5599	6131	7875	6311	5500
59	8040	2562	5543	6248	7958	6004	5585
0.60	1.8221	0.2606	0.5488	0.6367	9.8039	1.5707	0.5669
61	8404	2649	5434	6485	8119	5419	5753
62	8589	2693	5379	6605	8199	5140	5837
63	8776	2736	5326	6725	8277	4870	5920
64	8965	2779	5273	6846	8354	4607	6003
0.65	1.9155	0.2823	0.5220	0.6967	9.8431	1.4352	0.6085
66	9348	2866	5169	7090	8506	4105	6167
67	9542	2910	5117	7213	8581	3865	6249
68	9739	2953	5066	7336	8655	3631	6329
69	1.9937	2997	5016	7461	8728	3404	6410
0.70	2.0138	0.3040	0.4966	0.7586	9.8800	1.3182	0.6490
71	0340	3083	4916	7712	8872	2967	6569
72	0544	3127	4868	7838	8942	2758	6648
73	0751	3170	4819	7966	9012	2554	6727
74	0959	3214	4771	8094	9082	2355	6805
0.75	2.1170	0.3257	0.4724	0.8223	9.9150	1.2161	0.6882
76	1383	3301	4677	8353	9218	1972	6959
77	1598	3344	4630	8484	9286	1787	7036
78	1815	3387	4584	8615	9353	1607	7112
79	2034	3431	4538	8748	9419	1431	7187
0.80	2.2255	0.3474	0.4493	0.8881	9.9485	1.1260	0.7262
81	2479	3518	4449	9015	9550	1092	7337
82	2705	3561	4404	9150	9614	0928	7411
83	2933	3605	4360	9286	9678	0768	7484
84	3164	3648	4317	9423	9742	0612	7557
0.85	2.3396	0.3692	0.4274	0.9561	9.9805	1.0459	0.7630
86	3632	3735	4232	9700	9868	0309	7702
87	3869	3778	4190	9840	9930	0163	7773
88	4109	3822	4148	0.9981	9.9992	1.0019	7844
89	4351	3865	4107	1.0122	10.0053	0.9879	7915
0.90	2.4596	0.3909	0.4066	1.0265	10.0114	0.9742	0.7985
91	4843	3952	4025	0409	0174	9607	8054
92	5093	3996	3985	0554	0234	9475	8123
93	5345	4039	3946	0700	0294	9346	8192
94	5600	4082	3906	0847	0353	9219	8260
0.95	2.5857	0.4126	0.3867	1.0995	10.0412	0.9095	0.8327
96	6117	4169	3829	1144	0470	8973	8395
97	6379	4213	3791	1294	0529	8854	8461
98	6645	4256	3753	1446	0586	8737	8527
0.99	6912	4300	3716	1598	0644	8622	8593
1.00	2.7183	0.4343	0.3679	1.1752	10.0701	0.8509	0.8658
x	e^x	log e^x	e^{-x}	tan gd x	log* tan gd x	cot gd x	gd x (rad.)

* Attach − 10 to entries in this column.

TABLE 8.—EXPONENTIAL AND HYPERBOLIC FUNCTIONS.—(*Continued*)

x	$\tanh x$	$\log^*$ $\tanh x$	$\coth x$	$\cosh x$	$\log$ $\cosh x$	$\mathrm{sech}\ x$	gd x		
							°	′	″
0.50	0.4621	9.6648	2.1640	1.1276	0.0522	0.8868	27	31	26
51	4699	6720	1279	1329	0542	8827	28	01	51
52	4777	6792	0934	1383	0562	8785	28	32	07
53	4854	6861	0602	1438	0583	8743	29	02	15
54	4930	6928	2.0284	1494	0605	8700	29	32	14
0.55	0.5005	9.6994	1.9979	1.1551	0.0626	0.8657	30	02	04
56	5080	7058	9686	1609	0648	8614	30	31	45
57	5154	7121	9404	1669	0670	8570	31	01	17
58	5227	7182	9133	1730	0693	8525	31	30	40
59	5299	7242	8872	1792	0716	8481	31	59	54
0.60	0.5370	9.7300	1.8620	1.1855	0.0739	0.8435	32	28	59
61	5441	7357	8378	1919	0762	8390	32	57	54
62	5511	7413	8145	1984	0786	8344	33	26	40
63	5581	7467	7919	2051	0810	8298	33	55	16
64	5649	7520	7702	2119	0835	8252	34	23	43
0.65	0.5717	9.7571	1.7493	1.2188	0.0859	0.8205	34	52	00
66	5784	7622	7290	2258	0884	8158	35	20	08
67	5850	7671	7095	2330	0910	8110	35	48	06
68	5915	7720	6906	2402	0935	8063	36	15	54
69	5980	7767	6723	2476	0961	8015	36	43	32
0.70	0.6044	9.7813	1.6546	1.2552	0.0987	0.7967	37	11	00
71	6107	7858	6375	2628	1013	7919	37	38	18
72	6169	7902	6210	2706	1040	7870	38	05	27
73	6231	7945	6050	2785	1067	7822	38	32	25
74	6291	7988	5895	2865	1094	7773	38	59	14
0.75	0.6351	9.8029	1.5744	1.2947	0.1122	0.7724	39	25	52
76	6411	8069	5599	3030	1149	7675	39	52	20
77	6469	8109	5458	3114	1177	7625	40	18	38
78	6527	8147	5321	3199	1206	7576	40	44	46
79	6584	8185	5188	3286	1234	7527	41	10	43
0.80	0.6640	9.8222	1.5059	1.3374	0.1263	0.7477	41	36	30
81	6696	8258	4935	3464	1292	7427	42	02	08
82	6751	8293	4813	3555	1321	7378	42	27	34
83	6805	8328	4696	3647	1350	7328	42	52	51
84	6858	8362	4581	3740	1380	7278	43	17	57
0.85	0.6911	9.8395	1.4470	1.3835	0.1410	0.7228	43	42	53
86	6963	8428	4362	3932	1440	7178	44	07	39
87	7014	8459	4258	4029	1470	7128	44	32	14
88	7064	8491	4156	4128	1501	7078	44	56	40
89	7114	8521	4057	4229	1532	7028	45	20	54
0.90	0.7163	9.8551	1.3961	1.4331	0.1563	0.6978	45	44	59
91	7211	8580	3867	4434	1594	6928	46	08	53
92	7259	8609	3776	4539	1625	6878	46	32	37
93	7306	8637	3687	4645	1657	6828	46	56	10
94	7352	8664	3601	4753	1689	6778	47	19	34
0.95	0.7398	9.8691	1.3517	1.4862	0.1721	0.6728	47	42	47
96	7443	8717	3436	4973	1753	6678	48	05	49
97	7487	8743	3356	5085	1785	6629	48	28	42
98	7531	8768	3279	5199	1818	6579	48	51	24
0.99	7574	8793	3204	5314	1851	6530	49	13	56
1.00	0.7616	9.8817	1.3130	1.5431	0.1884	0.6480	49	36	18
x	sin gd x	$\log^*$ sin gd x	csc gd x	sec gd x	$\log$ sec gd x	cos gd x	gd x		

* Attach −10 to entries in this column.

TABLE 8.—EXPONENTIAL AND HYPERBOLIC FUNCTIONS.—(*Continued*)

x	e^x	$\log e^x$	e^{-x}	$\sinh x$	$\log \sinh x$	$\operatorname{csch} x$	gd x (rad.)
1.00	2.7183	0.4343	0.3679	1.1752	0.0701	0.8509	0.8658
01	7456	4386	3642	1907	0758	8398	8722
02	7732	4430	3606	2063	0815	8290	8786
03	8011	4473	3570	2220	0871	8183	8850
04	8292	4517	3535	2379	0927	8078	8913
1.05	2.8577	0.4560	0.3499	1.2539	0.0982	0.7975	0.8976
06	8864	4604	3465	2700	1038	7874	9038
07	9154	4647	3430	2862	1093	7775	9099
08	9447	4690	3396	3025	1148	7677	9160
09	2.9743	4734	3362	3190	1203	7581	9221
1.10	3.0042	0.4777	0.3329	1.3356	0.1257	0.7487	0.9281
11	0344	4821	3296	3524	1311	7394	9341
12	0649	4864	3263	3693	1365	7303	9400
13	0957	4908	3230	3863	1419	7213	9459
14	1268	4951	3198	4035	1472	7125	9517
1.15	3.1582	0.4994	0.3166	1.4208	0.1525	0.7038	0.9575
16	1899	5038	3135	4382	1578	6953	9632
17	2220	5081	3104	4558	1631	6869	9689
18	2544	5125	3073	4735	1684	6786	9746
19	2871	5168	3042	4914	1736	6705	9801
1.20	3.3201	0.5212	0.3012	1.5095	0.1788	0.6625	0.9857
21	3535	5255	2982	5276	1840	6546	9912
22	3872	5298	2952	5460	1892	6468	0.9966
23	4212	5342	2923	5645	1944	6392	1.0021
24	4556	5385	2894	5831	1995	6317	0074
1.25	3.4903	0.5429	0.2865	1.6019	0.2046	0.6243	1.0127
26	5254	5472	2837	6209	2098	6170	0180
27	5609	5516	2808	6400	2148	6098	0232
28	5966	5559	2780	6593	2199	6027	0284
29	6328	5602	2753	6788	2250	5957	0336
1.30	3.6693	0.5646	0.2725	1.6984	0.2300	0.5888	1.0387
31	7062	5689	2698	7182	2351	5820	0437
32	7434	5733	2671	7381	2401	5753	0487
33	7810	5776	2645	7583	2451	5687	0537
34	8190	5820	2618	7786	2501	5622	0586
1.35	3.8574	0.5863	0.2592	1.7991	0.2551	0.5558	1.0635
36	8962	5906	2567	8198	2600	5495	0683
37	9354	5950	2541	8406	2650	5433	0731
38	3.9749	5993	2516	8617	2699	5372	0779
39	4.0149	6037	2491	8829	2748	5311	0826
1.40	4.0552	0.6080	0.2466	1.9043	0.2797	0.5251	1.0872
41	0960	6124	2441	9259	2846	5192	0919
42	1371	6167	2417	9477	2895	5134	0965
43	1787	6210	2393	9697	2944	5077	1010
44	2207	6254	2369	1.9919	2993	5020	1055
1.45	4.2631	0.6297	0.2346	2.0143	0.3041	0.4965	1.1100
46	3060	6341	2322	0369	3090	4910	1144
47	3492	6384	2299	0597	3138	4855	1188
48	3929	6428	2276	0827	3186	4802	1231
49	4371	6471	2254	1059	3234	4749	1275
1.50	4.4817	0.6514	0.2231	2.1293	0.3282	0.4696	1.1317
x	e^x	$\log e^x$	e^{-x}	tan gd x	log tan gd x	cot gd x	gd x (rad.)

TABLE 8.—EXPONENTIAL AND HYPERBOLIC FUNCTIONS.—(*Continued*)

x	tanh x	log* tanh x	coth x	cosh x	log cosh x	sech x	gd x		
							°	′	″
1.00	0.7616	9.8817	1.3130	1.5431	0.1884	0.6480	49	36	18
01	7658	8841	3059	5549	1917	6431	49	58	29
02	7699	8864	2989	5669	1950	6382	50	20	31
03	7739	8887	2921	5790	1984	6333	50	42	22
04	7779	8909	2855	5913	2018	6284	51	04	03
1.05	0.7818	9.8931	1.2791	1.6038	0.2051	0.6235	51	25	35
06	7857	8952	2728	6164	2086	6187	51	46	56
07	7895	8973	2667	6292	2120	6138	52	08	07
08	7932	8994	2607	6421	2154	6090	52	29	08
09	7969	9014	2549	6552	2189	6041	52	49	59
1.10	0.8005	9.9034	1.2492	1.6685	0.2223	0.5993	53	10	40
11	8041	9053	2437	6820	2258	5945	53	31	11
12	8076	9072	2383	6956	2293	5898	53	51	33
13	8110	9090	2330	7093	2328	5850	54	11	44
14	8144	9108	2279	7233	2364	5803	54	31	46
1.15	0.8178	9.9126	1.2229	1.7374	0.2399	0.5756	54	51	38
16	8210	9144	2180	7517	2435	5709	55	11	20
17	8243	9161	2132	7662	2470	5662	55	30	53
18	8275	9177	2085	7808	2506	5615	55	50	16
19	8306	9194	2040	7957	2542	5569	56	09	30
1.20	0.8337	9.9210	1.1995	1.8107	0.2578	0.5523	56	28	34
21	8367	9226	1952	8258	2615	5477	56	47	28
22	8397	9241	1910	8412	2651	5431	57	06	13
23	8426	9256	1868	8568	2688	5386	57	24	49
24	8455	9271	1828	8725	2724	5340	57	43	15
1.25	0.8483	9.9285	1.1789	1.8884	0.2761	0.5295	58	01	32
26	8511	9300	1750	9045	2798	5251	58	19	39
27	8538	9314	1712	9208	2835	5206	58	37	38
28	8565	9327	1676	9373	2872	5162	58	55	27
29	8591	9341	1640	9540	2909	5118	59	13	07
1.30	0.8617	9.9354	1.1605	1.9709	0.2947	0.5074	59	30	38
31	8643	9367	1570	1.9880	2984	5030	59	48	00
32	8668	9379	1537	2.0053	3022	4987	60	05	13
33	8692	9391	1504	0228	3059	4944	60	22	17
34	8717	9404	1472	0404	3097	4901	60	39	13
1.35	0.8741	9.9415	1.1441	2.0583	0.3135	0.4858	60	55	59
36	8764	9427	1410	0764	3173	4816	61	12	37
37	8787	9438	1381	0947	3211	4774	61	29	06
38	8810	9450	1351	1132	3249	4732	61	45	26
39	8832	9460	1323	1320	3288	4690	62	01	38
1.40	0.8854	9.9471	1.1295	2.1509	0.3326	0.4649	62	17	41
41	8875	9482	1268	1700	3365	4608	62	33	36
42	8896	9492	1241	1894	3403	4567	62	49	22
43	8917	9502	1215	2090	3442	4527	63	05	00
44	8937	9512	1189	2288	3481	4487	63	20	30
1.45	0.8957	9.9522	1.1165	2.2488	0.3520	0.4447	63	35	51
46	8977	9531	1140	2691	3559	4407	63	51	04
47	8996	9540	1116	2896	3598	4368	64	06	09
48	9015	9550	1093	3103	3637	4328	64	21	06
49	9033	9558	1070	3312	3676	4290	64	35	55
1.50	0.9051	9.9567	1.1048	2.3524	0.3715	0.4251	64	50	36
x	sin gd x	log* sin gd x	csc gd x	sec gd x	log sec gd x	cos gd x	gd x		

* Attach −10 to entries in this column.

TABLE 8.—EXPONENTIAL AND HYPERBOLIC FUNCTIONS.—(*Continued*)

x	e^x	$\log e^x$	e^{-x}	$\sinh x$	$\dfrac{\log}{\sinh x}$	$\operatorname{csch} x$	gd x (rad.)
1.50	4.4817	0.6514	0.2231	2.1293	0.3282	0.4696	1.1317
51	5267	6558	2209	1529	3330	4645	1360
52	5722	6601	2187	1768	3378	4594	1402
53	6182	6645	2165	2008	3426	4544	1443
54	6646	6688	2144	2251	3474	4494	1484
1.55	4.7115	0.6732	0.2122	2.2496	0.3521	0.4445	1.1525
56	7588	6775	2101	2743	3569	4397	1566
57	8066	6818	2080	2993	3616	4349	1606
58	8550	6862	2060	3245	3663	4302	1645
59	9037	6905	2039	3499	3711	4255	1685
1.60	4.9530	0.6949	0.2019	2.3756	0.3758	0.4209	1.1724
61	5.0028	6992	1999	4015	3805	4164	1762
62	0531	7036	1979	4276	3852	4119	1800
63	1039	7079	1959	4540	3899	4075	1838
64	1552	7122	1940	4806	3946	4031	1876
1.65	5.2070	0.7166	0.1920	2.5075	0.3992	0.3988	1.1913
66	2593	7209	1901	5346	4039	3945	1950
67	3122	7253	1882	5620	4086	3903	1987
68	3656	7296	1864	5896	4132	3862	2023
69	4195	7340	1845	6175	4179	3820	2059
1.70	5.4739	0.7383	0.1827	2.6456	0.4225	0.3780	1.2094
71	5290	7426	1809	6740	4272	3740	2129
72	5845	7470	1791	7027	4318	3700	2164
73	6407	7513	1773	7317	4364	3661	2199
74	6973	7557	1755	7609	4411	3622	2233
1.75	5.7546	0.7600	0.1738	2.7904	0.4457	0.3584	1.2267
76	8124	7644	1720	8202	4503	3546	2300
77	8709	7687	1703	8503	4549	3508	2334
78	9299	7730	1686	8806	4595	3471	2367
79	5.9895	7774	1670	9112	4641	3435	2399
1.80	6.0496	0.7817	0.1653	2.9422	0.4687	0.3399	1.2432
81	1104	7861	1637	2.9734	4733	3363	2464
82	1719	7904	1620	3.0049	4778	3328	2495
83	2339	7948	1604	0367	4824	3293	2527
84	2965	7991	1588	0689	4870	3258	2558
1.85	6.3598	0.8034	0.1572	3.1013	0.4915	0.3224	1.2589
86	4237	8078	1557	1340	4961	3191	2619
87	4883	8121	1541	1671	5007	3157	2650
88	5535	8165	1526	2005	5052	3125	2680
89	6194	8208	1511	2341	5098	3092	2709
1.90	6.6859	0.8252	0.1496	3.2682	0.5143	0.3060	1.2739
91	7531	8295	1481	3025	5188	3028	2768
92	8210	8338	1466	3372	5234	2997	2797
93	8895	8382	1451	3722	5279	2965	2825
94	6.9588	8425	1437	4075	5324	2935	2853
1.95	7.0287	0.8469	0.1423	3.4432	0.5370	0.2904	1.2881
96	0993	8512	1409	4792	5415	2874	2909
97	1707	8556	1395	5156	5460	2844	2937
98	2427	8599	1381	5523	5505	2815	2964
1.99	3155	8642	1367	5894	5550	2786	2991
2.00	7.3891	0.8686	0.1353	3.6269	0.5595	0.2757	1.3018
x	e^x	$\log e^x$	e^{-x}	tan gd x	log tan gd x	cot gd x	gd x (rad.)

TABLE 8.—EXPONENTIAL AND HYPERBOLIC FUNCTIONS.—(*Continued*)

x	tanh x	$\dfrac{\log^*}{\tanh x}$	coth x	cosh x	$\dfrac{\log}{\cosh x}$	sech x	gd x
							° ′ ″
1.50	0.9051	9.9567	1.1048	2.3524	0.3715	0.4251	64 50 36
51	9069	9576	1026	3738	3754	4213	65 05 09
52	9087	9584	1005	3955	3794	4175	65 19 34
53	9104	9592	0984	4174	3833	4137	65 33 51
54	9121	9601	0963	4395	3873	4099	65 48 00
1.55	0.9138	9.9608	1.0943	2.4619	0.3913	0.4062	66 02 02
56	9154	9616	0924	4845	3952	4025	66 15 56
57	9170	9624	0905	5073	3992	3988	66 29 42
58	9186	9631	0886	5305	4032	3952	66 43 21
59	9201	9639	0868	5538	4072	3916	66 56 53
1.60	0.9217	9.9646	1.0850	2.5775	0.4112	0.3880	67 10 16
61	9232	9653	0832	6013	4152	3844	67 23 33
62	9246	9660	0815	6255	4192	3809	67 36 42
63	9261	9666	0798	6499	4232	3774	67 49 44
64	9275	9673	0782	6746	4273	3739	68 02 39
1.65	0.9289	9.9679	1.0766	2.6995	0.4313	0.3704	68 15 27
66	9302	9686	0750	7247	4353	3670	68 28 07
67	9316	9692	0735	7502	4394	3636	68 40 41
68	9329	9698	0720	7760	4434	3602	68 53 07
69	9341	9704	0705	8020	4475	3569	69 05 27
1.70	0.9354	9.9710	1.0691	2.8283	0.4515	0.3536	69 17 40
71	9366	9716	0676	8549	4556	3503	69 29 45
72	9379	9721	0663	8818	4597	3470	69 41 45
73	9391	9727	0649	9090	4637	3438	69 53 38
74	9402	9732	0636	9364	4678	3405	70 05 23
1.75	0.9414	9.9738	1.0623	2.9642	0.4719	0.3374	70 17 02
76	9425	9743	0610	2.9922	4760	3342	70 28 34
77	9436	9748	0598	3.0206	4801	3311	70 40 01
78	9447	9753	0585	0492	4842	3279	70 51 20
79	9458	9758	0574	0782	4883	3249	71 02 33
1.80	0.9468	9.9763	1.0562	3.1075	0.4924	0.3218	71 13 40
81	9478	9767	0550	1371	4965	3188	71 24 41
82	9488	9772	0539	1669	5006	3158	71 35 35
83	9498	9776	0528	1972	5048	3128	71 46 24
84	9508	9781	0518	2277	5089	3098	71 57 06
1.85	0.9517	9.9785	1.0507	3.2585	0.5130	0.3069	72 07 42
86	9527	9789	0497	2897	5172	3040	72 18 12
87	9536	9794	0487	3212	5213	3011	72 28 36
88	9545	9798	0477	3530	5254	2982	72 38 54
89	9554	9802	0467	3852	5296	2954	72 49 06
1.90	0.9562	9.9806	1.0458	3.4177	0.5337	0.2926	72 59 13
91	9571	9810	0448	4506	5379	2898	73 09 13
92	9579	9813	0439	4838	5421	2870	73 19 08
93	9587	9817	0430	5173	5462	2843	73 28 57
94	9595	9821	0422	5512	5504	2816	73 38 41
1.95	0.9603	9.9824	1.0413	3.5855	0.5545	0.2789	73 48 19
96	9611	9828	0405	6201	5587	2762	73 57 52
97	9618	9831	0397	6551	5629	2736	74 07 19
98	9626	9834	0389	6904	5671	2710	74 16 40
1.99	9633	9838	0381	7261	5713	2684	74 25 56
2.00	0.9640	9.9841	1.0373	3.7622	0.5754	0.2658	74 35 07
x	$\dfrac{\sin}{\text{gd } x}$	$\dfrac{\log^*}{\sin \text{ gd } x}$	$\dfrac{\csc}{\text{gd } x}$	$\dfrac{\sec}{\text{gd } x}$	$\dfrac{\log}{\sec \text{ gd } x}$	$\dfrac{\cos}{\text{gd } x}$	gd x

* Attach −10 to entries in this column.

TABLE 8.—EXPONENTIAL AND HYPERBOLIC FUNCTIONS.—(Continued)

x	e^x	log e^x	e^{-x}	sinh x	log sinh x	csch x	gd x (rad.)
2.00	7.3891	0.8686	0.1353	3.6269	0.5595	0.2757	1.3018
01	4633	8729	1340	6647	5640	2729	3044
02	5383	8773	1327	7028	5685	2701	3070
03	6141	8816	1313	7414	5730	2673	3096
04	6906	8860	1300	7803	5775	2645	3122
2.05	7.7679	0.8903	0.1287	3.8196	0.5820	0.2618	1.3147
06	8460	8946	1275	8593	5865	2591	3173
07	7.9248	8990	1262	8993	5910	2565	3198
08	8.0045	9033	1249	9398	5955	2538	3222
09	0849	9077	1237	3.9806	6000	2512	3247
2.10	8.1662	0.9120	0.1225	4.0219	0.6044	0.2486	1.3271
11	2482	9164	1212	0635	6089	2461	3295
12	3311	9207	1200	1056	6134	2436	3319
13	4149	9250	1188	1480	6178	2411	3342
14	4994	9294	1177	1909	6223	2386	3366
2.15	8.5849	0.9337	0.1165	4.2342	0.6268	0.2362	1.3389
16	6711	9381	1153	2779	6312	2338	3412
17	7583	9424	1142	3221	6357	2314	3434
18	8463	9468	1130	3666	6401	2290	3457
19	8.9352	9511	1119	4116	6446	2267	3479
2.20	9.0250	0.9554	0.1108	4.4571	0.6491	0.2244	1.3501
21	1157	9598	1097	5030	6535	2221	3523
22	2073	9641	1086	5494	6580	2198	3544
23	2999	9685	1075	5962	6624	2176	3566
24	3933	9728	1065	6434	6668	2154	3587
2.25	9.4877	0.9772	0.1054	4.6912	0.6713	0.2132	1.3608
26	5831	9815	1044	7394	6757	2110	3628
27	6794	9858	1033	7880	6802	2089	3649
28	7767	9902	1023	8372	6846	2067	3669
29	8749	9945	1013	8868	6890	2046	3690
2.30	9.9742	0.9989	0.1003	4.9370	0.6935	0.2026	1.3709
31	10.074	1.0032	0993	4.9876	6979	2005	3729
32	176	0076	0983	5.0387	7023	1985	3749
33	278	0119	0973	0903	7067	1965	3768
34	381	0162	0963	1425	7112	1945	3787
2.35	10.486	1.0206	0.0954	5.1951	0.7156	0.1925	1.3806
36	591	0249	0944	2483	7200	1905	3825
37	697	0293	0935	3020	7244	1886	3844
38	805	0336	0926	3562	7289	1867	3862
39	10.913	0380	0916	4109	7333	1848	3880
2.40	11.023	1.0423	0.0907	5.4662	0.7377	0.1829	1.3899
41	134	0466	0898	5221	7421	1811	3916
42	246	0510	0889	5785	7465	1793	3934
43	359	0553	0880	6354	7509	1775	3952
44	473	0597	0872	6929	7553	1757	3969
2.45	11.588	1.0640	0.0863	5.7510	0.7597	0.1739	1.3986
46	705	0684	0854	8097	7642	1721	4003
47	822	0727	0846	8689	7686	1704	4020
48	11.941	0771	0837	9288	7730	1687	4037
49	12.061	0814	0.829	5.9892	7774	1670	4054
2.50	12.182	1.0857	0.0821	6.0502	0.7818	0.1653	1.4070

x	e^x	log e^x	e^{-x}	tan gd x	log tan gd x	cot gd x	gd x (rad.)

TABLE 8.—EXPONENTIAL AND HYPERBOLIC FUNCTIONS.—(*Continued*)

x	tanh x	log* tanh x	coth x	cosh x	log cosh x	sech x	gd x
							° ′ ″
2.00	0.9640	9.9841	1.0373	3.7622	0.5754	0.2658	74 35 07
01	9647	9844	0366	7987	5796	2633	74 44 13
02	9654	9847	0358	8355	5838	2607	74 53 13
03	9661	9850	0351	8727	5880	2582	75 02 09
04	9667	9853	0344	9103	5922	2557	75 10 59
2.05	0.9674	9.9856	1.0337	3.9483	0.5964	0.2533	75 19 44
06	9680	9859	0330	3.9867	6006	2508	75 28 23
07	9687	9862	0324	4.0255	6048	2484	75 36 58
08	9693	9864	0317	0647	6090	2460	75 45 28
09	9699	9867	0311	1043	6132	2436	75 53 53
2.10	0.9705	9.9870	1.0304	4.1443	0.6175	0.2413	76 02 13
11	9710	9872	0298	1847	6217	2390	76 10 29
12	9716	9875	0292	2256	6259	2367	76 18 39
13	9721	9877	0286	2669	6301	2344	76 26 45
14	9727	9880	0281	3085	6343	2321	76 34 46
2.15	0.9732	9.9882	1.0275	4.3507	0.6386	0.2298	76 42 42
16	9737	9884	0270	3932	6428	2276	76 50 34
17	9743	9887	0264	4362	6470	2254	76 58 21
18	9748	9889	0259	4797	6512	2232	77 06 04
19	9753	9891	0254	5236	6555	2211	77 13 42
2.20	0.9757	9.9893	1.0249	4.5679	0.6597	0.2189	77 21 16
21	9762	9895	0244	6127	6640	2168	77 28 45
22	9767	9898	0239	6580	6682	2147	77 36 10
23	9771	9900	0234	7037	6724	2126	77 43 31
24	9776	9902	0229	7499	6767	2105	77 50 48
2.25	0.9780	9.9904	1.0225	4.7966	0.6809	0.2085	77 58 00
26	9785	9905	0220	8437	6852	2064	78 05 08
27	9789	9907	0216	8914	6894	2044	78 12 11
28	9793	9909	0211	9395	6937	2025	78 19 11
29	9797	9911	0207	4.9881	6979	2005	78 26 07
2.30	0.9801	9.9913	1.0203	5.0372	0.7022	0.1985	78 32 58
31	9805	9914	0199	0868	7064	1966	78 39 45
32	9809	9916	0195	1370	7107	1947	78 46 29
33	9812	9918	0191	1876	7150	1928	78 53 09
34	9816	9919	0187	2388	7192	1909	78 59 44
2.35	0.9820	9.9921	1.0184	5.2905	0.7235	0.1890	79 06 16
36	9823	9923	0180	3427	7278	1872	79 12 44
37	9827	9924	0176	3954	7320	1853	79 19 08
38	9830	9926	0173	4487	7363	1835	79 25 29
39	9833	9927	0169	5026	7406	1817	79 31 45
2.40	0.9837	9.9929	1.0166	5.5569	0.7448	0.1800	79 37 58
41	9840	9930	0163	6119	7491	1782	79 44 08
42	9843	9931	0159	6674	7534	1765	79 50 13
43	9846	9933	0156	7235	7577	1747	79 56 16
44	9849	9934	0153	7801	7619	1730	80 02 14
2.45	0.9852	9.9935	1.0150	5.8373	0.7662	0.1713	80 08 09
46	9855	9937	0147	8951	7705	1696	80 14 01
47	9858	9938	0144	5.9535	7748	1680	80 19 49
48	9861	9939	0141	6.0125	7791	1663	80 25 34
49	9863	9940	0138	0721	7833	1647	80 31 15
2.50	0.9866	9.9941	1.0136	6.1323	0.7876	0.1631	80 36 53
x	sin gd x	log* sin gd x	csc gd x	sec gd x	log sec gd x	cos gd x	gd x

* Attach −10 to entries in this column.

TABLE 8.—EXPONENTIAL AND HYPERBOLIC FUNCTIONS.—(*Continued*)

x	e^x	$\log e^x$	e^{-x}	$\sinh x$	$\log \sinh x$	$\operatorname{csch} x$	gd x (rad.)
2.50	12.182	1.0857	0.0821	6.0502	0.7818	0.1653	1.4070
51	305	0901	0813	1118	7862	1636	4086
52	429	0944	0805	1741	7906	1620	4102
53	554	0988	0797	2369	7950	1603	4118
54	680	1031	0789	3004	7994	1587	4134
2.55	12.807	1.1075	0.0781	6.3645	0.8038	0.1571	1.4149
56	12.936	1118	0773	4293	8082	1555	4165
57	13.066	1161	0765	4946	8126	1540	4180
58	197	1205	0758	5607	8169	1524	4195
59	330	1248	0750	6274	8213	1509	4210
2.60	13.464	1.1292	0.0743	6.6947	0.8257	0.1494	1.4225
61	599	1335	0735	7628	8301	1479	4240
62	736	1379	0728	8315	8345	1464	4254
63	13.874	1422	0721	9008	8389	1449	4269
64	14.013	1465	0714	6.9709	8433	1435	4283
2.65	14.154	1.1509	0.0707	7.0417	0.8477	0.1420	1.4297
66	296	1552	0699	1132	8521	1406	4311
67	440	1596	0693	1854	8564	1392	4325
68	585	1639	0686	2583	8608	1378	4339
69	732	1683	0679	3319	8652	1364	4352
2.70	14.880	1.1726	0.0672	7.4063	0.8696	0.1350	1.4366
71	15.029	1769	0665	4814	8740	1337	4379
72	180	1813	0659	5572	8784	1323	4392
73	333	1856	0652	6338	8827	1310	4405
74	487	1900	0646	7112	8871	1297	4418
2.75	15.643	1.1943	0.0639	7.7894	0.8915	0.1284	1.4431
76	800	1987	0633	8683	8959	1271	4444
77	15.959	2030	0627	7.9480	9003	1258	4456
78	16.119	2073	0620	8.0285	9046	1246	4469
79	281	2117	0614	1098	9090	1233	4481
2.80	16.445	1.2160	0.0608	8.1919	0.9134	0.1221	1.4493
81	610	2204	0602	2749	9178	1208	4505
82	777	2247	0596	3586	9221	1196	4517
83	16.945	2291	0590	4432	9265	1184	4529
84	17.116	2334	0584	5287	9309	1173	4541
2.85	17.288	1.2377	0.0578	8.6150	0.9353	0.1161	1.4552
86	462	2421	0573	7021	9396˙	1149	4564
87	637	2464	0567	7902	9440	1138	4575
88	814	2508	0561	8791	9484	1126	4586
89	17.993	2551	0556	8.9689	9527	1115	4598
2.90	18.174	1.2595	0.0550	9.0596	0.9571	0.1104	1.4609
91	357	2638	0545	1512	9615	1093	4620
92	541	2681	0539	2437	9658	1082	4630
93	728	2725	0534	3371	9702	1071	4641
94	18.916	2768	0529	4315	9746	1060	4652
2.95	19.106	1.2812	0.0523	9.5268	0.9789	0.1050	1.4662
96	298	2855	0518	6231	9833	1039	4673
97	492	2899	0513	7203	9877	1029	4683
98	688	2942	0508	8185	9920	1018	4693
2.99	19.886	2985	0503	9.9177	0.9964	1008	4703
3.00	20.086	1.3029	0.0498	10.0179	1.0008	0.0998	1.4713
x	e^x	$\log e^x$	e^{-x}	tan gd x	log tan gd x	cot gd x	gd x (rad.)

TABLE 8.—EXPONENTIAL AND HYPERBOLIC FUNCTIONS.—(*Continued*)

x	tanh x	log* tanh x	coth x	cosh x	log cosh x	sech x	gd x
							° ′ ″
2.50	0.9866	9.9941	1.0136	6.1323	0.7876	0.1631	80 36 53
51	9869	9943	0133	1931	7919	1615	80 42 28
52	9871	9944	0130	2545	7962	1599	80 47 59
53	9874	9945	0128	3166	8005	1583	80 53 28
54	9876	9946	0125	3793	8048	1568	80 58 52
2.55	0.9879	9.9947	1.0123	6.4426	0.8091	0.1552	81 04 14
56	9881	9948	0120	5066	8134	1537	81 09 33
57	9884	9949	0118	5712	8176	1522	81 14 48
58	9886	9950	0115	6365	8219	1507	81 20 01
59	9888	9951	0113	7024	8262	1492	81 25 10
2.60	0.9890	9.9952	1.0111	6.7690	0.8305	0.1477	81 30 16
61	9892	9953	0109	8363	8348	1463	81 35 19
62	9895	9954	0107	9043	8391	1448	81 40 20
63	9897	9955	0104	6.9729	8434	1434	81 45 17
64	9899	9956	0102	7.0423	8477	1420	81 50 11
2.65	0.9901	9.9957	1.0100	7.1123	0.8520	0.1406	81 55 03
66	9903	9958	0098	1831	8563	1392	81 59 51
67	9905	9958	0096	2546	8606	1378	82 04 37
68	9906	9959	0094	3268	8649	1365	82 09 20
69	9908	9960	0093	3998	8692	1351	82 14 00
2.70	0.9910	9.9961	1.0091	7.4735	0.8735	0.1338	82 18 37
71	9912	9962	0089	5479	8778	1325	82 23 12
72	9914	9962	0087	6231	8821	1312	82 27 44
73	9915	9963	0085	6991	8864	1299	82 32 13
74	9917	9964	0084	7758	8907	1286	82 36 40
2.75	0.9919	9.9965	1.0082	7.8533	0.8951	0.1273	82 41 04
76	9920	9965	0080	7.9316	8994	1261	82 45 25
77	9922	9966	0079	8.0106	9037	1248	82 49 44
78	9923	9967	0077	0905	9080	1236	82 54 00
79	9925	9967	0076	1712	9123	1224	82 58 14
2.80	0.9926	9.9968	1.0074	8.2527	0.9166	0.1212	83 02 25
81	9928	9969	0073	3351	9209	1200	83 06 34
82	9929	9969	0071	4182	9252	1188	83 10 40
83	9931	9970	0070	5022	9295	1176	83 14 44
84	9932	9970	0069	5871	9338	1165	83 18 45
2.85	0.9933	9.9971	1.0067	8.6728	0.9382	0.1153	83 22 44
86	9935	9972	0066	7594	9425	1142	83 26 41
87	9936	9972	0065	8469	9468	1130	83 30 35
88	9937	9973	0063	8.9352	9511	1119	83 34 27
89	9938	9973	0062	9.0244	9554	1108	83 38 17
2.90	0.9940	9.9974	1.0061	9.1146	0.9597	0.1097	83 42 04
91	9941	9974	0060	2056	9641	1086	83 45 49
92	9942	9975	0058	2976	9684	1076	83 49 32
93	9943	9975	0057	3905	9727	1065	83 53 13
94	9944	9976	0056	4844	9770	1054	83 56 52
2.95	0.9945	9.9976	1.0055	9.5791	0.9813	0.1044	84 00 28
96	9946	9977	0054	6749	9856	1034	84 04 02
97	9947	9977	0053	7716	9900	1023	84 07 34
98	9949	9978	0052	8693	9943	1013	84 11 04
2.99	9950	9978	0051	9.9680	0.9986	1003	84 14 32
3.00	0.9951	9.9978	1.0050	10.0677	1.0029	0.0993	84 17 58
x	sin gd x	log* sin gd x	csc gd x	sec gd x	log sec gd x	cos gd x	gd x

* Attach −10 to entries in this column.

TABLE 8.—EXPONENTIAL AND HYPERBOLIC FUNCTIONS.—(*Continued*)

x	e^x	log e^x	e^{-x}	sinh x	log sinh x	csch x	gd x (rad.)
3.00	20.086	1.3029	0.0498	10.018	1.0008	0.0998	1.4713
01	287	3072	0493	119	0051	0988	4723
02	491	3116	0488	221	0095	0978	4733
03	697	3159	0483	324	0139	0969	4742
04	20.905	3203	0478	429	0182	0959	4752
3.05	21.115	1.3246	0.0474	10.534	1.0226	0.0949	1.4761
06	328	3289	0469	640	0270	0940	4771
07	542	3333	0464	748	0313	0930	4780
08	758	3376	0460	856	0357	0921	4789
09	21.977	3420	0455	10.966	0400	0912	4799
3.10	22.198	1.3463	0.0450	11.076	1.0444	0.0903	1.4808
11	421	3507	0446	188	0488	0894	4817
12	646	3550	0442	301	0531	0885	4825
13	22.874	3593	0437	415	0575	0876	4834
14	23.104	3637	0433	530	0618	0867	4843
3.15	23.336	1.3680	0.0429	11.647	1.0662	0.0859	1.4851
16	571	3724	0424	764	0706	0850	4860
17	23.807	3767	0420	11.883	0749	0842	4868
18	24.047	3811	0416	12.003	0793	0833	4877
19	288	3854	0412	124	0836	0825	4885
3.20	24.533	1.3897	0.0408	12.246	1.0880	0.0817	1.4893
21	24.779	3941	0404	369	0923	0808	4901
22	25.028	3984	0400	494	0967	0800	4909
23	280	4028	0396	620	1011	0792	4917
24	534	4071	0392	747	1054	0784	4925
3.25	25.790	1.4115	0.0388	12.876	1.1098	0.0777	1.4933
26	26.050	4158	0384	13.006	1141	0769	4941
27	311	4201	0380	137	1185	0761	4948
28	576	4245	0376	269	1228	0754	4956
29	26.843	4288	0373	403	1272	0746	4963
3.30	27.113	1.4332	0.0369	13.538	1.1316	0.0739	1.4971
31	385	4375	0365	674	1359	0731	4978
32	660	4419	0362	812	1403	0724	4985
33	27.938	4462	0358	13.951	1446	0717	4992
34	28.219	4505	0354	14.092	1490	0710	5000
3.35	28.503	1.4549	0.0351	14.234	1.1533	0.0703	1.5007
36	28.789	4592	0347	377	1577	0696	5014
37	29.079	4636	0344	522	1620	0689	5020
38	371	4679	0340	668	1664	0682	5027
39	666	4723	0337	816	1707	0675	5034
3.40	29.964	1.4766	0.0334	14.965	1.1751	0.0668	1.5041
41	30.265	4809	0330	15.116	1794	0662	5047
42	569	4853	0327	268	1838	0655	5054
43	30.877	4896	0324	422	1881	0648	5060
44	31.187	4940	0321	577	1925	0642	5067
3.45	31.500	1.4983	0.0317	15.734	1.1968	0.0636	1.5073
46	31.817	5027	0314	15.893	2012	0629	5080
47	32.137	5070	0311	16.053	2056	0623	5086
48	460	5113	0308	214	2099	0617	5092
49	32.786	5157	0305	378	2143	0611	5098
3.50	33.115	1.5200	0.0302	16.543	1.2186	0.0604	1.5104
x	e^x	log e^x	e^{-x}	tan gd x	log tan gd x	cot gd x	gd x (rad.)

TABLE 8.—EXPONENTIAL AND HYPERBOLIC FUNCTIONS.—(*Continued*)

x	tanh x	log* tanh x	coth x	cosh x	log cosh x	sech x	gd x
							° ′ ″
3.00	0.9951	9.9978	1.0050	10.068	1.0029	0.0993	84 17 58
01	9952	9979	0049	168	0073	0983	84 21 22
02	9952	9979	0048	270	0116	0974	84 24 44
03	9953	9980	0047	373	0159	0964	84 28 04
04	9954	9980	0046	477	0202	0955	84 31 22
3.05	0.9955	9.9981	1.0045	10.581	1.0245	0.0945	84 34 38
06	9956	9981	0044	687	0289	0936	84 37 52
07	9957	9981	0043	794	0332	0926	84 41 04
08	9958	9982	0042	10.902	0375	0917	84 44 14
09	9959	9982	0041	11.011	0418	0908	84 47 22
3.10	0.9959	9.9982	1.0041	11.122	1.0462	0.0899	84 50 28
11	9960	9983	0040	233	0505	0890	84 53 33
12	9961	9983	0039	345	0548	0881	84 56 36
13	9962	9983	0038	459	0591	0873	84 59 37
14	9963	9984	0038	574	0635	0864	85 02 36
3.15	0.9963	9.9984	1.0037	11.689	1.0678	0.0855	85 05 33
16	9964	9984	0036	807	0721	0847	85 08 29
17	9965	9985	0035	11.925	0764	0839	85 11 22
18	9965	9985	0035	12.044	0808	0830	85 14 15
19	9966	9985	0034	165	0851	0822	85 17 05
3.20	0.9967	9.9986	1.0033	12.287	1.0894	0.0814	85 19 54
21	9967	9986	0033	410	0938	0806	85 22 41
22	9968	9986	0032	534	0981	0798	85 25 26
23	9969	9986	0031	660	1024	0790	85 28 10
24	9969	9987	0031	786	1067	0782	85 30 52
3.25	0.9970	9.9987	1.0030	12.915	1.1111	0.0774	85 33 32
26	9971	9987	0030	13.044	1154	0767	85 36 11
27	9971	9987	0029	175	1197	0759	85 38 49
28	9972	9988	0028	307	1241	0752	85 41 25
29	9972	9988	0028	440	1284	0744	85 43 59
3.30	0.9973	9.9988	1.0027	13.575	1.1327	0.0737	85 46 32
31	9973	9988	0027	711	1371	0729	85 49 03
32	9974	9989	0026	848	1414	0722	85 51 32
33	9974	9989	0026	13.987	1457	0715	85 54 01
34	9975	9989	0025	14.127	1501	0708	85 56 27
3.35	0.9975	9.9989	1.0025	14.269	1.1544	0.0701	85 58 53
36	9976	9990	0024	412	1587	0694	86 01 16
37	9976	9990	0024	556	1631	0687	86 03 39
38	9977	9990	0023	702	1674	0680	86 06 00
39	9977	9990	0023	850	1717	0673	86 08 19
3.40	0.9978	9.9990	1.0022	14.999	1.1761	0.0667	86 10 38
41	9978	9991	0022	15.149	1804	0660	86 12 54
42	9979	9991	0021	301	1847	0654	86 15 10
43	9979	9991	0021	455	1891	0647	86 17 24
44	9979	9991	0021	610	1934	0641	86 19 37
3.45	0.9980	9.9991	1.0020	15.766	1.1977	0.0634	86 21 48
46	9980	9991	0020	15.924	2021	0628	86 23 59
47	9981	9992	0019	16.084	2064	0622	86 26 07
48	9981	9992	0019	245	2107	0616	86 28 15
49	9981	9992	0019	408	2151	0609	86 30 21
3.50	0.9982	9.9992	1.0018	16.573	1.2194	0.0603	86 32 26
x	sin gd x	log* sin gd x	csc gd x	sec gd x	log sec gd x	cos gd x	gd x

* Attach −10 to entries in this column.

TABLE 8.—EXPONENTIAL AND HYPERBOLIC FUNCTIONS.—(*Continued*)

x	e^x	log e^x	e^{-x}	sinh x	log sinh x	csch x	gd x (rad.)
3.50	33.115	1.5200	0.03020	16.543	1.2186	0.06045	1.5104
55	34.813	5417	02872	17.392	2404	05750	5134
60	36.598	5635	02732	18.285	2621	05469	5162
65	38.475	5852	02599	19.224	2839	05202	5188
70	40.447	6069	02472	20.211	3056	04948	5214
3.75	42.521	1.6286	0.02352	21.249	1.3273	0.04706	1.5238
80	44.701	6503	02237	22.339	3491	04476	5261
85	46.993	6720	02128	23.486	3708	04258	5282
90	49.402	6937	02024	24.691	3925	04050	5303
3.95	51.935	7155	01925	25.958	4143	03852	5323
4.00	54.598	1.7372	0.01832	27.290	1.4360	0.03664	1.5342
05	57.397	7589	01742	28.690	4577	03486	5360
10	60.340	7806	01657	30.162	4795	03315	5377
15	63.434	8023	01576	31.709	5012	03154	5393
20	66.686	8240	01500	33.336	5229	03000	5408
4.25	70.105	1.8458	0.01426	35.046	1.5446	0.02853	1.5423
30	73.700	8675	01357	36.843	5664	02714	5437
35	77.478	8892	01291	38.733	5881	02582	5450
40	81.451	9109	01228	40.719	6098	02456	5462
45	85.627	9326	01168	42.808	6315	02336	5474
4.50	90.017	1.9543	0.01111	45.003	1.6532	0.02222	1.5486
55	94.632	9760	01057	47.311	6750	02114	5497
60	99.484	1.9978	01005	49.737	6967	02011	5507
65	104.58	2.0195	00956	52.288	7184	01912	5517
70	109.95	0412	00910	54.969	7401	01819	5526
4.75	115.58	2.0629	0.00865	57.788	1.7618	0.01730	1.5535
80	121.51	0846	00823	60.751	7836	01646	5543
85	127.74	1063	00783	63.866	8053	01566	5551
90	134.29	1280	00745	67.141	8270	01489	5559
4.95	141.17	1498	00708	70.584	8487	01417	5566
5.00	148.41	2.1715	0.00674	74.203	1.8704	0.01348	1.5573
05	156.02	1932	00641	78.008	8921	01282	5580
10	164.02	2149	00610	82.008	9139	01219	5586
15	172.43	2366	00580	86.213	9356	01160	5592
20	181.27	2583	00552	90.633	9573	01103	5598
5.25	190.57	2.2800	0.00525	95.281	1.9790	0.01050	1.5603
30	200.34	3018	00499	100.17	2.0007	00998	5608
35	210.61	3235	00475	105.30	0224	00950	5613
40	221.41	3452	00452	110.70	0442	00903	5618
45	232.76	3669	00430	116.38	0659	00859	5622
5.50	244.69	2.3886	0.00409	122.34	2.0876	0.00817	1.5626
55	257.24	4103	00389	128.62	1093	00778	5630
60	270.43	4320	00370	135.21	1310	00740	5634
65	284.29	4538	00352	142.14	1527	00704	5638
70	298.87	4755	00335	149.43	1744	00669	5641
5.75	314.19	2.4972	0.00318	157.09	2.1962	0.00637	1.5644
80	330.30	5189	00303	165.15	2179	00606	5647
85	347.23	5406	00288	173.62	2396	00576	5650
90	365.04	5623	00274	182.52	2613	00548	5653
5.95	383.75	5841	00261	191.88	2830	00521	5656
6.00	403.43	2.6058	0.00248	201.71	2.3047	0.00496	1.5658
x	e^x	log e^x	e^{-x}	tan gd x	log tan gd x	cot gd x	gd x (rad.)

TABLE 8.—EXPONENTIAL AND HYPERBOLIC FUNCTIONS.—(*Continued*)

x	tanh x	log* tanh x	coth x	cosh x	log cosh x	sech x	gd x
							° ′ ″
3.50	0.9982	9.9992	1.0018	16.573	1.2194	0.06034	86 32 26
55	9984	9993	0017	17.421	2411	05740	86 42 33
60	9985	9994	0015	18.313	2628	05461	86 52 11
65	9986	9994	0014	19.250	2844	05195	87 01 20
70	9988	9995	0012	20.236	3061	04942	87 10 03
3.75	0.9989	9.9995	1.0011	21.272	1.3278	0.04701	87 18 20
80	9990	9996	0010	22.362	3495	04472	87 26 13
85	9991	9996	0009	23.507	3712	04254	87 33 43
90	9992	9996	0008	24.711	3929	04047	87 40 51
3.95	9993	9997	0007	25.977	4146	03850	87 47 38
4.00	0.9993	9.9997	1.0007	27.308	1.4363	0.03662	87 54 05
05	9994	9997	0006	28.707	4580	03483	88 00 13
10	9995	9998	0005	30.178	4797	03314	88 06 04
15	9995	9998	0005	31.725	5014	03152	88 11 37
20	9996	9998	0004	33.351	5231	02998	88 16 54
4.25	0.9996	9.9998	1.0004	35.060	1.5448	0.02852	88 21 56
30	9996	9998	0004	36.857	5665	02713	88 26 43
35	9997	9999	0003	38.746	5882	02581	88 31 16
40	9997	9999	0003	40.732	6099	02455	88 35 35
45	9997	9999	0003	42.819	6316	02335	88 39 42
4.50	0.9998	9.9999	1.0002	45.014	1.6533	0.02222	88 43 37
55	9998	9999	0002	47.321	6751	02113	88 47 21
60	9998	9999	0002	49.747	6968	02010	88 50 53
65	9998	9999	0002	52.297	7185	01912	88 54 16
70	9998	9999	0002	54.978	7402	01819	88 57 28
4.75	0.9999	9.9999	1.0001	57.796	1.7619	0.01730	89 00 31
80	9999	9999	0001	60.759	7836	01646	89 03 25
85	9999	9.9999	0001	63.874	8053	01566	89 06 11
90	9999	10.0000	0001	67.149	8270	01489	89 08 48
4.95	9999	0000	0001	70.591	8487	01417	89 11 18
5.00	0.9999	10.0000	1.0001	74.210	1.8705	0.01348	89 13 40
05	9999	0000	0001	78.014	8922	01282	89 15 56
10	9999	0000	0001	82.014	9139	01219	89 18 05
15	9999	0000	0001	86.219	9356	01160	89 20 08
20	9999	0000	0001	90.639	9573	01103	89 22 04
5.25	0.9999	10.0000	1.0001	95.286	1.9790	0.01049	89 23 55
30	1.0000	0000	0000	100.17	2.0007	00998	89 25 41
35	0000	0000	0000	105.31	0225	00950	89 27 21
40	0000	0000	0000	110.71	0442	00903	89 28 57
45	0000	0000	0000	116.38	0659	00859	89 30 28
5.50	1.0000	10.0000	1.0000	122.35	2.0876	0.00817	89 31 54
55	0000	0000	0000	128.62	1093	00777	89 33 16
60	0000	0000	0000	135.22	1310	00740	89 34 35
65	0000	0000	0000	142.15	1527	00703	89 35 49
70	0000	0000	0000	149.44	1745	00669	89 37 00
5.75	1.0000	10.0000	1.0000	157.10	2.1962	0.00637	89 38 07
80	0000	0000	0000	165.15	2179	00606	89 39 11
85	0000	0000	0000	173.62	2396	00576	89 40 12
90	0000	0000	0000	182.52	2613	00548	89 41 10
5.95	0000	0000	0000	191.88	2830	00521	89 42 05
6.00	1.0000	10.0000	1.0000	201.72	2.3047	0.00496	89 42 57
x	sin gd x	log* sin gd x	csc gd x	sec gd x	log sec gd x	cos gd x	gd x

* Attach −10 to entries in this column.

TABLE 8.—EXPONENTIAL AND HYPERBOLIC FUNCTIONS.—(*Concluded*)

x	e^x	$\log e^x$	e^{-x}	x	e^x	$\log e^x$	e^{-x}
6.0	403.43	2.6058	0.00248	**8.0**	2981.0	3.4744	0.00034
6.1	445.86	6492	00224	8.1	3294.5	5178	00030
6.2	492.75	6926	00203	8.2	3641.0	5612	00027
6.3	544.57	7361	00184	8.3	4023.9	6046	00025
6.4	601.85	7795	00166	8.4	4447.1	6481	00022
6.5	665.14	2.8229	0.00150	8.5	4914.8	3.6915	0.00020
6.6	735.10	8663	00136	8.6	5431.7	7349	00018
6.7	812.41	9098	00123	8.7	6002.9	7784	00017
6.8	897.85	9532	00111	8.8	6634.2	8218	00015
6.9	992.27	2.9966	00101	8.9	7332.0	8652	00014
7.0	1096.6	3.0401	0.00091	**9.0**	8103.1	3.9087	0.00012
7.1	1212.0	0835	00083	9.1	8955.3	3.9521	00011
7.2	1339.4	1269	00075	9.2	9897.1	3.9955	00010
7.3	1480.3	1703	00068	9.3	10938	4.0389	00009
7.4	1636.0	2138	00061	9.4	12088	4.0824	00008
7.5	1808.0	3.2572	0.00055	9.5	13360	4.1258	0.00007
7.6	1998.2	3006	00050	9.6	14765	1692	00007
7.7	2208.3	3441	00045	9.7	16318	2127	00006
7.8	2440.6	3875	00041	9.8	18034	2561	00006
7.9	2697.3	4309	00037	9.9	19930	2995	00005
8.0	2981.0	3.4744	0.00034	**10.0**	22026	4.3429	0.00005
$\pi/6$	1.6881	0.2274	0.59238	$3\pi/4$	10.551	1.0233	0.09478
$\pi/4$	2.1933	3411	45594	$5\pi/6$	13.708	1.1370	07295
$\pi/3$	2.8497	4548	35092	π	23.141	1.3644	04321
$\pi/2$	4.8105	6822	20788	$3\pi/2$	111.32	2.0466	00898
$2\pi/3$	8.1205	0.9096	0.12314	2π	535.49	2.7288	0.00187
x	e^x	$\log e^x$	e^{-x}	x	e^x	$\log e^x$	e^{-x}

For $x > 6$, the following approximations may be used:

$$\sinh x = \cosh x = \tfrac{1}{2}e^x.$$
$$\log \sinh x = \log \cosh x = \log e^x - 0.3010.$$
$$\tanh x = \coth x = 1.0000.$$
$$\operatorname{csch} x = \operatorname{sech} x = 2e^{-x}.$$

TABLE 9.—COMMON LOGARITHMS OF POWERS OF e

x	$\log e^x$	$\log e^{\frac{x}{10}}$	$\log e^{\frac{x}{100}}$	$\log e^{\frac{x}{1000}}$	$\log e^{\frac{x}{10000}}$
1	0.434294	0.043429	0.004343	0.000434	0.000043
2	0.868589	086859	008686	000869	000087
3	1.302883	130288	013029	001303	000130
4	1.737178	0.173718	0.017372	0.001737	0.000174
5	2.171472	217147	021715	002171	000217
6	2.605767	260577	026058	002606	000261
7	3.040061	0.304006	0.030401	0.003040	0.000304
8	3.474356	347436	034744	003474	000347
9	3.908650	0.390865	0.039087	0.003909	0.000391

$\log e = 0.43429\ 44819\ 03252.$

TABLE 10.—INVERSE TRIGONOMETRIC AND HYPERBOLIC FUNCTIONS

x	$\sin^{-1} x$	$\cos^{-1} x$	$\tan^{-1} x$	$\text{gd}^{-1} x$	$\sinh^{-1} x$	$\text{sech}^{-1} x$	$\tanh^{-1} x$	x
0.00	0.0000	1.5708	0.0000	0.0000	0.0000	∞	0.0000	**0.00**
01	0100	5608	0100	0100	0100	5.2983	0100	01
02	0200	5508	0200	0200	0200	4.6051	0200	02
03	0300	5408	0300	0300	0300	4.1995	0300	03
04	0400	5308	0400	0400	0400	3.9116	0400	04
0.05	0.0500	1.5208	0.0500	0.0500	0.0500	3.6883	0.0500	**0.05**
06	0600	5108	0599	0600	0600	5057	0601	06
07	0701	5007	0699	0701	0699	3512	0701	07
08	0801	4907	0798	0801	0799	2173	0802	08
09	0901	4807	0898	0901	0899	3.0991	0902	09
0.10	0.1002	1.4706	0.0997	0.1002	0.0998	2.9932	0.1003	**0.10**
11	1102	4606	1096	1102	1098	8974	1104	11
12	1203	4505	1194	1203	1197	8098	1206	12
13	1304	4404	1293	1304	1296	7291	1307	13
14	1405	4303	1391	1405	1395	6543	1409	14
0.15	0.1506	1.4202	0.1489	0.1506	0.1494	2.5846	0.1511	0.15
16	1607	4101	1587	1607	1593	5193	1614	16
17	1708	4000	1684	1708	1692	4578	1717	17
18	1810	3898	1781	1810	1790	3997	1820	18
19	1912	3796	1878	1912	1889	3447	1923	19
0.20	0.2014	1.3694	0.1974	0.2013	0.1987	2.2924	0.2027	**0.20**
21	2116	3592	2070	2116	2085	2426	2132	21
22	2218	3490	2166	2218	2183	1949	2237	22
23	2321	3387	2261	2321	2280	1493	2342	23
24	2424	3284	2355	2423	2378	1055	2448	24
0.25	0.2527	1.3181	0.2450	0.2526	0.2475	2.0634	0.2554	0.25
26	2630	3078	2544	2630	2572	2.0229	2661	26
27	2734	2974	2637	2733	2668	1.9837	2769	27
28	2838	2870	2730	2837	2765	9459	2877	28
29	2942	2766	2823	2942	2861	9093	2986	29
0.30	0.3047	1.2661	0.2915	0.3046	0.2957	1.8738	0.3095	**0.30**
31	3152	2556	3006	3151	3052	8394	3205	31
32	3257	2451	3097	3256	3148	8059	3316	32
33	3363	2345	3187	3362	3243	7734	3428	33
34	3469	2239	3277	3467	3338	7417	3541	34
0.35	0.3576	1.2132	0.3367	0.3574	0.3432	1.7108	0.3654	0.35
36	3683	2025	3456	3680	3526	6807	3769	36
37	3790	1918	3544	3787	3620	6513	3884	37
38	3898	1810	3631	3895	3714	6225	4001	38
39	4006	1702	3719	4003	3807	5944	4118	39
0.40	0.4115	1.1593	0.3805	0.4111	0.3900	1.5668	0.4236	**0.40**
41	4225	1483	3891	4220	3993	5398	4356	41
42	4334	1374	3976	4329	4085	5133	4477	42
43	4445	1263	4061	4439	4177	4873	4599	43
44	4556	1152	4145	4549	4269	4618	4722	44
0.45	0.4668	1.1040	0.4229	0.4660	0.4360	1.4367	0.4847	0.45
46	4780	0928	4311	4771	4452	4120	4973	46
47	4893	0815	4394	4883	4542	3877	5101	47
48	5007	0701	4475	4996	4633	3638	5230	48
49	5121	0587	4556	5109	4722	3402	5361	49
0.50	0.5236	1.0472	0.4636	0.5222	0.4812	1.3170	0.5493	**0.50**
x	$\sin^{-1} x$	$\cos^{-1} x$	$\tan^{-1} x$	$\text{gd}^{-1} x$	$\sinh^{-1} x$	$\text{sech}^{-1} x$	$\tanh^{-1} x$	x

TABLE 10.—INVERSE TRIGONOMETRIC AND HYPERBOLIC FUNCTIONS.—
(*Continued*)

x	$\sin^{-1} x$	$\cos^{-1} x$	$\tan^{-1} x$	$\mathrm{gd}^{-1} x$	$\sinh^{-1} x$	$\operatorname{sech}^{-1} x$	$\tanh^{-1} x$	x
0.50	0.5236	1.0472	0.4636	0.5222	0.4812	1.3170	0.5493	0.50
51	5352	0356	4716	5337	4901	2940	5627	51
52	5469	0239	4795	5452	4990	2714	5763	52
53	5586	0122	4874	5567	5079	2490	5901	53
54	5704	1.0004	4951	5683	5167	2269	6042	54
0.55	0.5824	0.9884	0.5028	0.5800	0.5255	1.2050	0.6184	0.55
56	5944	9764	5105	5918	5342	1833	6328	56
57	6065	9643	5181	6036	5429	1619	6475	57
58	6187	9521	5256	6156	5516	1406	6625	58
59	6311	9397	5330	6275	5602	1195	6777	59
0.60	0.6435	0.9273	0.5404	0.6396	0.5688	1.0986	0.6931	0.60
61	6561	9147	5477	6518	5774	0779	7089	61
62	6687	9021	5550	6640	5859	0572	7250	62
63	6816	8892	5622	6764	5944	0367	7414	63
64	6945	8763	5693	6888	6028	1.0163	7582	64
0.65	0.7076	0.8632	0.5764	0.7013	0.6112	0.9961	0.7753	0.65
66	7208	8500	5834	7139	6196	9759	7928	66
67	7342	8366	5903	7266	6279	9557	8107	67
68	7478	8230	5972	7394	6362	9356	8291	68
69	7615	8093	6040	7523	6445	9156	8480	69
0.70	0.7754	0.7954	0.6107	0.7654	0.6527	0.8956	0.8673	0.70
71	7895	7813	6174	7785	6608	8756	8872	71
72	8038	7670	6240	7917	6690	8556	9076	72
73	8183	7525	6306	8051	6771	8356	9287	73
74	8331	7377	6371	8186	6851	8155	9505	74
0.75	0.8481	0.7227	0.6435	0.8322	0.6931	0.7954	0.9730	0.75
76	8633	7075	6499	8459	7011	7752	0.9962	76
77	8788	6920	6562	8598	7091	7549	1.0203	77
78	8947	6761	6624	8738	7170	7344	0454	78
79	9108	6600	6686	8879	7248	7139	0714	79
0.80	0.9273	0.6435	0.6747	0.9022	0.7327	0.6931	1.0986	0.80
81	9442	6266	6808	9166	7405	6722	1270	81
82	9614	6094	6868	9312	7482	6510	1568	82
83	9791	5917	6928	9459	7559	6296	1881	83
84	0.9973	5735	6987	9608	7636	6078	2212	84
0.85	1.0160	0.5548	0.7045	0.9759	0.7712	0.5857	1.2562	0.85
86	0353	5355	7103	0.9911	7788	5631	2933	86
87	0552	5156	7160	1.0065	7864	5401	3331	87
88	0759	4949	7217	0221	7939	5165	3758	88
89	0973	4735	7273	0379	8014	4922	4219	89
0.90	1.1198	0.4510	0.7328	1.0539	0.8089	0.4671	1.4722	0.90
91	1433	4275	7383	0701	8163	4412	5275	91
92	1681	4027	7438	0865	8237	4141	5890	92
93	1944	3764	7491	1031	8310	3856	6584	93
94	2226	3482	7545	1200	8383	3554	7380	94
0.95	1.2532	0.3176	0.7598	1.1370	0.8456	0.3230	1.8318	0.95
96	2870	2838	7650	1544	8528	2877	1.9459	96
97	3252	2456	7702	1719	8600	2481	2.0923	97
98	3705	2003	7753	1897	8672	2017	2.2976	98
99	4293	1415	7804	2078	8743	1420	2.6467	99
1.00	1.5708	0.0000	0.7854	1.2262	0.8814	0.0000	∞	1.00
x	$\sin^{-1} x$	$\cos^{-1} x$	$\tan^{-1} x$	$\mathrm{gd}^{-1} x$	$\sinh^{-1} x$	$\operatorname{sech}^{-1} x$	$\tanh^{-1} x$	x

TABLE 10.—INVERSE TRIGONOMETRIC AND HYPERBOLIC FUNCTIONS.—
(*Continued*)

x	csc⁻¹ x	sec⁻¹ x	tan⁻¹ x	gd⁻¹ x*	sinh⁻¹ x	cosh⁻¹ x	coth⁻¹ x	x
1.00	1.5708	0.0000	0.7854	1.2262	0.8814	0.0000	∞	**1.00**
01	4300	1408	7904	2449	8884	1413	2.6516	01
02	3724	1984	7953	2638	8954	1997	3076	02
03	3289	2419	8002	2831	9024	2443	2.1073	03
04	2925	2782	8050	3027	9094	2819	1.9659	04
1.05	1.2610	0.3098	0.8098	1.3226	0.9163	0.3149	1.8568	**1.05**
06	2327	3381	8145	3428	9232	3447	7681	06
07	2071	3637	8192	3635	9300	3720	6934	07
08	1835	3873	8238	3845	9368	3974	6290	08
09	1616	4092	8284	4059	9436	4211	5726	09
1.10	1.1411	0.4297	0.8330	1.4278	0.9503	0.4436	1.5223	**1.10**
11	1218	4490	8375	4500	9571	4648	4770	11
12	1037	4671	8419	4727	9637	4851	4358	12
13	0864	4844	8464	4959	9704	5045	3982	13
14	0700	5008	8507	5196	9770	5232	3635	14
1.15	1.0543	0.5165	0.8551	1.5438	0.9836	0.5411	1.3313	**1.15**
16	0393	5315	8593	5686	9901	5584	3013	16
17	0250	5458	8636	5939	0.9966	5751	2733	17
18	1.0112	5596	8678	6199	1.0031	5913	2471	18
19	0.9979	5729	8719	6465	0096	6071	2223	19
1.20	0.9851	0.5857	0.8761	1.6737	1.0160	0.6224	1.1989	**1.20**
21	9728	5980	8801	7017	0224	6372	1768	21
22	9608	6100	8842	7304	0287	6517	1558	22
23	9493	6215	8882	7599	0350	6659	1358	23
24	9381	6327	8921	7902	0413	6797	1168	24
1.25	0.9273	0.6435	0.8961	1.8215	1.0476	0.6931	1.0986	**1.25**
26	9168	6540	8999	8537	0538	7063	0812	26
27	9066	6642	9038	8869	0600	7192	0646	27
28	8967	6741	9076	9212	0662	7319	0486	28
29	8870	6838	9114	9567	0723	7443	0332	29
1.30	0.8776	0.6932	0.9151	1.9934	1.0785	0.7564	1.0184	**1.30**
31	8685	7023	9188	2.0315	0845	7684	1.0042	31
32	8596	7112	9225	0710	0906	7801	0.9905	32
33	8509	7199	9261	1121	0966	7916	9773	33
34	8424	7284	9297	1549	1026	8029	9645	34
1.35	0.8342	0.7366	0.9332	2.1996	1.1086	0.8140	0.9521	**1.35**
36	8261	7447	9368	2463	1145	8249	9402	36
37	8182	7526	9403	2952	1204	8357	9286	37
38	8105	7603	9437	3467	1263	8463	9173	38
39	8030	7678	9472	4008	1322	8567	9065	39
1.40	0.7956	0.7752	0.9505	2.4580	1.1380	0.8670	0.8959	**1.40**
41	7884	7824	9539	5186	1438	8771	8856	41
42	7813	7895	9572	5831	1496	8871	8756	42
43	7744	7964	9605	6519	1553	8970	8659	43
44	7676	8031	9638	7258	1610	9067	8565	44
1.45	0.7610	0.8098	0.9670	2.8056	1.1667	0.9163	0.8473	**1.45**
46	7545	8163	9703	8922	1724	9258	8383	46
47	7481	8227	9734	2.9869	1780	9351	8296	47
48	7419	8289	9766	3.0916	1836	9443	8211	48
49	7357	8350	9797	2084	1892	9534	8128	49
1.50	0.7297	0.8411	0.9828	3.3407	1.1948	0.9624	0.8047	**1.50**
x	csc⁻¹ x	sec⁻¹ x	tan⁻¹ x	gd⁻¹ x*	sinh⁻¹ x	cosh⁻¹ x	coth⁻¹ x	x

* For $x > 1.50$ use $gd^{-1}x = \ln 200 - \ln 100(\pi/2 - x)$, to 3 decimal places.

TABLE 10.—INVERSE TRIGONOMETRIC AND HYPERBOLIC FUNCTIONS.—
(Continued)

x	$\csc^{-1} x$	$\sec^{-1} x$	$\tan^{-1} x$	$\cot^{-1} x$	$\sinh^{-1} x$	$\cosh^{-1} x$	$\coth^{-1} x$	x
1.50	0.7297	0.8411	0.9828	0.5880	1.1948	0.9624	0.8047	**1.50**
55	7012	8696	0.9978	5730	2222	1.0059	7670	55
60	6751	8957	1.0122	5586	2490	0470	7332	60
65	6511	9197	0259	5449	2752	0860	7027	65
70	6289	9419	0391	5317	3008	1232	6750	70
1.75	0.6082	0.9626	1.0517	0.5191	1.3259	1.1588	0.6496	1.75
80	5890	9818	0637	5071	3504	1929	6264	80
85	5711	0.9997	0752	4956	3745	2257	6049	85
90	5543	1.0165	0863	4845	3980	2572	5850	90
95	5385	0323	0969	4739	4210	2876	5665	95
2.00	0.5236	1.0472	1.1071	0.4636	1.4436	1.3170	0.5493	**2.00**
05	5096	0612	1170	4538	4658	3454	5332	05
10	4963	0745	1264	4444	4875	3729	5180	10
15	4838	0870	1354	4354	5088	3995	5038	15
20	4719	0989	1442	4266	5297	4254	4904	20
2.25	0.4606	1.1102	1.1526	0.4182	1.5502	1.4506	0.4778	2.25
30	4498	1210	1607	4101	5703	4750	4658	30
35	4395	1312	1685	4023	5900	4989	4544	35
40	4298	1410	1760	3948	6094	5221	4437	40
45	4204	1504	1833	3875	6285	5447	4334	45
2.50	0.4115	1.1593	1.1903	0.3805	1.6472	1.5668	0.4236	**2.50**
55	4030	1678	1971	3737	6656	5884	4143	55
60	3948	1760	2036	3672	6837	6094	4055	60
65	3869	1839	2100	3608	7015	6300	3970	65
70	3794	1914	2161	3547	7191	6502	3889	70
2.75	0.3722	1.1986	1.2220	0.3488	1.7363	1.6699	0.3811	2.75
80	3652	2056	2278	3430	7532	6892	3736	80
85	3585	2123	2333	3375	7699	7082	3664	85
90	3521	2187	2387	3321	7863	7267	3596	90
95	3458	2250	2440	3268	8025	7449	3529	95
3.00	0.3398	1.2310	1.2490	0.3218	1.8184	1.7627	0.3466	**3.00**
10	3285	2423	2588	3120	8496	7975	3345	10
20	3178	2530	2679	3029	8799	8309	3233	20
30	3079	2629	2766	2942	9093	8633	3129	30
40	2985	2723	2847	2861	9379	8946	3031	40
3.50	0.2898	1.2810	1.2925	0.2783	1.9657	1.9248	0.2939	3.50
60	2815	2893	2998	2709	1.9928	9542	2853	60
70	2737	2971	3068	2640	2.0193	1.9827	2772	70
80	2663	3045	3135	2573	0450	2.0104	2695	80
90	2593	3115	3198	2510	0702	0373	2623	90
4.00	0.2527	1.3181	1.3258	0.2450	2.0947	2.0634	0.2554	**4.00**
10	2464	3244	3316	2392	1187	0889	2489	10
20	2404	3304	3371	2337	1421	1137	2428	20
30	2347	3361	3423	2285	1650	1380	2369	30
40	2293	3415	3473	2235	1874	1616	2313	40
4.50	0.2241	1.3467	1.3521	0.2187	2.2093	2.1846	0.2260	4.50
60	2191	3517	3567	2141	2308	2072	2209	60
70	2144	3564	3612	2096	2518	2292	2161	70
80	2099	3609	3654	2054	2724	2507	2114	80
90	2055	3653	3695	2013	2926	2718	2070	90
5.00	0.2014	1.3694	1.3734	0.1974	2.3124	2.2924	0.2027	**5.00**
x	$\csc^{-1} x$	$\sec^{-1} x$	$\tan^{-1} x$	$\cot^{-1} x$	$\sinh^{-1} x$	$\cosh^{-1} x$	$\coth^{-1} x$	x

TABLE 10.—Inverse Trigonometric and Hyperbolic Functions.—
(*Concluded*)

x	$\csc^{-1} x$	$\sec^{-1} x$	$\tan^{-1} x$	$\cot^{-1} x$	$\sinh^{-1} x$	$\cosh^{-1} x$	$\coth^{-1} x$	x
5.0	0.2014	1.3694	1.3734	0.1974	2.3124	2.2924	0.2027	**5.0**
1	1974	3734	3772	1936	3319	3126	1987	1
2	1935	3773	3808	1900	3509	3324	1947	2
3	1898	3810	3843	1865	3696	3518	1910	3
4	1863	3845	3877	1831	3880	3709	1873	4
5.5	0.1828	1.3880	1.3909	0.1799	2.4061	2.3895	0.1839	5.5
6	1795	3913	3941	1767	4238	4078	1805	6
7	1764	3944	3971	1737	4412	4258	1773	7
8	1733	3975	4001	1707	4584	4435	1742	8
9	1703	4005	4029	1679	4752	4608	1711	9
6.0	0.1674	1.4033	1.4056	0.1651	2.4918	2.4779	0.1682	**6.0**
2	1620	4088	4109	1599	5241	5111	1627	2
4	1569	4139	4158	1550	5555	5433	1575	4
6	1521	4187	4204	1504	5859	5744	1527	6
8	1476	4232	4248	1460	6154	6046	1481	8
7.0	0.1433	1.4274	1.4289	0.1419	2.6441	2.6339	0.1438	**7.0**
2	1393	4315	4328	1380	6720	6624	1398	2
4	1355	4352	4365	1343	6992	6900	1360	4
6	1320	4388	4400	1308	7256	7169	1323	6
8	1286	4422	4433	1275	7514	7431	1289	8
8.0	0.1253	1.4455	1.4464	0.1244	2.7765	2.7687	0.1257	**8.0**
2	1223	4485	4494	1214	8010	7935	1226	2
4	1193	4515	4523	1185	8249	8178	1196	4
6	1165	4543	4550	1158	8483	8415	1168	6
8	1139	4569	4576	1132	8711	8647	1141	8
9.0	0.1113	1.4595	1.4601	0.1107	2.8934	2.8873	0.1116	**9.0**
2	1089	4619	4625	1083	9153	9094	1091	2
4	1066	4642	4648	1060	9367	9310	1068	4
6	1044	4664	4670	1038	9576	9522	1045	6
8	1022	4686	4691	1017	9781	9729	1024	8
10.0	0.1002	1.4706	1.4711	0.0997	2.9982	2.9929	0.1003	**10.0**
10.5	0954	4754	4758	0950	3.0468	3.0422	0955	10.5
11.0	0910	4798	4801	0907	0931	0890	0912	11.0
11.5	0871	4837	4841	0867	1374	1336	0872	11.5
12.0	0834	4874	4877	0831	1798	1763	0835	12.0
12.5	0.0801	1.4907	1.4910	0.0798	3.2205	3.2173	0.0802	12.5
13.0	0770	4938	4940	0768	2596	2566	0771	13.0
13.5	0741	4967	4969	0739	2972	2945	0742	13.5
14.0	0715	4993	4995	0713	3335	3309	0716	14.0
14.5	0690	5018	5019	0689	3685	3661	0691	14.5
15	0.0667	1.5041	1.5042	0.0666	3.4023	3.4001	0.0668	**15**
16	0625	5083	5084	0624	4667	4648	0626	16
17	0589	5119	5120	0588	5272	5255	0589	17
18	0556	5152	5153	0555	5843	5827	0556	18
19	0527	5181	5182	0526	6383	6369	0527	19
20	0.0500	1.52C8	1.5208	0.0500	3.6895	3.6883	0.0500	**20**
21	0476	5232	5232	0476	7382	7371	0477	21
22	0455	5253	5254	0454	7847	7837	0455	22
23	0435	5273	5273	0435	8291	8282	0435	23
24	0417	5291	5292	0416	8716	8708	0417	24
25	0.0400	1.5308	1.5308	0.0400	3.9124	3.9116	0.0400	**25**

x	$\csc^{-1} x$	$\sec^{-1} x$	$\tan^{-1} x$	$\cot^{-1} x$	$\sinh^{-1} x$	$\cosh^{-1} x$	$\coth^{-1} x$	x

Table 11.—Squares, Cubes, Square Roots, Cube Roots, Reciprocals and
Natural Logarithms
(1–50)

N	N²	N³	$\sqrt{N}$	$\sqrt[3]{N}$	1000/N	ln N	N
1	1	1	1.0000	1.0000	1000.000	0.0000	1
2	4	8	1.4142	1.2599	500.000	0.6931	2
3	9	27	1.7321	1.4422	333.333	1.0986	3
4	16	64	2.0000	1.5874	250.000	1.3863	4
5	25	125	2.2361	1.7100	200.000	1.6094	5
6	36	216	2.4495	1.8171	166.667	1.7918	6
7	49	343	2.6458	1.9129	142.857	1.9459	7
8	64	512	2.8284	2.0000	125.000	2.0794	8
9	81	729	3.0000	2.0801	111.111	2.1972	9
10	1 00	1 000 ·	3.1623	2.1544	100.0000	2.3026	10
11	1 21	1 331	3166	2240	90.9091	3979	11
12	1 44	1 728	4641	2894	83.3333	4849	12
13	1 69	2 197	6056	3513	76.9231	5649	13
14	1 96	2 744	7417	4101	71.4286	6391	14
15	2 25	3 375	3.8730	2.4662	66.6667	2.7081	15
16	2 56	4 096	4.0000	5198	62.5000	7726	16
17	2 89	4 913	1231	5713	58.8235	8332	17
18	3 24	5 832	2426	6207	55.5556	8904	18
19	3 61	6 859	3589	6684	52.6316	9444	19
20	4 00	8 000	4.4721	2.7144	50.0000	2.9957	20
21	4 41	9 261	5826	7589	47.6190	3.0445	21
22	4 84	10 648	6904	8020	45.4545	0910	22
23	5 29	12 167	7958	8439	43.4783	1355	23
24	5 76	13 824	4.8990	8845	41.6667	1781	24
25	6 25	15 625	5.0000	2.9240	40.0000	3.2189	25
26	6 76	17 576	0990	2.9625	38.4615	2531	26
27	7 29	19 683	1962	3.0000	37.0370	2958	27
28	7 84	21 952	2915	0366	35.7143	3322	28
29	8 41	24 389	3852	0723	34.4828	3673	29
30	9 00	27 000	5.4772	3.1072	33.3333	3.4012	30
31	9 61	29 791	5678	1414	32.2581	4340	31
32	10 24	32 768	6569	1748	31.2500	4657	32
33	10 89	35 937	7446	2075	30.3030	4965	33
34	11 56	39 304	8310	2396	29.4118	5264	34
35	12 25	42 875	5.9161	3.2711	28.5714	3.5553	35
36	12 96	46 656	6.0000	3019	27.7778	5835	36
37	13 69	50 653	0828	3322	27.0270	6109	37
38	14 44	54 872	1644	3620	26.3158	6376	38
39	15 21	59 319	2450	3912	25.6410	6636	39
40	16 00	64 000	6.3246	3.4200	25.0000	3.6889	40
41	16 81	68 921	4031	4482	24.3902	7136	41
42	17 64	74 088	4807	4760	23.8095	7377	42
43	18 49	79 507	5574	5034	23.2558	7612	43
44	19 36	85 184	6332	5303	22.7273	7842	44
45	20 25	91 125	6.7082	3.5569	22.2222	3.8067	45
46	21 16	97 336	7823	5830	21.7391	8286	46
47	22 09	103 823	8557	6088	21.2766	8501	47
48	23 04	110 592	6.9282	6342	20.8333	8712	48
49	24 01	117 649	7.0000	6593	20.4082	8918	49
50	25 00	125 000	7.0711	3.6840	20.0000	3.9120	50
N	N²	N³	$\sqrt{N}$	$\sqrt[3]{N}$	1000/N	ln N	N

TABLE 11.—SQUARES, CUBES, SQUARE ROOTS, CUBE ROOTS, RECIPROCALS AND NATURAL LOGARITHMS.—(Continued)
(50–100)

N	N^2	N^3	$\sqrt{N}$	$\sqrt[3]{N}$	1000/N	ln N	N
50	25 00	125 000	7.0711	3.6840	20.0000	3.9120	**50**
51	26 01	132 651	1414	7084	19.6078	9318	51
52	27 04	140 608	2111	7325	19.2308	9512	52
53	28 09	148 877	2801	7563	18.8679	9703	53
54	29 16	157 464	3485	7798	18.5185	3.9890	54
55	30 25	166 375	7.4162	3.8030	18.1818	4.0073	55
56	31 36	175 616	4833	8259	17.8571	0254	56
57	32 49	185 193	5498	8485	5439	0431	57
58	33 64	195 112	6158	8709	17.2414	0604	58
59	34 81	205 379	6811	8930	16.9492	0775	59
60	36 00	216 000	7.7460	3.9149	16.6667	4.0943	**60**
61	37 21	226 981	8102	9365	3934	1109	61
62	38 44	238 328	8740	9579	16.1290	1271	62
63	39 69	250 047	7.9373	3.9791	15.8730	1431	63
64	40 96	262 144	8.0000	4.0000	6250	1589	64
65	42 25	274 625	8.0623	4.0207	15.3846	4.1744	65
66	43 56	287 496	1240	0412	15.1515	1897	66
67	44 89	300 763	1854	0615	14.9254	2047	67
68	46 24	314 432	2462	0817	7059	2195	68
69	47 61	328 509	3066	1016	4928	2341	69
70	49 00	343 000	8.3666	4.1213	14.2857	4.2485	**70**
71	50 41	357 911	4261	1408	14.0845	2627	71
72	51 84	373 248	4853	1602	13.8889	2767	72
73	53 29	389 017	5440	1793	6986	2905	73
74	54 76	405 224	6023	1983	5135	3041	74
75	56 25	421 875	8.6603	4.2172	13.3333	4.3175	75
76	57 76	438 976	7178	2358	13.1579	3307	76
77	59 29	456 533	7750	2543	12.9870	3438	77
78	60 84	474 552	8318	2727	8205	3567	78
79	62 41	493 039	8882	2908	6582	3694	79
80	64 00	512 000	8.9443	4.3089	12.5000	4.3820	**80**
81	65 61	531 441	9.0000	3267	3457	3944	81
82	67 24	551 368	0554	3445	1951	4067	82
83	68 89	571 787	1104	3621	12.0482	4188	83
84	70 56	592 704	1652	3795	11.9048	4308	84
85	72 25	614 125	9.2195	4.3968	11.7647	4.4427	85
86	73 96	636 056	2736	4140	6279	4543	86
87	75 69	658 503	3274	4310	4943	4659	87
88	77 44	681 472	3808	4480	3636	4773	88
89	79 21	704 969	4340	4647	2360	4886	89
90	81 00	729 000	9.4868	4.4814	11.1111	4.4998	**90**
91	82 81	753 571	5394	4979	10.9890	5109	91
92	84 64	778 688	5917	5144	8696	5218	92
93	86 49	804 357	6437	5307	7527	5326	93
94	88 36	830 584	6954	5468	6383	5433	94
95	90 25	857 375	9.7468	4.5629	10.5263	4.5539	95
96	92 16	884 736	7980	5789	4167	5643	96
97	94 09	912 673	8489	5947	3093	5747	97
98	96 04	941 192	8995	6104	2041	5850	98
99	98 01	970 299	9.9499	6261	1010	5951	99
100	1 00 00	1 000 000	10.0000	4.6416	10.00000	4.6052	**100**
N	N^2	N^3	$\sqrt{N}$	$\sqrt[3]{N}$	1000/N	ln N	N

TABLE 11.—SQUARES, CUBES, SQUARE ROOTS, CUBE ROOTS, RECIPROCALS AND
NATURAL LOGARITHMS.—(*Continued*)
(100–150)

N	N^2	N^3	$\sqrt{N}$	$\sqrt[3]{N}$	$1000/N$	$\ln N$	N
100	1 00 00	1 000 000	10.0000	4.6416	10.00000	4.6052	**100**
101	1 02 01	1 030 301	0499	6570	9.90099	6151	101
102	1 04 04	1 061 208	0995	6723	80392	6250	102
103	1 06 09	1 092 727	1489	6875	70874	6347	103
104	1 08 16	1 124 864	1980	7027	61538	6444	104
105	1 10 25	1 157 625	10.2470	4.7177	9.52381	4.6540	105
106	1 12 36	1 191 016	2956	7326	43396	6634	106
107	1 14 49	1 225 043	3441	7475	34579	6728	107
108	1 16 64	1 259 712	3923	7622	25926	6821	108
109	1 18 81	1 295 029	4403	7769	17431	6913	109
110	1 21 00	1 331 000	10.4881	4.7914	9.09091	4.7005	**110**
111	1 23 21	1 367 631	5357	8059	9.00901	7095	111
112	1 25 44	1 404 928	5830	8203	8.92857	7185	112
113	1 27 69	1 442 897	6301	8346	84956	7274	113
114	1 29 96	1 481 544	6771	8488	77193	7362	114
115	1 32 25	1 520 875	10.7238	4.8629	8.69565	4.7449	115
116	1 34 56	1 560 896	7703	8770	62069	7536	116
117	1 36 89	1 601 613	8167	8910	54701	7622	117
118	1 39 24	1 643 032	8628	9049	47458	7707	118
119	1 41 61	1 685 159	9087	9187	40336	7791	119
120	1 44 00	1 728 000	10.9545	4.9324	8.33333	4.7875	**120**
121	1 46 41	1 771 561	11.0000	9461	26446	7958	121
122	1 48 84	1 815 848	0454	9597	19672	8040	122
123	1 51 29	1 860 867	0905	9732	13008	8122	123
124	1 53 76	1 906 624	1355	4.9866	06452	8203	124
125	1 56 25	1 953 125	11.1803	5.0000	8.00000	4.8283	125
126	1 58 76	2 000 376	2250	0133	7.93651	8363	126
127	1 61 29	2 048 383	2694	0265	87402	8442	127
128	1 63 84	2 097 152	3137	0397	81250	8520	128
129	1 66 41	2 146 689	3578	0528	75194	8598	129
130	1 69 00	2 197 000	11.4018	5.0658	7.69231	4.8675	**130**
131	1 71 61	2 248 091	4455	0788	63359	8752	131
132	1 74 24	2 299 968	4891	0916	57576	8828	132
133	1 76 89	2 352 637	5326	1045	51880	8903	133
134	1 79 56	2 406 104	5758	1172	46269	8978	134
135	1 82 25	2 460 375	11.6190	5.1299	7.40741	4.9053	135
136	1 84 96	2 515 456	6619	1426	35294	9127	136
137	1 87 69	2 571 353	7047	1551	29927	9200	137
138	1 90 44	2 628 072	7473	1676	24638	9273	138
139	1 93 21	2 685 619	7898	1801	19424	9345	139
140	1 96 00	2 744 000	11.8322	5.1925	7.14286	4.9416	**140**
141	1 98 81	2 803 221	8743	2048	09220	9488	141
142	2 01 64	2 863 288	9164	2171	7.04225	9558	142
143	2 04 49	2 924 207	11.9583	2293	6.99301	9628	143
144	2 07 36	2 985 984	12.0000	2415	94444	9698	144
145	2 10 25	3 048 625	12.0416	5.2536	6.89655	4.9767	145
146	2 13 16	3 112 136	0830	2656	84932	9836	146
147	2 16 09	3 176 523	1244	·2776	80272	9904	147
148	2 19 04	3 241 792	1655	2896	75676	4.9972	148
149	2 22 01	3 307 949	2066	3015	71141	5.0039	149
150	2 25 00	3 375 000	12.2474	5.3133	6.66667	5.0106	**150**
N	N^2	N^3	$\sqrt{N}$	$\sqrt[3]{N}$	$1000/N$	$\ln N$	N

TABLE 11.—SQUARES, CUBES, SQUARE ROOTS, CUBE ROOTS, RECIPROCALS AND NATURAL LOGARITHMS.—(*Continued*)
(150–200)

N	N^2	N^3	$\sqrt{N}$	$\sqrt[3]{N}$	1000/N	ln N	N
150	2 25 00	3 375 000	12.2474	5.3133	6.66667	5.0106	150
151	2 28 01	3 442 951	2882	3251	62252	0173	151
152	2 31 04	3 511 808	3288	3368	57895	0239	152
153	2 34 09	3 581 577	3693	3485	53595	0304	153
154	2 37 16	3 652 264	4097	3601	49351	0370	154
155	2 40 25	3 723 875	12.4499	5.3717	6.45161	5.0434	155
156	2 43 36	3 796 416	4900	3832	41026	0499	156
157	2 46 49	3 869 893	5300	3947	36943	0562	157
158	2 49 64	3 944 312	5698	4061	32911	0626	158
159	2 52 81	4 019 679	6095	4175	28931	0689	159
160	2 56 00	4 096 000	12.6491	5.4288	6.25000	5.0752	160
161	2 59 21	4 173 281	6886	4401	21118	0814	161
162	2 62 44	4 251 528	7279	4514	17284	0876	162
163	2 65 69	4 330 747	7671	4626	13497	0938	163
164	2 68 96	4 410 944	8062	4737	09756	0999	164
165	2 72 25	4 492 125	12.8452	5.4848	6.06061	5.1059	165
166	2 75 56	4 574 296	8841	4959	6.02410	1120	166
167	2 78 89	4 657 463	9228	5069	5.98802	1180	167
168	2 82 24	4 741 632	12.9615	5178	95238	1240	168
169	2 85 61	4 826 809	13.0000	5288	91716	1299	169
170	2 89 00	4 913 000	13.0384	5.5397	5.88235	5.1358	170
171	2 92 41	5 000 211	0767	5505	84795	1417	171
172	2 95 84	5 088 448	1149	5613	81395	1475	172
173	2 99 29	5 177 717	1529	5721	78035	1533	173
174	3 02 76	5 268 024	1909	5828	74713	1591	174
175	3 06 25	5 359 375	13.2288	5.5934	5.71429	5.1648	175
176	3 09 76	5 451 776	2665	6041	68182	1705	176
177	3 13 29	5 545 233	3041	6147	64972	1761	177
178	3 16 84	5 639 752	3417	6252	61798	1818	178
179	3 20 41	5 735 339	3791	6357	58659	1874	179
180	3 24 00	5 832 000	13.4164	5.6462	5.55556	5.1930	180
181	3 27 61	5 929 741	4536	6567	52486	1985	181
182	3 31 24	6 028 568	4907	6671	49451	2040	182
183	3 34 89	6 128 487	5277	6774	46448	2095	183
184	3 38 56	6 229 504	5647	6877	43478	2149	184
185	3 42 25	6 331 625	13.6015	5.6980	5.40541	5.2204	185
186	3 45 96	6 434 856	6382	7083	37634	2257	186
187	3 49 69	6 539 203	6748	7185	34759	2311	187
188	3 53 44	6 644 672	7113	7287	31915	2364	188
189	3 57 21	6 751 269	7477	7388	29101	2417	189
190	3 61 00	6 859 000	13.7840	5.7489	5.26316	5.2470	190
191	3 64 81	6 967 871	8203	7590	23560	2523	191
192	3 68 64	7 077 888	8564	7690	20833	2575	192
193	3 72 49	7 189 057	8924	7790	18135	2627	193
194	3 76 36	7 301 384	9284	7890	15464	2679	194
195	3 80 25	7 414 875	13.9642	5.7989	5.12821	5.2730	195
196	3 84 16	7 529 536	14.0000	8088	10204	2781	196
197	3 88 09	7 645 373	0357	8186	07614	2832	197
198	3 92 04	7 762 392	0712	8285	05051	2883	198
199	3 96 01	7 880 599	1067	8383	02513	2933	199
200	4 00 00	8 000 000	14.1421	5.8480	5.00000	5.2983	200
N	N^2	N^3	$\sqrt{N}$	$\sqrt[3]{N}$	1000/N	ln N	N

TABLE 11.—SQUARES, CUBES, SQUARE ROOTS, CUBE ROOTS, RECIPROCALS AND
NATURAL LOGARITHMS.—*(Continued)*
(200–250)

N	N^2	N^3	$\sqrt{N}$	$\sqrt[3]{N}$	$1000/N$	$\ln N$	N
200	4 00 00	8 000 000	14.1421	5.8480	5.00000	5.2983	**200**
201	4 04 01	8 120 601	1774	8578	4.97512	3033	201
202	4 08 04	8 242 408	2127	8675	95050	3083	202
203	4 12 09	8 365 427	2478	8771	92611	3132	203
204	4 16 16	8 489 664	2829	8868	90196	3181	204
205	4 20 25	8 615 125	14.3178	5.8964	4.87805	5.3230	205
206	4 24 36	8 741 816	3527	9059	85437	3279	206
207	4 28 49	8 869 743	3875	9155	83092	3327	207
208	4 32 64	8 998 912	4222	9250	80769	3375	208
209	4 36 81	9 129 329	4568	9345	78469	3423	209
210	4 41 00	9 261 000	14.4914	5.9439	4.76190	5.3471	**210**
211	4 45 21	9 393 931	5258	9533	73934	3519	211
212	4 49 44	9 528 128	5602	9627	71698	3566	212
213	4 53 69	9 663 597	5945	9721	69484	3613	213
214	4 57 96	9 800 344	6287	9814	67290	3660	214
215	4 62 25	9 938 375	14.6629	5.9907	4.65116	5.3706	215
216	4 66 56	10 077 696	6969	6.0000	62963	3753	216
217	4 70 89	10 218 313	7309	0092	60829	3799	217
218	4 75 24	10 360 232	7648	0185	58716	3845	218
219	4 79 61	10 503 459	7986	0277	56621	3891	219
220	4 84 00	10 648 000	14.8324	6.0368	4.54545	5.3936	**220**
221	4 88 41	10 793 861	8661	0459	52489	3982	221
222	4 92 84	10 941 048	8997	0550	50450	4027	222
223	4 97 29	11 089 567	9332	0641	48430	4072	223
224	5 01 76	11 239 424	14.9666	0732	46429	4116	224
225	5 06 25	11 390 625	15.0000	6.0822	4.44444	5.4161	225
226	5 10 76	11 543 176	0333	0912	42478	4205	226
227	5 15 29	11 697 083	0665	1002	40529	4250	227
228	5 19 84	11 852 352	0997	1091	38596	4293	228
229	5 24 41	12 008 989	1327	1180	36681	4337	229
230	5 29 00	12 167 000	15.1658	6.1269	4.34783	5.4381	**230**
231	5 33 61	12 326 391	1987	1358	32900	4424	231
232	5 38 24	12 487 168	2315	1446	31034	4467	232
233	5 42 89	12 649 337	2643	1534	29185	4510	233
234	5 47 56	12 812 904	2971	1622	27350	4553	234
235	5 52 25	12 977 875	15.3297	6.1710	4.25532	5.4596	235
236	5 56 96	13 144 256	3623	1797	23729	4638	236
237	5 61 69	13 312 053	3948	1885	21941	4681	237
238	5 66 44	13 481 272	4272	1972	20168	4723	238
239	5 71 21	13 651 919	4596	2058	18410	4765	239
240	5 76 00	13 824 000	15.4919	6.2145	4.16667	5.4806	**240**
241	5 80 81	13 997 521	5242	2231	14938	4848	241
242	5 85 64	14 172 488	5563	2317	13223	4889	242
243	5 90 49	14 348 907	5885	2403	11523	4931	243
244	5 95 36	14 526 784	6205	2488	09836	4972	244
245	6 00 25	14 706 125	15.6525	6.2573	4.08163	5.5013	245
246	6 05 16	14 886 936	6844	2658	06504	5053	246
247	6 10 09	15 069 223	7162	2743	04858	5094	247
248	6 15 04	15 252 992	7480	2828	03226	5134	248
249	6 20 01	15 438 249	7797	2912	01606	5175	249
250	6 25 00	15 625 000	15.8114	6.2996	4.00000	5.5215	**250**
N	N^2	N^3	$\sqrt{N}$	$\sqrt[3]{N}$	$1000/N$	$\ln N$	N

TABLE 11.—SQUARES, CUBES, SQUARE ROOTS, CUBE ROOTS, RECIPROCALS AND NATURAL LOGARITHMS.—(*Continued*)
(250–300)

N	N^2	N^3	$\sqrt{N}$	$\sqrt[3]{N}$	1000/N	ln N	N
250	6 25 00	15 625 000	15.8114	6.2996	4.00000	5.5215	**250**
251	6 30 01	15 813 251	8430	3080	3.98406	5255	251
252	6 35 04	16 003 008	8745	3164	96825	5294	252
253	6 40 09	16 194 277	9060	3247	95257	5334	253
254	6 45 16	16 387 064	9374	3330	93701	5373	254
255	6 50 25	16 581 375	15.9687	6.3413	3.92157	5.5413	255
256	6 55 36	16 777 216	16.0000	3496	90625	5452	256
257	6 60 49	16 974 593	0312	3579	89105	5491	257
258	6 65 64	17 173 512	0624	3661	87597	5530	258
259	6 70 81	17 373 979	0935	3743	86100	5568	259
260	6 76 00	17 576 000	16.1245	6.3825	3.84615	5.5607	**260**
261	6 81 21	17 779 581	1555	3907	83142	5645	261
262	6 86 44	17 984 728	1864	3988	81679	5683	262
263	6 91 69	18 191 447	2173	4070	80228	5722	263
264	6 96 96	18 399 744	2481	4151	78788	5759	264
265	7 02 25	18 609 625	16.2788	6.4232	3.77358	5.5797	265
266	7 07 56	18 821 096	3095	4312	75940	5835	266
267	7 12 89	19 034 163	3401	4393	74532	5872	267
268	7 18 24	19 248 832	3707	4473	73134	5910	268
269	7 23 61	19 465 109	4012	4553	71747	5947	269
270	7 29 00	19 683 000	16.4317	6.4633	3.70370	5.5984	**270**
271	7 34 41	19 902 511	4621	4713	69004	6021	271
272	7 39 84	20 123 648	4924	4792	67647	6058	272
273	7 45 29	20 346 417	5227	4872	66300	6095	273
274	7 50 76	20 570 824	5529	4951	64964	6131	274
275	7 56 25	20 796 875	16.5831	6.5030	3.63636	5.6168	275
276	7 61 76	21 024 576	6132	5108	62319	6204	276
277	7 67 29	21 253 933	6433	5187	61011	6240	277
278	7 72 84	21 484 952	6733	5265	59712	6276	278
279	7 78 41	21 717 639	7033	5343	58423	6312	279
280	7 84 00	21 952 000	16.7332	6.5421	3.57143	5.6348	**280**
281	7 89 61	22 188 041	7631	5499	55872	6384	281
282	7 95 24	22 425 768	7929	5577	54610	6419	282
283	8 00 89	22 665 187	8226	5654	53357	6454	283
284	8 06 56	22 906 304	8523	5731	52113	6490	284
285	8 12 25	23 149 125	16.8819	6.5808	3.50877	5.6525	285
286	8 17 96	23 393 656	9115	5885	49650	6560	286
287	8 23 69	23 639 903	9411	5962	48432	6595	287
288	8 29 44	23 887 872	16.9706	6039	47222	6630	288
289	8 35 21	24 137 569	17.0000	6115	46021	6664	289
290	8 41 00	24 389 000	17.0294	6.6191	3.44828	5.6699	**290**
291	8 46 81	24 642 171	0587	6267	43643	6733	291
292	8 52 64	24 897 088	0880	6343	42466	6768	292
293	8 58 49	25 153 757	1172	6419	41297	6802	293
294	8 64 36	25 412 184	1464	6494	40136	6836	294
295	8 70 25	25 672 375	17.1756	6.6569	3.38983	5.6870	295
296	8 76 16	25 934 336	2047	6644	37838	6904	296
297	8 82 09	26 198 073	2337	6719	36700	6937	297
298	8 88 04	26 463 592	2627	6794	35570	6971	298
299	8 94 01	26 730 899	2916	6869	34448	7004	299
300	9 00 00	27 000 000	17.3205	6.6943	3.33333	5.7038	**300**
N	N^2	N^3	$\sqrt{N}$	$\sqrt[3]{N}$	1000/N	ln N	N

TABLE 11.—SQUARES, CUBES, SQUARE ROOTS, CUBE ROOTS, RECIPROCALS AND
NATURAL LOGARITHMS.—(Continued)
(300–350)

N	N²	N³	√N	∛N	1000/N	ln N	N
300	9 00 00	27 000 000	17.3205	6.6943	3.33333	5.7038	**300**
301	9 06 01	27 270 901	3494	7018	32226	7071	301
302	9 12 04	27 543 608	3781	7092	31126	7104	302
303	9 18 09	27 818 127	4069	7166	30033	7137	303
304	9 24 16	28 094 464	4356	7240	28947	7170	304
305	9 30 25	28 372 625	17.4642	6.7313	3.27869	5.7203	305
306	9 36 36	28 652 616	4929	7387	26797	7236	306
307	9 42 49	28 934 443	5214	7460	25733	7268	307
308	9 48 64	29 218 112	5499	7533	24675	7301	308
309	9 54 81	29 503 629	5784	7606	23625	7333	309
310	9 61 00	29 791 000	17.6068	6.7679	3.22581	5.7366	**310**
311	9 67 21	30 080 231	6352	7752	21543	7398	311
312	9 73 44	30 371 328	6635	7824	20513	7430	312
313	9 79 69	30 664 297	6918	7897	19489	7462	313
314	9 85 96	30 959 144	7200	7969	18471	7494	314
315	9 92 25	31 255 875	17.7482	6.8041	3.17460	5.7526	315
316	9 98 56	31 554 496	7764	8113	16456	7557	316
317	10 04 89	31 855 013	8045	8185	15457	7589	317
318	10 11 24	32 157 432	8326	8256	14465	7621	318
319	10 17 61	32 461 759	8606	8328	13480	7652	319
320	10 24 00	32 768 000	17.8885	6.8399	3.12500	5.7683	**320**
321	10 30 41	33 076 161	9165	8470	11526	7714	321
322	10 36 84	33 386 248	9444	8541	10559	7746	322
323	10 43 29	33 698 267	17.9722	8612	09598	7777	323
324	10 49 76	34 012 224	18.0000	8683	08642	7807	324
325	10 56 25	34 328 125	18.0278	6.8753	3.07692	5.7838	325
326	10 62 76	34 645 976	0555	8824	06748	7869	326
327	10 69 29	34 965 783	0831	8894	05810	7900	327
328	10 75 84	35 287 552	1108	8964	04878	7930	328
329	10 82 41	35 611 289	1384	9034	03951	7961	329
330	10 89 00	35 937 000	18.1659	6.9104	3.03030	5.7991	**330**
331	10 95 61	36 264 691	1934	9174	02115	8021	331
332	11 02 24	36 594 368	2209	9244	01205	8051	332
333	11 08 89	36 926 037	2483	9313	3.00300	8081	333
334	11 15 56	37 259 704	2757	9382	2.99401	8111	334
335	11 22 25	37 595 375	18.3030	6.9451	2.98507	5.8141	335
336	11 28 96	37 933 056	3303	9521	97619	8171	336
337	11 35 69	38 272 753	3576	9589	96736	8201	337
338	11 42 44	38 614 472	3848	9658	95858	8230	338
339	11 49 21	38 958 219	4120	9727	94985	8260	339
340	11 56 00	39 304 000	18.4391	6.9795	2.94118	5.8289	**340**
341	11 62 81	39 651 821	4662	9864	93255	8319	341
342	11 69 64	40 001 688	4932	6.9932	92398	8348	342
343	11 76 49	40 353 607	5203	7.0000	91545	8377	343
344	11 83 36	40 707 584	5472	0068	90698	8406	344
345	11 90 25	41 063 625	18.5742	7.0136	2.89855	5.8435	345
346	11 97 16	41 421 736	6011	0203	89017	8464	346
347	12 04 09	41 781 923	6279	0271	88184	8493	347
348	12 11 04	42 144 192	6548	0338	87356	8522	348
349	12 18 01	42 508 549	6815	0406	86533	8551	349
350	12 25 00	42 875 000	18.7083	7.0473	2.85714	5.8579	**350**
N	N²	N³	√N	∛N	1000/N	ln N	N

TABLE 11.—SQUARES, CUBES, SQUARE ROOTS, CUBE ROOTS, RECIPROCALS AND
NATURAL LOGARITHMS.—(*Continued*)
(350–400)

N	N^2	N^3	$\sqrt{N}$	$\sqrt[3]{N}$	1000/N	ln N	N
350	12 25 00	42 875 000	18.7083	7.0473	2.85714	5.8579	**350**
351	12 32 01	43 243 551	7350	0540	84900	8608	351
352	12 39 04	43 614 208	7617	0607	84091	8636	352
353	12 46 09	43 986 977	7883	0674	83286	8665	353
354	12 53 16	44 361 864	8149	0740	82486	8693	354
355	12 60 25	44 738 875	18.8414	7.0807	2.81690	5.8721	355
356	12 67 36	45 118 016	8680	0873	80899	8749	356
357	12 74 49	45 499 293	8944	0940	80112	8777	357
358	12 81 64	45 882 712	9209	1006	79330	8805	358
359	12 88 81	46 268 279	9473	1072	78552	8833	359
360	12 96 00	46 656 000	18.9737	7.1138	2.77778	5.8861	**360**
361	13 03 21	47 045 881	19.0000	1204	77008	8889	361
362	13 10 44	47 437 928	0263	1269	76243	8916	362
363	13 17 69	47 832 147	0526	1335	75482	8944	363
364	13 24 96	48 228 544	0788	1400	74725	8972	364
365	13 32 25	48 627 125	19.1050	7.1466	2.73973	5.8999	365
366	13 39 56	49 027 896	1311	1531	73224	9026	366
367	13 46 89	49 430 863	1572	1596	72480	9054	367
368	13 54 24	49 836 032	1833	1661	71739	9081	368
369	13 61 61	50 243 409	2094	1726	71003	9108	369
370	13 69 00	50 653 000	19.2354	7.1791	2.70270	5.9135	**370**
371	13 76 41	51 064 811	2614	1855	69542	9162	371
372	13 83 84	51 478 848	2873	1920	68817	9189	372
373	13 91 29	51 895 117	3132	1984	68097	9216	373
374	13 98 76	52 313 624	3391	2048	67380	9243	374
375	14 06 25	52 734 375	19.3649	7.2112	2.66667	5.9269	375
376	14 13 76	53 157 376	3907	2177	65957	9296	376
377	14 21 29	53 582 633	4165	2240	65252	9322	377
378	14 28 84	54 010 152	4422	2304	64550	9349	378
379	14 36 41	54 439 939	4679	2368	63852	9375	379
380	14 44 00	54 872 000	19.4936	7.2432	2.63158	5.9402	**380**
381	14 51 61	55 306 341	5192	2495	62467	9428	381
382	14 59 24	55 742 968	5448	2558	61780	9454	382
383	14 66 89	56 181 887	5704	2622	61097	9480	383
384	14 74 56	56 623 104	5959	2685	60417	9506	384
385	14 82 25	57 066 625	19.6214	7.2748	2.59740	5.9532	385
386	14 89 96	57 512 456	6469	2811	59067	9558	386
387	14 97 69	57 960 603	6723	2874	58398	9584	387
388	15 05 44	58 411 072	6977	2936	57732	9610	388
389	15 13 21	58 863 869	7231	2999	57069	9636	389
390	15 21 00	59 319 000	19.7484	7.3061	2.56410	5.9661	**390**
391	15 28 81	59 776 471	7737	3124	55754	9687	391
392	15 36 64	60 236 288	7990	3186	55102	9713	392
393	15 44 49	60 698 457	8242	3248	54453	9738	393
394	15 52 36	61 162 984	8494	3310	53807	9764	394
395	15 60 25	61 629 875	19.8746	7.3372	2.53165	5.9789	395
396	15 68 16	62 099 136	8997	3434	52525	9814	396
397	15 76 09	62 570 773	9249	3496	51889	9839	397
398	15 84 04	63 044 792	9499	3558	51256	9865	398
399	15 92 01	63 521 199	19.9750	3619	50627	9890	399
400	16 00 00	64 000 000	20.0000	7.3681	2.50000	5.9915	**400**
N	N^2	N^3	$\sqrt{N}$	$\sqrt[3]{N}$	1000/N	ln N	N

TABLE 11.—SQUARES, CUBES, SQUARE ROOTS, CUBE ROOTS, RECIPROCALS AND
NATURAL LOGARITHMS.—(*Continued*)
(400–450)

N	N²	N³	$\sqrt{N}$	$\sqrt[3]{N}$	1000/N	ln N	N
400	16 00 00	64 000 000	20.0000	7.3681	2.50000	5.9915	**400**
401	16 08 01	64 481 201	0250	3742	49377	9940	401
402	16 16 04	64 964 808	0499	3803	48756	9965	402
403	16 24 09	65 450 827	0749	3864	48139	5.9989	403
404	16 32 16	65 939 264	0998	3925	47525	6.0014	404
405	16 40 25	66 430 125	20.1246	7.3986	2.46914	6.0039	405
406	16 48 36	66 923 416	1494	4047	46305	0064	406
407	16 56 49	67 419 143	1742	4108	45700	0088	407
408	16 64 64	67 917 312	1990	4169	45098	0113	408
409	16 72 81	68 417 929	2237	4229	44499	0137	409
410	16 81 00	68 921 000	20.2485	7.4290	2.43902	6.0162	**410**
411	16 89 21	69 426 531	2731	4350	43309	0186	411
412	16 97 44	69 934 528	2978	4410	42718	0210	412
413	17 05 69	70 444 997	3224	4470	42131	0234	413
414	17 13 96	70 957 944	3470	4530	41546	0259	414
415	17 22 25	71 473 375	20.3715	7.4590	2.40964	6.0283	415
416	17 30 56	71 991 296	3961	4650	40385	0307	416
417	17 38 89	72 511 713	4206	4710	39808	0331	417
418	17 47 24	73 034 632	4450	4770	39234	0355	418
419	17 55 61	73 560 059	4695	4829	38663	0379	419
420	17 64 00	74 088 000	20.4939	7.4889	2.38095	6.0403	**420**
421	17 72 41	74 618 461	5183	4948	37530	0426	421
422	17 80 84	75 151 448	5426	5007	36967	0450	422
423	17 89 29	75 686 967	5670	5067	36407	0474	423
424	17 97 76	76 225 024	5913	5126	35849	0497	424
425	18 06 25	76 765 625	20.6155	7.5185	2.35294	6.0521	425
426	18 14 76	77 308 776	6398	5244	34742	0544	426
427	18 23 29	77 854 483	6640	5302	34192	0568	427
428	18 31 84	78 402 752	6882	5361	33645	0591	428
429	18 40 41	78 953 589	7123	5420	33100	0615	429
430	18 49 00	79 507 000	20.7364	7.5478	2.32558	6.0638	**430**
431	18 57 61	80 062 991	7605	5537	32019	0661	431
432	18 66 24	80 621 568	7846	5595	31481	0684	432
433	18 74 89	81 182 737	8087	5654	30947	0707	433
434	18 83 56	81 746 504	8327	5712	30415	0730	434
435	18 92 25	82 312 875	20.8567	7.5770	2.29885	6.0753	435
436	19 00 96	82 881 856	8806	5828	29358	0776	436
437	19 09 69	83 453 453	9045	5886	28833	0799	437
438	19 18 44	84 027 672	9284	5944	28311	0822	438
439	19 27 21	84 604 519	9523	6001	27790	0845	439
440	19 36 00	85 184 000	20.9762	7.6059	2.27273	6.0868	**440**
441	19 44 81	85 766 121	21.0000	6117	26757	0890	441
442	19 53 64	86 350 888	0238	6174	26244	0913	442
443	19 62 49	86 938 307	0476	6232	25734	0936	443
444	19 71 36	87 528 384	0713	6289	25225	0958	444
445	19 80 25	88 121 125	21.0950	7.6346	2.24719	6.0981	445
446	19 89 16	88 716 536	1187	6403	24215	1003	446
447	19 98 09	89 314 623	1424	6460	23714	1026	447
448	20 07 04	89 915 392	1660	6517	23214	1048	448
449	20 16 01	90 518 849	1896	6574	22717	1070	449
450	20 25 00	91 125 000	21.2132	7.6631	2.22222	6.1092	**450**
N	N²	N³	$\sqrt{N}$	$\sqrt[3]{N}$	1000/N	ln N	N

TABLE 11.—SQUARES, CUBES, SQUARE ROOTS, CUBE ROOTS, RECIPROCALS AND NATURAL LOGARITHMS.—(*Continued*)
(450–500)

N	N^2	N^3	$\sqrt{N}$	$\sqrt[3]{N}$	1000/N	ln N	N
450	20 25 00	91 125 000	21.2132	7.6631	2.22222	6.1092	**450**
451	20 34 01	91 733 851	2368	6688	21729	1115	451
452	20 43 04	92 345 408	2603	6744	21239	1137	452
453	20 52 09	92 959 677	2838	6801	20751	1159	453
454	20 61 16	93 576 664	3073	6857	20264	1181	454
455	20 70 25	94 196 375	21.3307	7.6914	2.19780	6.1203	455
456	20 79 36	94 818 816	3542	6970	19298	1225	456
457	20 88 49	95 443 993	3776	7026	18818	1247	457
458	20 97 64	96 071 912	4009	7082	18341	1269	458
459	21 06 81	96 702 579	4243	7138	17865	1291	459
460	21 16 00	97 336 000	21.4476	7.7194	2.17391	6.1312	**460**
461	21 25 21	97 972 181	4709	7250	16920	1334	461
462	21 34 44	98 611 128	4942	7306	16450	1356	462
463	21 43 69	99 252 847	5174	7362	15983	1377	463
464	21 52 96	99 897 344	5407	7418	15517	1399	464
465	21 62 25	100 544 625	21.5639	7.7473	2.15054	6.1420	465
466	21 71 56	101 194 696	5870	7529	14592	1442	466
467	21 80 89	101 847 563	6102	7584	14133	1463	467
468	21 90 24	102 503 232	6333	7639	13675	1485	468
469	21 99 61	103 161 709	6564	7695	13220	1506	469
470	22 09 00	103 823 000	21.6795	7.7750	2.12766	6.1527	**470**
471	22 18 41	104 487 111	7025	7805	12314	1549	471
472	22 27 84	105 154 048	7256	7860	11864	1570	472
473	22 37 29	105 823 817	7486	7915	11416	1591	473
474	22 46 76	106 496 424	7715	7970	10970	1612	474
475	22 56 25	107 171 875	21.7945	7.8025	2.10526	6.1633	475
476	22 65 76	107 850 176	8174	8079	10084	1654	476
477	22 75 29	108 531 333	8403	8134	09644	1675	477
478	22 84 84	109 215 352	8632	8188	09205	1696	478
479	22 94 41	109 902 239	8861	8243	08768	1717	479
480	23 04 00	110 592 000	21.9089	7.8297	2.08333	6.1738	**480**
481	23 13 61	111 284 641	9317	8352	07900	1759	481
482	23 23 24	111 980 168	9545	8406	07469	1779	482
483	23 32 89	112 678 587	21.9773	8460	07039	1800	483
484	23 42 56	113 379 904	22.0000	8514	06612	1821	484
485	23 52 25	114 084 125	22.0227	7.8568	2.06186	6.1841	485
486	23 61 96	114 791 256	0454	8622	05761	1862	486
487	23 71 69	115 501 303	0681	8676	05339	1883	487
488	23 81 44	116 214 272	0907	8730	04918	1903	488
489	23 91 21	116 930 169	1133	8784	04499	1924	489
490	24 01 00	117 649 000	22.1359	7.8837	2.04082	6.1944	**490**
491	24 10 81	118 370 771	1585	8891	03666	1964	491
492	24 20 64	119 095 488	1811	8944	03252	1985	492
493	24 30 49	119 823 157	2036	8998	02840	2005	493
494	24 40 36	120 553 784	2261	9051	02429	2025	494
495	24 50 25	121 287 375	22.2486	7.9105	2.02020	6.2046	495
496	24 60 16	122 023 936	2711	9158	01613	2066	496
497	24 70 09	122 763 473	2935	9211	01207	2086	497
498	24 80 04	123 505 992	3159	9264	00803	2106	498
499	24 90 01	124 251 499	3383	9317	00401	2126	499
500	25 00 00	125 000 000	22.3607	7.9370	2.00000	6.2146	**500**
N	N^2	N^3	$\sqrt{N}$	$\sqrt[3]{N}$	1000/N	ln N	N

TABLE 11.—SQUARES, CUBES, SQUARE ROOTS, CUBE ROOTS, RECIPROCALS AND
NATURAL LOGARITHMS.—(*Continued*)
(500–550)

N	N²	N³	$\sqrt{N}$	$\sqrt[3]{N}$	1000/N	ln N	N
500	25 00 00	125 000 000	22.3607	7.9370	2.00000	6.2146	**500**
501	25 10 01	125 751 501	3830	9423	1.99601	2166	501
502	25 20 04	126 506 008	4054	9476	99203	2186	502
503	25 30 09	127 263 527	4277	9528	98807	2206	503
504	25 40 16	128 024 064	4499	9581	98413	2226	504
505	25 50 25	128 787 625	22.4722	7.9634	1.98020	6.2246	505
506	25 60 36	129 554 216	4944	9686	97628	2265	506
507	25 70 49	130 323 843	5167	9739	97239	2285	507
508	25 80 64	131 096 512	5389	9791	96850	2305	508
509	25 90 81	131 872 229	5610	9843	96464	2324	509
510	26 01 00	132 651 000	22.5832	7.9896	1.96078	6.2344	**510**
511	26 11 21	133 432 831	6053	7.9948	95695	2364	511
512	26 21 44	134 217 728	6274	8.0000	95312	2383	512
513	26 31 69	135 005 697	6495	0052	94932	2403	513
514	26 41 96	135 796 744	6716	0104	94553	2422	514
515	26 52 25	136 590 875	22.6936	8.0156	1.94175	6.2442	515
516	26 62 56	137 388 096	7156	0208	93798	2461	516
517	26 72 89	138 188 413	7376	0260	93424	2480	517
518	26 83 24	138 991 832	7596	0311	93050	2500	518
519	26 93 61	139 798 359	7816	0363	92678	2519	519
520	27 04 00	140 608 000	22.8035	8.0415	1.92308	6.2538	**520**
521	27 14 41	141 420 761	8254	0466	91939	2558	521
522	27 24 84	142 236 648	8473	0517	91571	2577	522
523	27 35 29	143 055 667	8692	0569	91205	2596	523
524	27 45 76	143 877 824	8910	0620	90840	2615	524
525	27 56 25	144 703 125	22.9129	8.0671	1.90476	6.2634	525
526	27 66 76	145 531 576	9347	0723	90114	2653	526
527	27 77 29	146 363 183	9565	0774	89753	2672	527
528	27 87 84	147 197 952	22.9783	0825	89394	2691	**528**
529	27 98 41	148 035 889	23.0000	0876	89036	2710	529
530	28 09 00	148 877 000	23.0217	8.0927	1.88679	6.2729	**530**
531	28 19 61	149 721 291	0434	0978	88324	2748	531
532	28 30 24	150 568 768	0651	1028	87970	2766	532
533	28 40 89	151 419 437	0868	1079	87617	2785	533
534	28 51 56	152 273 304	1084	1130	87266	2804	534
535	28 62 25	153 130 375	23.1301	8.1180	1.86916	6.2823	535
536	28 72 96	153 990 656	1517	1231	86567	2841	536
537	28 83 69	154 854 153	1733	1281	86220	2860	537
538	28 94 44	155 720 872	1948	1332	85874	2879	538
539	29 05 21	156 590 819	2164	1382	85529	2897	539
540	29 16 00	157 464 000	23.2379	8.1433	1.85185	6.2916	**540**
541	29 26 81	158 340 421	2594	1483	84843	2934	541
542	29 37 64	159 220 088	2809	1533	84502	2953	542
543	29 48 49	160 103 007	3024	1583	84162	2971	543
544	29 59 36	160 989 184	3238	1633	83824	2989	544
545	29 70 25	161 878 625	23.3452	8.1683	1.83486	6.3008	545
546	29 81 16	162 771 336	3666	1733	83150	3026	546
547	29 92 09	163 667 323	3880	1783	82815	3044	547
548	30 03 04	164 566 592	4094	1833	82482	3063	548
549	30 14 01	165 469 149	4307	1882	82149	3081	549
550	30 25 00	166 375 000	23.4521	8.1932	1.81818	6.3099	**550**
N	N²	N³	$\sqrt{N}$	$\sqrt[3]{N}$	1000/N	ln N	N

TABLE 11.—SQUARES, CUBES, SQUARE ROOTS, CUBE ROOTS, RECIPROCALS AND NATURAL LOGARITHMS.—(*Continued*)
(550–600)

N	N^2	N^3	$\sqrt{N}$	$\sqrt[3]{N}$	1000/N	ln N	N
550	30 25 00	166 375 000	23.4521	8.1932	1.81818	6.3099	**550**
551	30 36 01	167 284 151	4734	1982	81488	3117	551
552	30 47 04	168 196 608	4947	2031	81159	3135	552
553	30 58 09	169 112 377	5160	2081	80832	3154	553
554	30 69 16	170 031 464	5372	2130	80505	3172	554
555	30 80 25	170 953 875	23.5584	8.2180	1.80180	6.3190	555
556	30 91 36	171 879 616	5797	2229	79856	3208	556
557	31 02 49	172 808 693	6008	2278	79533	3226	557
558	31 13 64	173 741 112	6220	2327	79211	3244	558
559	31 24 81	174 676 879	6432	2377	78891	3261	559
560	31 36 00	175 616 000	23.6643	8.2426	1.78571	6.3279	**560**
561	31 47 21	176 558 481	6854	2475	78253	3297	561
562	31 58 44	177 504 328	7065	2524	77936	3315	562
563	31 69 69	178 453 547	7276	2573	77620	3333	563
564	31 80 96	179 406 144	7487	2621	77305	3351	564
565	31 92 25	180 362 125	23.7697	8.2670	1.76991	6.3368	565
566	32 03 56	181 321 496	7908	2719	76678	3386	566
567	32 14 89	182 284 263	8118	2768	76367	3404	567
568	32 26 24	183 250 432	8328	2816	76056	3421	568
569	32 37 61	184 220 009	8537	2865	75747	3439	569
570	32 49 00	185 193 000	23.8747	8.2913	1.75439	6.3456	**570**
571	32 60 41	186 169 411	8956	2962	75131	3474	571
572	32 71 84	187 149 248	9165	3010	74825	3491	572
573	32 83 29	188 132 517	9374	3059	74520	3509	573
574	32 94 76	189 119 224	9583	3107	74216	3526	574
575	33 06 25	190 109 375	23.9792	8.3155	1.73913	6.3544	575
576	33 17 76	191 102 976	24.0000	3203	73611	3561	576
577	33 29 29	192 100 033	0208	3251	73310	3578	577
578	33 40 84	193 100 552	0416	3300	73010	3596	578
579	33 52 41	194 104 539	0624	3348	72712	3613	579
580	33 64 00	195 112 000	24.0832	8.3396	1.72414	6.3630	**580**
581	33 75 61	196 122 941	1039	3443	72117	3648	581
582	33 87 24	197 137 368	1247	3491	71821	3665	582
583	33 98 89	198 155 287	1454	3539	71527	3682	583
584	34 10 56	199 176 704	1661	3587	71233	3699	584
585	34 22 25	200 201 625	24.1868	8.3634	1.70940	6.3716	585
586	34 33 96	201 230 056	2074	3682	70648	3733	586
587	34 45 69	202 262 003	2281	3730	70358	3750	587
588	34 57 44	203 297 472	2487	3777	70068	3767	588
589	34 69 21	204 336 469	2693	3825	69779	3784	589
590	34 81 00	205 379 000	24.2899	8.3872	1.69492	6.3801	**590**
591	34 92 81	206 425 071	3105	3919	69205	3818	591
592	35 04 64	207 474 688	3311	3967	68919	3835	592
593	35 16 49	208 527 857	3516	4014	68634	3852	593
594	35 28 36	209 584 584	3721	4061	68350	3869	594
595	35 40 25	210 644 875	24.3926	8.4108	1.68067	6.3886	595
596	35 52 16	211 708 736	4131	4155	67785	3902	596
597	35 64 09	212 776 173	4336	4202	67504	3919	597
598	35 76 04	213 847 192	4540	4249	67224	3936	598
599	35 88 01	214 921 799	4745	4296	66945	3953	599
600	36 00 00	216 000 000	24.4949	8.4343	1.66667	6.3969	**600**
N	N^2	N^3	$\sqrt{N}$	$\sqrt[3]{N}$	1000/N	ln N	N

TABLE 11.—Squares, Cubes, Square Roots, Cube Roots, Reciprocals and Natural Logarithms.—(*Continued*)
(600–650)

N	N²	N³	√N	∛N	1000/N	ln N	N
600	36 00 00	216 000 000	24.4949	8.4343	1.66667	6.3969	**600**
601	36 12 01	217 081 801	5153	4390	66389	3986	601
602	36 24 04	218 167 208	5357	4437	66113	4003	602
603	36 36 09	219 256 227	5561	4484	65837	4019	603
604	36 48 16	220 348 864	5764	4530	65563	4036	604
605	36 60 25	221 445 125	24.5967	8.4577	1.65289	6.4052	605
606	36 72 36	222 545 016	6171	4623	65017	4069	606
607	36 84 49	223 648 543	6374	4670	64745	4085	607
608	36 96 64	224 755 712	6577	4716	64474	4102	608
609	37 08 81	225 866 529	6779	4763	64204	4118	609
610	37 21 00	226 981 000	24.6982	8.4809	1.63934	6.4135	**610**
611	37 33 21	228 099 131	7184	4856	63666	4151	611
612	37 45 44	229 220 928	7386	4902	63399	4167	612
613	37 57 69	230 346 397	7588	4948	63132	4184	613
614	37 69 96	231 475 544	7790	4994	62866	4200	614
615	37 82 25	232 608 375	24.7992	8.5040	1.62602	6.4216	615
616	37 94 56	233 744 896	8193	5086	62338	4232	616
617	38 06 89	234 885 113	8395	5132	62075	4249	617
618	38 19 24	236 029 032	8596	5178	61812	4265	618
619	38 31.61	237 176 659	8797	5224	61551	4281	619
620	38 44 00	238 328 000	24.8998	8.5270	1.61290	6.4297	**620**
621	38 56 41	239 483 061	9199	5316	61031	4313	621
622	38 68 84	240 641 848	9399	5362	60772	4329	622
623	38 81 29	241 804 367	9600	5408	60514	4345	623
624	38 93 76	242 970 624	24.9800	5453	60256	4362	624
625	39 06 25	244 140 625	25.0000	8.5499	1.60000	6.4378	625
626	39 18 76	245 314 376	0200	5544	59744	4394	626
627	39 31 29	246 491 883	0400	5590	59490	4409	627
628	39 43 84	247 673 152	0599	5635	59236	4425	628
629	39 56 41	248 858 189	0799	5681	58983	4441	629
630	39 69 00	250 047 000	25.0998	8.5726	1.58730	6.4457	**630**
631	39 81 61	251 239 591	1197	5772	58479	4473	631
632	39 94 24	252 435 968	1396	5817	58228	4489	632
633	40 06 89	253 636 137	1595	5862	57978	4505	633
634	40 19 56	254 840 104	1794	5907	57729	4520	634
635	40 32 25	256 047 875	25.1992	8.5952	1.57480	6.4536	·635
636	40 44 96	257 259 456	2190	5997	57233	4552	636
637	40 57 69	258 474 853	2389	6043	56986	4568	637
638	40 70 44	259 694 072	2587	6088	56740	4583	638
639	40 83 21	260 917 119	2784	6132	56495	4599	639
640	40 96 00	262 144 000	25.2982	8.6177	1.56250	6.4615	**640**
641	41 08 81	263 374 721	3180	6222	56006	4630	641
642	41 21 64	264 609 288	3377	6267	55763	4646	642
643	41 34 49	265 847 707	3574	6312	55521	4661	643
644	41 47 36	267 089 984	3772	6357	55280	4677	644
645	41 60 25	268 336 125	25.3969	8.6401	1.55039	6.4693	645
646	41 73 16	269 586 136	4165	6446	54799	4708	646
647	41 86 09	270 840 023	4362	6490	54560	4723	647
648	41 99 04	272 097 792	4558	6535	54321	4739	648
649	42 12 01	273 359 449	4755	6579	54083	4754	649
650	42 25 00	274 625 000	25.4951	8.6624	1.53846	6.4770	**650**
N	N²	N³	√N	∛N	1000/N	ln N	N

TABLE 11.—SQUARES, CUBES, SQUARE ROOTS, CUBE ROOTS, RECIPROCALS AND
NATURAL LOGARITHMS.—(*Continued*)
(650–700)

N	N²	N³	√N	∛N	1000/N	ln N	N
650	42 25 00	274 625 000	25.4951	8.6624	1.53846	6.4770	**650**
651	42 38 01	275 894 451	5147	6668	53610	4785	651
652	42 51 04	277 167 808	5343	6713	53374	4800	652
653	42 64 09	278 445 077	5539	6757	53139	4816	653
654	42 77 16	279 726 264	5734	6801	52905	4831	654
655	42 90 25	281 011 375	25.5930	8.6845	1.52672	6.4846	655
656	43 03 36	282 300 416	6125	6890	52439	4862	656
657	43 16 49	283 593 393	6320	6934	52207	4877	657
658	43 29 64	284 890 312	6515	6978	51976	4892	658
659	43 42 81	286 191 179	6710	7022	51745	4907	659
660	43 56 00	287 496 000	25.6905	8.7066	1.51515	6.4922	**660**
661	43 69 21	288 804 781	7099	7110	51286	4938	661
662	43 82 44	290 117 528	7294	7154	51057	4953	662
663	43 95 69	291 434 247	7488	7198	50830	4968	663
664	44 08 96	292 754 944	7682	7241	50602	4983	664
665	44 22 25	294 079 625	25.7876	8.7285	1.50376	6.4998	665
666	44 35 56	295 408 296	8070	7329	50150	5013	666
667	44 48 89	296 740 963	8263	7373	49925	5028	667
668	44 62 24	298 077 632	8457	7416	49701	5043	668
669	44 75 61	299 418 309	8650	7460	49477	5058	669
670	44 89 00	300 763 000	25.8844	8.7503	1.49254	6.5073	**670**
671	45 02 41	302 111 711	9037	7547	49031	5088	671
672	45 15 84	303 464 448	9230	7590	48810	5103	672
673	45 29 29	304 821 217	9422	7634	48588	5117	673
674	45 42 76	306 182 024	9615	7677	48368	5132	674
675	45 56 25	307 546 875	25.9808	8.7721	1.48148	6.5147	675
676	45 69 76	308 915 776	26.0000	7764	47929	5162	676
677	45 83 29	310 288 733	0192	7807	47710	5177	677
678	45 96 84	311 665 752	0384	7850	47493	5191	678
679	46 10 41	313 046 839	0576	7893	47275	5206	679
680	46 24 00	314 432 000	26.0768	8.7937	1.47059	6.5221	**680**
681	46 37 61	315 821 241	0960	7980	46843	5236	681
682	46 51 24	317 214 568	1151	8023	46628	5250	682
683	46 64 89	318 611 987	1343	8066	46413	5265	683
684	46 78 56	320 013 504	1534	8109	46199	5280	684
685	46 92 25	321 419 125	26.1725	8.8152	1.45985	6.5294	685
686	47 05 96	322 828 856	1916	8194	45773	5309	686
687	47 19 69	324 242 703	2107	8237	45560	5323	687
688	47 33 44	325 660 672	2298	8280	45349	5338	688
689	47 47 21	327 082 769	2488	8323	45138	5352	689
690	47 61 00	328 509 000	26.2679	8.8366	1.44928	6.5367	**690**
691	47 74 81	329 939 371	2869	8408	44718	5381	691
692	47 88 64	331 373 888	3059	8451	44509	5396	692
693	48 02 49	332 812 557	3249	8493	44300	5410	693
694	48 16 36	334 255 384	3439	8536	44092	5425	694
695	48 30 25	335 702 375	26.3629	8.8578	1.43885	6.5439	695
696	48 44 16	337 153 536	3818	8621	43678	5453	696
697	48 58 09	338 608 873	4008	8663	43472	5468	697
698	48 72 04	340 068 392	4197	8706	43266	5482	698
699	48 86 01	341 532 099	4386	8748	43062	5497	699
700	49 00 00	343 000 000	26.4575	8.8790	1.42857	6.5511	**700**
N	N²	N³	√N	∛N	1000/N	ln N	N

TABLE 11.—SQUARES, CUBES, SQUARE ROOTS, CUBE ROOTS, RECIPROCALS AND
NATURAL LOGARITHMS.—*(Continued)*
(700–750)

N	N²	N³	$\sqrt{N}$	$\sqrt[3]{N}$	1000/N	ln N	N
700	49 00 00	343 000 000	26.4575	8.8790	1.42857	6.5511	**700**
701	49 14 01	344 472 101	4764	8833	42653	5525	701
702	49 28 04	345 948 408	4953	8875	42450	5539	702
703	49 42 09	347 428 927	5141	8917	42248	5554	703
704	49 56 16	348 913 664	5330	8959	42045	5568	704
705	49 70 25	350 402 625	26.5518	8.9001	1.41844	6.5582	705
706	49 84 36	351 895 816	5707	9043	41643	5596	706
707	49 98 49	353 393 243	5895	9085	41443	5610	707
708	50 12 64	354 894 912	6083	9127	41243	5624	708
709	50 26 81	356 400 829	6271	9169	41044	5639	709
710	50 41 00	357 911 000	26.6458	8.9211	1.40845	6.5653	**710**
711	50 55 21	359 425 431	6646	9253	40647	5667	711
712	50 69 44	360 944 128	6833	9295	40449	5681	712
713	50 83 69	362 467 097	7021	9337	40252	5695	713
714	50 97 96	363 994 344	7208	9378	40056	5709	714
715	51 12 25	365 525 875	26.7395	8.9420	1.39860	6.5723	715
716	51 26 56	367 061 696	7582	9462	39665	5737	716
717	51 40 89	368 601 813	7769	9503	39470	5751	717
718	51 55 24	370 146 232	7955	9545	39276	5765	718
719	51 69 61	371 694 959	8142	9587	39082	5779	719
720	51 84 00	373 248 000	26.8328	8.9628	1.38889	6.5793	**720**
721	51 98 41	374 805 361	8514	9670	38696	5806	721
722	52 12 84	376 367 048	8701	9711	38504	5820	722
723	52 27 29	377 933 067	8887	9752	38313	5834	723
724	52 41 76	379 503 424	9072	9794	38122	5848	724
725	52 56 25	381 078 125	26.9258	8.9835	1.37931	6.5862	725
726	52 70 76	382 657 176	9444	9876	37741	5876	726
727	52 85 29	384 240 583	9629	9918	37552	5889	727
728	52 99 84	385 828 352	26.9815	8.9959	37363	5903	728
729	53 14 41	387 420 489	27.0000	9.0000	37174	5917	729
730	53 29 00	389 017 000	27.0185	9.0041	1.36986	6.5930	**730**
731	53 43 61	390 617 891	0370	0082	36799	5944	731
732	53 58 24	392 223 168	0555	0123	36612	5958	732
733	53 72 89	393 832 837	0740	0164	36426	5971	733
734	53 87 56	395 446 904	0924	0205	36240	5985	734
735	54 02 25	397 065 375	27.1109	9.0246	1.36054	6.5999	735
736	54 16 96	398 688 256	1293	0287	35870	6012	736
737	54 31 69	400 315 553	1477	0328	35685	6026	737
738	54 46 44	401 947 272	1662	0369	35501	6039	738
739	54 61 21	403 583 419	1846	0410	35318	6053	739
740	54 76 00	405 224 000	27.2029	9.0450	1.35135	6.6067	**740**
741	54 90 81	406 869 021	2213	0491	34953	6080	741
742	55 05 64	408 518 488	2397	0532	34771	6093	742
743	55 20 49	410 172 407	2580	0572	34590	6107	743
744	55 35 36	411 830 784	2764	0613	34409	6120	744
745	55 50 25	413 493 625	27.2947	9.0654	1.34228	6.6134	745
746	55 65 16	415 160 936	3130	0694	34048	6147	746
747	55 80 09	416 832 723	3313	0735	33869	6161	747
748	55 95 04	418 508 992	3496	0775	33690	6174	748
749	56 10 01	420 189 749	3679	0816	33511	6187	749
750	56 25 00	421 875 000	27.3861	9.0856	1.33333	6.6201	**750**
N	N²	N³	$\sqrt{N}$	$\sqrt[3]{N}$	1000/N	ln N	N

TABLE 11.—SQUARES, CUBES, SQUARE ROOTS, CUBE ROOTS, RECIPROCALS AND
NATURAL LOGARITHMS.—(Continued)
(750–800)

N	N²	N³	$\sqrt{N}$	$\sqrt[3]{N}$	1000/N	ln N	N
750	56 25 00	421 875 000	27.3861	9.0856	1.33333	6.6201	750
751	56 40 01	423 564 751	4044	0896	33156	6214	751
752	56 55 04	425 259 008	4226	0937	32979	6227	752
753	56 70 09	426 957 777	4408	0977	32802	6241	753
754	56 85 16	428 661 064	4591	1017	32626	6254	754
755	57 00 25	430 368 875	27.4773	9.1057	1.32450	6.6267	755
756	57 15 36	432 081 216	4955	1098	32275	6280	756
757	57 30 49	433 798 093	5136	1138	32100	6294	757
758	57 45 64	435 519 512	5318	1178	31926	6307	758
759	57 60 81	437 245 479	5500	1218	31752	6320	759
760	57 76 00	438 976 000	27.5681	9.1258	1.31579	6.6333	760
761	57 91 21	440 711 081	5862	1298	31406	6346	761
762	58 06 44	442 450 728	6043	1338	31234	6359	762
763	58 21 69	444 194 947	6225	1378	31062	6373	763
764	58 36 96	445 943 744	6405	1418	30890	6386	764
765	58 52 25	447 697 125	27.6586	9.1458	1.30719	6.6399	765
766	58 67 56	449 455 096	6767	1498	30548	6412	766
767	58 82 89	451 217 663	6948	1537	30378	6425	767
768	58 98 24	452 984 832	7128	1577	30208	6438	768
769	59 13 61	454 756 609	7308	1617	30039	6451	769
770	59 29 00	456 533 000	27.7489	9.1657	1.29870	6.6464	770
771	59 44 41	458 314 011	7669	1696	29702	6477	771
772	59 59 84	460 099 648	7849	1736	29534	6490	772
773	59 75 29	461 889 917	8029	1775	29366	6503	773
774	59 90 76	463 684 824	8209	1815	29199	6516	774
775	60 06 25	465 484 375	27.8388	9.1855	1.29032	6.6529	775
776	60 21 76	467 288 576	8568	1894	28866	6542	776
777	60 37 29	469 097 433	8747	1933	28700	6554	777
778	60 52 84	470 910 952	8927	1973	28535	6567	778
779	60 68 41	472 729 139	9106	2012	28370	6580	779
780	60 84 00	474 552 000	27.9285	9.2052	1.28205	6.6593	780
781	60 99 61	476 379 541	9464	2091	28041	6606	781
782	61 15 24	478 211 768	9643	2130	27877	6619	782
783	61 30 89	480 048 687	27.9821	2170	27714	6631	783
784	61 46 56	481 890 304	28.0000	2209	27551	6644	784
785	61 62 25	483 736 625	28.0179	9.2248	1.27389	6.6657	785
786	61 77 96	485 587 656	0357	2287	27226	6670	786
787	61 93 69	487 443 403	0535	2326	27065	6682	787
788	62 09 44	489 303 872	0713	2365	26904	6695	788
789	62 25 21	491 169 069	0891	2404	26743	6708	789
790	62 41 00	493 039 000	28.1069	9.2443	1.26582	6.6720	790
791	62 56 81	494 913 671	1247	2482	26422	6733	791
792	62 72 64	496 793 088	1425	2521	26263	6746	792
793	62 88 49	498 677 257	1603	2560	26103	6758	793
794	63 04 36	500 566 184	1780	2599	25945	6771	794
795	63 20 25	502 459 875	28.1957	9.2638	1.25786	6.6783	795
796	63 36 16	504 358 336	2135	2677	25628	6796	796
797	63 52 09	506 261 573	2312	2716	25471	6809	797
798	63 68 04	508 169 592	2489	2754	25313	6821	798
799	63 84 01	510 082 399	2666	2793	25156	6834	799
800	64 00 00	512 000 000	28.2843	9.2832	1.25000	6.6846	800
N	N²	N³	$\sqrt{N}$	$\sqrt[3]{N}$	1000/N	ln N	N

TABLE 11.—SQUARES, CUBES, SQUARE ROOTS, CUBE ROOTS, RECIPROCALS AND
NATURAL LOGARITHMS.—(Continued)
(800–850)

N	N²	N³	√N	∛N	1000/N	ln N	N
800	64 00 00	512 000 000	28.2843	9.2832	1.25000	6.6846	**800**
801	64 16 01	513 922 401	3019	2870	24844	6859	801
802	64 32 04	515 849 608	3196	2909	24688	6871	802
803	64 48 09	517 781 627	3373	2948	24533	6884	803
804	64 64 16	519 718 464	3549	2986	24378	6896	804
805	64 80 25	521 660 125	28.3725	9.3025	1.24224	6.6908	805
806	64 96 36	523 606 616	3901	3063	24069	6921	806
807	65 12 49	525 557 943	4077	3102	23916	6933	807
808	65 28 64	527 514 112	4253	3140	23762	6946	808
809	65 44 81	529 475 129	4429	3179	23609	6958	809
810	65 61 00	531 441 000	28.4605	9.3217	1.23457	6.6970	**810**
811	65 77 21	533 411 731	4781	3255	23305	6983	811
812	65 93 44	535 387 328	4956	3294	23153	6995	812
813	66 09 69	537 367 797	5132	3332	23001	7007	813
814	66 25 96	539 353 144	5307	3370	22850	7020	814
815	66 42 25	541 343 375	28.5482	9.3408	1.22699	6.7032	815
816	66 58 56	543 338 496	5657	3447	22549	7044	816
817	66 74 89	545 338 513	5832	3485	22399	7056	817
818	66 91 24	547 343 432	6007	3523	22249	7069	818
819	67 07 61	549 353 259	6182	3561	22100	7081	819
820	67 24 00	551 368 000	28.6356	9.3599	1.21951	6.7093	**820**
821	67 40 41	553 387 661	6531	3637	21803	7105	821
822	67 56 84	555 412 248	6705	3675	21655	7117	822
823	67 73 29	557 441 767	6880	3713	21507	7130	823
824	67 89 76	559 476 224	7054	3751	21359	7142	824
825	68 06 25	561 515 625	28.7228	9.3789	1.21212	6.7154	825
826	68 22 76	563 559 976	7402	3827	21065	7166	826
827	68 39 29	565 609 283	7576	3865	20919	7178	827
828	68 55 84	567 663 552	7750	3902	20773	7190	828
829	68 72 41	569 722 789	7924	3940	20627	7202	829
830	68 89 00	571 787 000	28.8097	9.3978	1.20482	6.7214	**830**
831	69 05 61	573 856 191	8271	4016	20337	7226	831
832	69 22 24	575 930 368	8444	4053	20192	7238	832
833	69 38 89	578 009 537	8617	4091	20048	7250	833
834	69 55 56	580 093 704	8791	4129	19904	7262	834
835	69 72 25	582 182 875	28.8964	9.4166	1.19760	6.7274	835
836	69 88 96	584 277 056	9137	4204	19617	7286	836
837	70 05 69	586 376 253	9310	4241	19474	7298	837
838	70 22 44	588 480 472	9482	4279	19332	7310	838
839	70 39 21	590 589 719	9655	4316	19190	7322	839
840	70 56 00	592 704 000	28.9828	9.4354	1.19048	6.7334	**840**
841	70 72 81	594 823 321	29.0000	4391	18906	7346	841
842	70 89 64	596 947 688	0172	4429	18765	7358	842
843	71 06 49	599 077 107	0345	4466	18624	7370	843
844	71 23 36	601 211 584	0517	4503	18483	7382	844
845	71 40 25	603 351 125	29.0689	9.4541	1.18343	6.7393	845
846	71 57 16	605 495 736	0861	4578	18203	7405	846
847	71 74 09	607 645 423	1033	4615	18064	7417	847
848	71 91 04	609 800 192	1204	4652	17925	7429	848
849	72 08 01	611 960 049	1376	4690	17786	7441	849
850	72 25 00	614 125 000	29.1548	9.4727	1.17647	6.7452	**850**
N	N²	N³	√N	∛N	1000/N	ln N	N

TABLE 11.—SQUARES, CUBES, SQUARE ROOTS, CUBE ROOTS, RECIPROCALS AND NATURAL LOGARITHMS.—(*Continued*)

(850–900)

N	N^2	N^3	$\sqrt{N}$	$\sqrt[3]{N}$	$1000/N$	$\ln N$	N
850	72 25 00	614 125 000	29.1548	9.4727	1.17647	6.7452	850
851	72 42 01	616 295 051	1719	4764	17509	7464	851
852	72 59 04	618 470 208	1890	4801	17371	7476	852
853	72 76 09	620 650 477	2062	4838	17233	7488	853
854	72 93 16	622 835 864	2233	4875	17096	7499	854
355	73 10 25	625 026 375	29.2404	9.4912	1.16959	6.7511	855
856	73 27 36	627 222 016	2575	4949	16822	7523	856
857	73 44 49	629 422 793	2746	4986	16686	7534	857
858	73 61 64	631 628 712	2916	5023	16550	7546	858
859	73 78 81	633 839 779	3087	5060	16414	7558	859
860	73 96 00	636 056 000	29.3258	9.5097	1.16279	6.7569	860
861	74 13 21	638 277 381	3428	5134	16144	7581	861
862	74 30 44	640 503 928	3598	5171	16009	7593	862
863	74 47 69	642 735 647	3769	5207	15875	7604	863
864	74 64 96	644 972 544	3939	5244	15741	7616	864
865	74 82 25	647 214 625	29.4109	9.5281	1.15607	6.7627	865
866	74 99 56	649 461 896	4279	5317	15473	7639	866
867	75 16 89	651 714 363	4449	5354	15340	7650	867
868	75 34 24	653 972 032	4618	5391	15207	7662	868
869	75 51 61	656 234 909	4788	5427	15075	7673	869
870	75 69 00	658 503 000	29.4958	9.5464	1.14943	6.7685	870
871	75 86 41	660 776 311	5127	5501	14811	7696	871
872	76 03 84	663 054 848	5296	5537	14679	7708	872
873	76 21 29	665 338 617	5466	5574	14548	7719	873
874	76 38 76	667 627 624	5635	5610	14416	7731	874
875	76 56 25	669 921 875	29.5804	9.5647	1.14286	6.7742	875
876	76 73 76	672 221 376	5973	5683	14155	7754	876
877	76 91 29	674 526 133	6142	5719	14025	7765	877
878	77 08 84	676 836 152	6311	5756	13895	7776	878
879	77 26 41	679 151 439	6479	5792	13766	7788	879
880	77 44 00	681 472 000	29.6648	9.5828	1.13636	6.7799	880
881	77 61 61	683 797 841	6816	5865	13507	7811	881
882	77 79 24	686 128 968	6985	5901	13379	7822	882
883	77 96 89	688 465 387	7153	5937	13250	7833	883
884	78 14 56	690 807 104	7321	5973	13122	7845	884
885	78 32 25	693 154 125	29.7489	9.6010	1.12994	6.7856	885
886	78 49 96	695 506 456	7658	6046	12867	7867	886
887	78 67 69	697 864 103	7825	6082	12740	7878	887
888	78 85 44	700 227 072	7993	6118	12613	7890	888
889	79 03 21	702 595 369	8161	6154	12486	7901	889
890	79 21 00	704 969 000	29.8329	9.6190	1.12360	6.7912	890
891	79 38 81	707 347 971	8496	6226	12233	7923	891
892	79 56 64	709 732 288	8664	6262	12108	7935	892
893	79 74 49	712 121 957	8831	6298	11982	7946	893
894	79 92 36	714 516 984	8998	6334	11857	7957	894
895	80 10 25	716 917 375	29.9166	9.6370	1.11732	6.7968	895
896	80 28 16	719 323 136	9333	6406	11607	7979	896
897	80 46 09	721 734 273	9500	6442	11483	7991	897
898	80 64 04	724 150 792	9666	6477	11359	8002	898
899	80 82 01	726 572 699	29.9833	6513	11235	8013	899
900	81 00 00	729 000 000	30.0000	9.6549	1.11111	6.8024	900
N	N^2	N^3	$\sqrt{N}$	$\sqrt[3]{N}$	$1000/N$	$\ln N$	N

TABLE 11.—Squares, Cubes, Square Roots, Cube Roots, Reciprocals and
Natural Logarithms.—(*Continued*)
(900–950)

N	N²	N³	$\sqrt{N}$	$\sqrt[3]{N}$	1000/N	ln N	N
900	81 00 00	729 000 000	30.0000	9.6549	1.11111	6.8024	**900**
901	81 18 01	731 432 701	0167	6585	10988	8035	901
902	81 36 04	733 870 808	0333	˙6620	10865	8046	902
903	81 54 09	736 314 327	0500	6656	10742	8057	903
904	81 72 16	738 763 264	0666	6692	10619	8068	904
905	81 90 25	741 217 625	30.0832	9.6727	1.10497	6.8079	905
906	82 08 36	743 677 416	0998	6763	10375	8090	906
907	82 26 49	746 142 643	1164	6799	10254	8101	907
908	82 44 64	748 613 312	1330	6834	10132	8112	908
909	82 62 81	751 089 429	1496	6870	10011	8123	909
910	82 81 00	753 571 000	30.1662	9.6905	1.09890	6.8134	**910**
911	82 99 21	756 058 031	1828	6941	09769	8145	911
912	83 17 44	758 550 528	1993	6976	09649	8156	912
913	83 35 69	761 048 497	2159	7012	09529	8167	913
914	83 53 96	763 551 944	2324	7047	09409	8178	914
915	83 72 25	766 060 875	30.2490	9.7082	1.09290	6.8189	915
916	83 90 56	768 575 296	2655	7118	09170	8200	916
917	84 08 89	771 095 213	2820	7153	09051	8211	917
918	84 27 24	773 620 632	2985	7188	08932	8222	918
919	84 45 61	776 151 559	3150	7224	08814	8233	919
920	84 64 00	778 688 000	30.3315	9.7259	1.08696	6.8244	**920**
921	84 82 41	781 229 961	3480	7294	08578	8255	921
922	85 00 84	783 777 448	3645	7329	08460	8265	922
923	85 19 29	786 330 467	3809	7364	08342	8276	923
924	85 37 76	788 889 024	3974	7400	08225	8287	924
925	85 56 25	791 453 125	30.4138	9.7435	1.08108	6.8298	925
926	85 74 76	794 022 776	4302	7470	07991	8309	926
927	85 93 29	796 597 983	4467	7505	07875	8320	927
928	86 11 84	799 178 752	4631	7540	07759	8330	928
929	86 30 41	801 765 089	4795	7575	07643	8341	929
930	86 49 00	804 357 000	30.4959	9.7610	1.07527	6.8352	**930**
931	86 67 61	806 954 491	5123	7645	07411	8363	931
932	86 86 24	809 557 568	5287	7680	07296	8373	932
933	87 04 89	812 166 237	5450	7715	07181	8384	933
934	87 23 56	814 780 504	5614	7750	07066	8395	934
935	87 42 25	817 400 375	30.5778	9.7785	1.06952	6.8405	935
936	87 60 96	820 025 856	5941	7819	06838	8416	936
937	87 79 69	822 656 953	6105	7854	06724	8427	937
938	87 98 44	825 293 672	6268	7889	06610	8437	938
939	88 17 21	827 936 019	6431	7924	06496	8448	939
940	88 36 00	830 584 000	30.6594	9.7959	1.06383	6.8459	**940**
941	88 54 81	833 237 621	6757	7993	06270	8469	941
942	88 73 64	835 896 888	6920	8028	06157	8480	942
943	88 92 49	838 561 807	7083	8063	06045	8491	943
944	89 11 36	841 232 384	7246	8097	05932	8501	944
945	89 30 25	843 908 625	30.7409	9.8132	1.05820	6.8512	945
946	89 49 16	846 590 536	7571	8167	05708	8522	946
947	89 68 09	849 278 123	7734	8201	05597	8533	947
948	89 87 04	851 971 392	7896	8236	05485	8544	948
949	90 06 01	854 670 349	8058	8270	05374	8554	949
950	90 25 00	857 375 000	30.8221	9.8305	1.05263	6.8565	**950**
N	N²	N³	$\sqrt{N}$	$\sqrt[3]{N}$	1000/N	ln N	N

TABLE 11.—SQUARES, CUBES, SQUARE ROOTS, CUBE ROOTS, RECIPROCALS AND NATURAL LOGARITHMS.—(*Concluded*)
(950–1000)

N	N^2	N^3	$\sqrt{N}$	$\sqrt[3]{N}$	1000/N	ln N	N
950	90 25 00	857 375 000	30.8221	9.8305	1.05263	6.8565	**950**
951	90 44 01	860 085 351	8383	8339	05152	8575	951
952	90 63 04	862 801 408	8545	8374	05042	8586	952
953	90 82 09	865 523 177	8707	8408	04932	8596	953
954	91 01 16	868 250 664	8869	8443	04822	8607	954
955	91 20 25	870 983 875	30.9031	9.8477	1.04712	6.8617	955
956	91 39 36	873 722 816	9192	8511	04603	8628	956
957	91 58 49	876 467 493	9354	8546	04493	8638	957
958	91 77 64	879 217 912	9516	8580	04384	8648	958
959	91 96 81	881 974 079	9677	8614	04275	8659	959
960	92 16 00	884 736 000	30.9839	9.8648	1.04167	6.8669	**960**
961	92 35 21	887 503 681	31.0000	8683	04058	8680	961
962	92 54 44	890 277 128	0161	8717	03950	8690	962
963	92 73 69	893 056 347	0322	8751	03842	8701	963
964	92 92 96	895 841 344	0483	8785	03734	8711	964
965	93 12 25	898 632 125	31.0644	9.8819	1.03627	6.8721	965
966	93 31 56	901 428 696	0805	8854	03520	8732	966
967	93 50 89	904 231 063	0966	8888	03413	8742	967
968	93 70 24	907 039 232	1127	8922	03306	8752	968
969	93 89 61	909 853 209	1288	8956	03199	8763	969
970	94 09 00	912 673 000	31.1448	9.8990	1.03093	6.8773	**970**
971	94 28 41	915 498 611	1609	9024	02987	8783	971
972	94 47 84	918 330 048	1769	9058	02881	8794	972
973	94 67 29	921 167 317	1929	9092	02775	8804	973
974	94 86 76	924 010 424	2090	9126	02669	8814	974
975	95 06 25	926 859 375	31.2250	9.9160	1.02564	6.8824	975
976	95 25 76	929 714 176	2410	9194	02459	8835	976
977	95 45 29	932 574 833	2570	9227	02354	8845	977
978	95 64 84	935 441 352	2730	9261	02249	8855	978
979	95 84 41	938 313 739	2890	9295	02145	8865	979
980	96 04 00	941 192 000	31.3050	9.9329	1.02041	6.8876	**980**
981	96 23 61	944 076 141	3209	9363	01937	8886	981
982	96 43 24	946 966 168	3369	9396	01833	8896	982
983	96 62 89	949 862 087	3528	9430	01729	8906	983
984	96 82 56	952 763 904	3688	9464	01626	8916	984
985	97 02 25	955 671 625	31.3847	9.9497	1.01523	6.8926	985
986	97 21 96	958 585 256	4006	9531	01420	8937	986
987	97 41 69	961 504 803	4166	9565	01317	8947	987
988	97 61 44	964 430 272	4325	9598	01215	8957	988
989	97 81 21	967 361 669	4484	9632	01112	8967	989
990	98 01 00	970 299 000	31.4643	9.9666	1.01010	6.8977	**990**
991	98 20 81	973 242 271	4802	9699	00908	8987	991
992	98 40 64	976 191 488	4960	9733	00806	8997	992
993	98 60 49	979 146 657	5119	9766	00705	9007	993
994	98 80 36	982 107 784	5278	9800	00604	9017	994
.995	99 00 25	985 074 875	31.5436	9.9833	1.00503	6.9027	995
996	99 20 16	988 047 936	5595	9866	00402	9037	996
997	99 40 09	991 026 973	5753	9900	00301	9048	997
998	99 60 04	994 011 992	5911	9933	00200	9058	998
999	99 80 01	997 002 999	6070	9.9967	00100	9068	999
1000			31.6228	10.0000	1.00000	6.9078	**1000**
N	N^2	N^3	$\sqrt{N}$	$\sqrt[3]{N}$	1000/N	ln N	N

TABLE 12.—SOME FUNCTIONS OF THE INTEGERS 1–100

n	πn	$\dfrac{\pi n^2}{4}$	$\dfrac{1}{\sqrt{n}}$	$\dfrac{0.6745}{\sqrt{n-1}}$	$\dfrac{0.6745}{\sqrt{n(n-1)}}$	$\log n!$	$\Gamma\!\left(1+\dfrac{n}{100}\right)$	n
1	3.1416	0.7854	1.0000			0.0000	0.9943	1
2	6.2832	3.1416	0.7071	0.6745	0.4769	0.3010	9888	2
3	9.4248	7.0686	5774	4769	2754	0.7782	9835	3
4	12.566	12.566	5000	3894	1947	1.3802	9784	4
5	15.708	19.635	0.4472	0.3372	0.1508	2.0792	0.9735	5
6	18.850	28.274	4082	3016	1231	2.8573	9687	6
7	21.991	38.485	3780	2754	1041	3.7024	9642	7
8	25.133	50.265	3536	2549	0901	4.6055	9597	8
9	28.274	63.617	3333	2385	0795	5.5598	9555	9
10	31.416	78.540	0.3162	0.2248	0.0711	6.5598	0.9514	10
11	34.558	95.033	3015	2133	0643	7.6012	9474	11
12	37.699	113.10	2887	2034	0587	8.6803	9436	12
13	40.841	132.73	2774	1947	0540	9.7943	9399	13
14	43.982	153.94	2673	1871	0500	10.9404	9364	14
15	47.124	176.71	0.2582	0.1803	0.0465	12.1165	0.9330	15
16	50.265	201.06	2500	1742	0435	13.3206	9298	16
17	53.407	226.98	2425	1686	0409	14.5511	9267	17
18	56.549	254.47	2357	1636	0386	15.8063	9237	18
19	59.690	283.53	2294	1590	0365	17.0851	9209	19
20	62.832	314.16	0.2236	0.1547	0.0346	18.3861	0.9182	20
21	65.973	346.36	2182	1508	0329	19.7083	9156	21
22	69.115	380.13	2132	1472	0314	21.0508	9131	22
23	72.257	415.48	2085	1438	0300	22.4125	9108	23
24	75.398	452.39	2041	1406	0287	23.7927	9085	24
25	78.540	490.87	0.2000	0.1377	0.0275	25.1906	0.9064	25
26	81.681	530.93	1961	1349	0265	26.6056	9044	26
27	84.823	572.56	1925	1323	0255	28.0370	9025	27
28	87.965	615.75	1890	1298	0245	29.4841	9007	28
29	91.106	660.52	1857	1275	0237	30.9465	8990	29
30	94.248	706.86	0.1826	0.1252	0.0229	32.4237	0.8975	30
31	97.389	754.77	1796	1231	0221	33.9150	8960	31
32	100.53	804.25	1768	1211	0214	35.4202	8946	32
33	103.67	855.30	1741	1192	0208	36.9387	8934	33
34	106.81	907.92	1715	1174	0201	38.4702	8922	34
35	109.96	962.11	0.1690	0.1157	0.0196	40.0142	0.8912	35
36	113.10	1017.9	1667	1140	0190	41.5705	8902	36
37	116.24	1075.2	1644	1124	0185	43.1387	8893	37
38	119.38	1134.1	1622	1109	0180	44.7185	8885	38
39	122.52	1194.6	1601	1094	0175	46.3096	8879	39
40	125.66	1256.6	0.1581	0.1080	0.0171	47.9116	0.8873	40
41	128.81	1320.3	1562	1066	0167	49.5244	8868	41
42	131.95	1385.4	1543	1053	0163	51.1477	8864	42
43	135.09	1452.2	1525	1041	0159	52.7811	8860	43
44	138.23	1520.5	1508	1029	0155	54.4246	8858	44
45	141.37	1590.4	0.1491	0.1017	0.0152	56.0778	0.8857	45
46	144.51	1661.9	1474	1005	0148	57.7406	8856	46
47	147.65	1734.9	1459	0994	0145	59.4127	8856	47
48	150.80	1809.6	1443	0984	0142	61.0939	8857	48
49	153.94	1885.7	1429	0974	0139	62.7841	8859	49
50	157.08	1963.5	0.1414	0.0964	0.0136	64.4831	0.8862	50
n	πn	$\dfrac{\pi n^2}{4}$	$\dfrac{1}{\sqrt{n}}$	$\dfrac{0.6745}{\sqrt{n-1}}$	$\dfrac{0.6745}{\sqrt{n(n-1)}}$	$\log n!$	$\Gamma\!\left(1+\dfrac{n}{100}\right)$	n

TABLE 12.—SOME FUNCTIONS OF THE INTEGERS 1–100.—(*Continued*)

n	πn	$\dfrac{\pi n^2}{4}$	$\dfrac{1}{\sqrt{n}}$	$\dfrac{0.6745}{\sqrt{n-1}}$	$\dfrac{0.6745}{\sqrt{n(n-1)}}$	$\log n!$	$\Gamma\left(1+\dfrac{n}{100}\right)$	n
50	157.08	1963.5	0.1414	0.0964	0.0136	64.4831	0.8862	50
51	160.22	2042.8	1400	0954	0134	66.1906	8866	51
52	163.36	2123.7	1387	0944	0131	67.9066	8870	52
53	166.50	2206.2	1374	0935	0128	69.6309	8876	53
54	169.65	2290.2	1361	0926	0126	71.3633	8882	54
55	172.79	2375.8	0.1348	0.0918	0.0124	73.1037	0.8889	55
56	175.93	2463.0	1336	0909	0122	74.8519	8896	56
57	179.07	2551.8	1325	0901	0119	76.6077	8905	57
58	182.21	2642.1	1313	0893	0117	78.3712	8914	58
59	185.35	2734.0	1302	0886	0115	80.1420	8924	59
60	188.50	2827.4	0.1291	0.0878	0.0113	81.9202	0.8935	60
61	191.64	2922.5	1280	0871	0111	83.7055	8947	61
62	194.78	3019.1	1270	0864	0110	85.4979	8959	62
63	197.92	3117.2	1260	0857	0108	87.2972	8972	63
64	201.06	3217.0	1250	0850	0106	89.1034	8986	64
65	204.20	3318.3	0.1240	0.0843	0.0105	90.9163	0.9001	65
66	207.35	3421.2	1231	0837	0103	92.7359	9017	66
67	210.49	3525.7	1222	0830	0101	94.5619	9033	67
68	213.63	3631.7	1213	0824	0100	96.3945	9050	68
69	216.77	3739.3	1204	0818	0098	98.2333	9068	69
70	219.91	3848.5	0.1195	0.0812	0.0097	100.0784	0.9086	70
71	223.05	3959.2	1187	0806	0096	101.9297	9106	71
72	226.19	4071.5	1179	0800	0094	103.7870	9126	72
73	229.34	4185.4	1170	0795	0093	105.6503	9147	73
74	232.48	4300.8	1162	0789	0092	107.5196	9168	74
75	235.62	4417.9	0.1155	0.0784	0.0091	109.3946	0.9191	75
76	238.76	4536.5	1147	0779	0089	111.2754	9214	76
77	241.90	4656.6	1140	0774	0088	113.1619	9238	77
78	245.04	4778.4	1132	0769	0087	115.0540	9262	78
79	248.19	4901.7	1125	0764	0086	116.9516	9288	79
80	251.33	5026.5	0.1118	0.0759	0.0085	118.8547	0.9314	80
81	254.47	5153.0	1111	0754	0084	120.7632	9341	81
82	257.61	5281.0	1104	0749	0083	122.6770	9368	82
83	260.75	5410.6	1098	0745	0082	124.5961	9397	83
84	263.89	5541.8	1091	0740	0081	126.5204	9426	84
85	267.04	5674.5	0.1085	0.0736	0.0080	128.4498	0.9456	85
86	270.18	5808.8	1078	0732	0079	130.3843	9487	86
87	273.32	5944.7	1072	0727	0078	132.3238	9518	87
88	276.46	6082.1	1066	0723	0077	134.2683	9551	88
89	279.60	6221.1	1060	0719	0076	136.2177	9584	89
90	282.74	6361.7	0.1054	0.0715	0.0075	138.1719	0.9618	90
91	285.88	6503.9	1048	0711	0075	140.1310	9652	91
92	289.03	6647.6	1043	0707	0074	142.0948	9688	92
93	292.17	6792.9	1037	0703	0073	144.0632	9724	93
94	295.31	6939.8	1031	0699	0072	146.0364	9761	94
95	298.45	7088.2	0.1026	0.0696	0.0071	148.0141	0.9799	95
96	301.59	7238.2	1021	0692	0071	149.9964	9837	96
97	304.73	7389.8	1015	0688	0070	151.9831	9877	97
98	307.88	7543.0	1010	0685	0069	153.9744	9917	98
99	311.02	7697.7	1005	0681	0068	155.9700	0.9958	99
100	314.16	7854.0	0.1000	0.0678	0.0068	157.9700	1.0000	100
n	πn	$\dfrac{\pi n^2}{4}$	$\dfrac{1}{\sqrt{n}}$	$\dfrac{0.6745}{\sqrt{n-1}}$	$\dfrac{0.6745}{\sqrt{n(n-1)}}$	$\log n!$	$\Gamma\left(1+\dfrac{n}{100}\right)$	n

TABLE 13.—AMOUNT OF 1 AT COMPOUND INTEREST

$$s = (1 + i)^n$$

n	$\frac{1}{2}\%$	$\frac{3}{4}\%$	1%	$1\frac{1}{4}\%$	$1\frac{1}{2}\%$	$1\frac{3}{4}\%$	n
1	1.0050	1.0075	1.0100	1.0125	1.0150	1.0175	**1**
2	0100	0151	0201	0252	0302	0353	2
3	0151	0227	0303	0380	0457	0534	3
4	0202	0303	0406	0509	0614	0719	4
5	1.0253	1.0381	1.0510	1.0641	1.0773	1.0906	5
6	0304	0459	0615	0774	0934	1097	6
7	0355	0537	0721	0909	1098	1291	7
8	0407	0616	0829	1045	1265	1489	8
9	0459	0696	0937	1183	1434	1690	9
10	1.0511	1.0776	1.1046	1.1323	1.1605	1.1894	**10**
11	0564	0857	1157	1464	1779	2103	11
12	0617	0938	1268	1608	1956	2314	12
13	0670	1020	1381	1753	2136	2530	13
14	0723	1103	1495	1900	2318	2749	14
15	1.0777	1.1186	1.1610	1.2048	1.2502	1.2972	15
16	0831	1270	1726	2199	2690	3199	16
17	0885	1354	1843	2351	2880	3430	17
18	0939	1440	1961	2506	3073	3665	18
19	0994	1525	2081	2662	3270	3904	19
20	1.1049	1.1612	1.2202	1.2820	1.3469	1.4148	**20**
22	1160	1787	2447	3143	3876	4647	22
24	1272	1964	2697	3474	4295	5164	24
26	1385	2144	2953	3812	4727	5700	26
28	1499	2327	3213	4160	5172	6254	28
30	1.1614	1.2513	1.3478	1.4516	1.5631	1.6828	30
32	1730	2701	3749	4881	6103	7422	32
34	1848	2892	4026	5256	6590	8037	34
36	1967	3086	4308	5639	7091	8674	36
38	2087	3283	4595	6033	7608	1.9333	38
40	1.2208	1.3483	1.4889	1.6436	1.8140	2.0016	**40**
44	2454	3893	5493	7274	1.9253	1454	44
48	2705	4314	6122	8154	2.0435	2996	48
52	2961	4748	6777	1.9078	1689	4648	52
56	3222	5196	7458	2.0050	3020	6420	56
60	1.3489	1.5657	1.8167	2.1072	2.4432	2.8318	60
64	3760	6132	8905	2145	5931	3.0353	64
68	4038	6621	1.9672	3274	7523	2534	68
72	4320	7126	2.0471	4459	2.9212	4872	72
76	4609	7645	2.1302	5705	3.1004	3.7378	76
80	1.4903	1.8180	2.2167	2.7015	3.2907	4.0064	**80**
84	5204	8732	3067	8391	4926	2943	84
88	5510	9300	4004	2.9838	7069	6029	88
92	5823	1.9886	4979	3.1358	3.9344	4.9336	92
96	6141	2.0489	5993	3.2955	4.1758	5.2882	96
100	1.6467	2.1111	2.7048	3.4634	4.4320	5.6682	**100**
n	$\frac{1}{2}\%$	$\frac{3}{4}\%$	1%	$1\frac{1}{4}\%$	$1\frac{1}{2}\%$	$1\frac{3}{4}\%$	n

TABLE 13.—AMOUNT OF 1 AT COMPOUND INTEREST.—(*Continued*)

$$s = (1 + i)^n$$

n	2%	2½%	3%	4%	5%	6%	n
1	1.0200	1.0250	1.0300	1.0400	1.0500	1.0600	1
2	0404	0506	0609	0816	1025	1236	2
3	0612	0769	0927	1249	1576	1910	3
4	0824	1038	1255	1699	2155	2625	4
5	1.1041	1.1314	1.1593	1.2167	1.2763	1.3382	5
6	1262	1597	1941	2653	3401	4185	6
7	1487	1887	2299	3159	4071	5036	7
8	1717	2184	2668	3686	4775	5938	8
9	1951	2489	3048	4233	5513	6895	9
10	1.2190	1.2801	1.3439	1.4802	1.6289	1.7908	10
11	2434	3121	3842	5395	7103	1.8983	11
12	2682	3449	4258	6010	7959	2.0122	12
13	2936	3785	4685	6651	8856	1329	13
14	3195	4130	5126	7317	1.9799	2609	14
15	1.3459	1.4483	1.5580	1.8009	2.0789	2.3966	15
16	3728	4845	6047	8730	1829	5404	16
17	4002	5216	6528	1.9479	2920	6928	17
18	4282	5597	7024	2.0258	4066	2.8543	18
19	4568	5987	7535	1068	5270	3.0256	19
20	1.4859	1.6386	1.8061	2.1911	2.6533	3.2071	20
21	5157	6796	8603	2788	7860	3996	21
22	5460	7216	9161	3699	2.9253	6035	22
23	5769	7646	1.9736	4647	3.0715	3.8197	23
24	6084	8087	2.0328	5633	2251	4.0489	24
25	1.6406	1.8539	2.0938	2.6658	3.3864	4.2919	25
26	6734	9003	1566	7725	5557	5494	26
27	7069	9478	2213	8834	7335	4.8223	27
28	7410	1.9965	2879	2.9987	3.9201	5.1117	28
29	7758	2.0464	3566	3.1187	4.1161	4184	29
30	1.8114	2.0976	2.4273	3.2434	4.3219	5.7435	30
31	8476	1500	5001	3731	5380	6.0881	31
32	8845	2038	5751	5081	4.7649	4534	32
33	9222	2589	6523	6484	5.0032	6.8406	33
34	9607	3153	7319	7943	2533	7.2510	34
35	1.9999	2.3732	2.8139	3.9461	5.5160	7.6861	35
36	2.0399	4325	8983	4.1039	5.7918	8.1473	36
37	0807	4933	2.9852	2681	6.0814	8.6361	37
38	1223	5557	3.0748	4388	6.3855	9.1543	38
39	1647	6196	1670	6164	6.7048	9.7035	39
40	2.2080	2.6851	3.2620	4.8010	7.0400	10.2857	40
42	2972	8210	4607	5.1928	7.7616	11.5570	42
44	3901	2.9638	6715	5.6165	8.5572	12.9855	44
46	4866	3.1139	3.8950	6.0748	9.4343	14.5905	46
48	5871	2715	4.1323	6.5705	10.4013	16.3939	48
50	2.6916	3.4371	4.3839	7.1067	11.4674	18.4202	50
n	2%	2½%	3%	4%	5%	6%	n

TABLE 14.—PRESENT VALUE OF 1 AT COMPOUND INTEREST

$$v^n = (1 + i)^{-n}$$

n	½%	¾%	1%	1¼%	1½%	1¾%	n
1	0.99502	0.99256	0.99010	0.98765	0.98522	0.98280	1
2	99007	98517	98030	97546	97066	96590	2
3	98515	97783	97059	96342	95632	94929	3
4	98025	97055	96098	95152	94218	93296	4
5	0.97537	0.96333	0.95147	0.93978	0.92826	0.91691	5
6	97052	95616	94205	92817	91454	90114	6
7	96569	94904	93272	91672	90103	88564	7
8	96089	94198	92348	90540	88771	87041	8
9	95610	93496	91434	89422	87459	85544	9
10	0.95135	0.92800	0.90529	0.88318	0.86167	0.84073	10
11	94661	92109	89632	87228	84893	82627	11
12	94191	91424	88745	86151	83639	81206	12
13	93722	90743	87866	85087	82403	79809	13
14	93256	90068	86996	84037	81185	78436	14
15	0.92792	0.89397	0.86135	0.82999	0.79985	0.77087	15
16	92330	88732	85282	81975	78803	75762	16
17	91871	88071	84438	80963	77639	74459	17
18	91414	87416	83602	79963	76491	73178	18
19	90959	86765	82774	78976	75361	71919	19
20	0.90506	0.86119	0.81954	0.78001	0.74247	0.70682	20
22	89608	84842	80340	76087	72069	68272	22
24	88719	83583	78757	74220	69954	65944	24
26	87838	82343	77205	72398	67902	63695	26
28	86966	81122	75684	70622	65910	61523	28
30	0.86103	0.79919	0.74192	0.68889	0.63976	0.59425	30
32	85248	78733	72730	67198	62099	57398	32
34	84402	77565	71297	65549	60277	55441	34
36	83564	76415	69892	63941	58509	53550	36
38	82735	75281	68515	62372	56792	51724	38
40	0.81914	0.74165	0.67165	0.60841	0.55126	0.49960	40
44	80296	71981	64545	57892	51939	46611	44
48	78710	69861	62026	55086	48936	43486	48
52	77155	67804	59606	52415	46107	40570	52
56	75631	65808	57280	49874	43441	37851	56
60	0.74137	0.63870	0.55045	0.47457	0.40930	0.35313	60
64	72673	61989	52897	45156	38563	32946	64
68	71237	60164	50833	42967	36334	30737	68
72	69830	58392	48850	40884	34233	28676	72
76	68451	56673	46944	38903	32254	26754	76
80	0.67099	0.55004	0.45112	0.37017	0.30389	0.24960	80
84	65773	53385	43352	35222	28632	23287	84
88	64474	51813	41660	33515	26977	21726	88
92	63201	50287	40034	31890	25417	20269	92
96	61952	48806	38472	30344	23947	18910	96
100	0.60729	0.47369	0.36971	0.28873	0.22563	0.17642	100
n	½%	¾%	1%	1¼%	1½%	1¾%	n

TABLE 14.—PRESENT VALUE OF 1 AT COMPOUND INTEREST.—(*Continued*)
$$v^n = (1 + i)^{-n}$$

n	2%	2½%	3%	4%	5%	6%	n
1	0.98039	0.97561	0.97087	0.96154	0.95238	0.94340	1
2	96117	95181	94260	92456	90703	89000	2
3	94232	92860	91514	88900	86384	83962	3
4	92385	90595	88849	85480	82270	79209	4
5	0.90573	0.88385	0.86261	0.82193	0.78353	0.74726	5
6	88797	86230	83748	79031	74622	70496	6
7	87056	84127	81309	75992	71068	66506	7
8	85349	82075	78941	73069	67684	62741	8
9	83676	80073	76642	70259	64461	59190	9
10	0.82035	0.78120	0.74409	0.67556	0.61391	0.55839	10
11	80426	76214	72242	64958	58468	52679	11
12	78849	74356	70138	62460	55684	49697	12
13	77303	72542	68095	60057	53032	46884	13
14	75788	70773	66112	57748	50507	44230	14
15	0.74301	0.69047	0.64186	0.55526	0.48102	0.41727	15
16	72845	67362	62317	53391	45811	39365	16
17	71416	65720	60502	51337	43630	37136	17
18	70016	64117	58739	49363	41552	35034	18
19	68643	62553	57029	47464	39573	33051	19
20	0.67297	0.61027	0.55368	0.45639	0.37689	0.31180	20
21	65978	59539	53755	43883	35894	29416	21
22	64684	58086	52189	42196	34185	27751	22
23	63416	56670	50669	40573	32557	26180	23
24	62172	55288	49193	39012	31007	24698	24
25	0.60953	0.53939	0.47761	0.37512	0.29530	0.23300	25
26	59758	52623	46369	36069	28124	21981	26
27	58586	51340	45019	34682	26785	20737	27
28	57437	50088	43708	33348	25509	19563	28
29	56311	48866	42435	32065	24295	18456	29
30	0.55207	0.47674	0.41199	0.30832	0.23138	0.17411	30
31	54125	46511	39999	29646	22036	16425	31
32	53063	45377	38834	28506	20987	15496	32
33	52023	44270	37703	27409	19987	14619	33
34	51003	43191	36604	26355	19035	13791	34
35	0.50003	0.42137	0.35538	0.25342	0.18129	0.13011	35
36	49022	41109	34503	24367	17266	12274	36
37	48061	40107	33498	23430	16444	11579	37
38	47119	39128	32523	22529	15661	10924	38
39	46195	38174	31575	21662	14915	10306	39
40	0.45289	0.37243	0.30656	0.20829	0.14205	0.09722	40
42	43530	35448	28896	19257	12884	08653	42
44	41840	33740	27237	17805	11686	07701	44
46	40215	32115	25674	16461	10600	06854	46
48	38654	30567	24200	15219	09614	06100	48
50	0.37153	0.29094	0.22811	0.14071	0.08720	0.05429	50
n	2%	2½%	3%	4%	5%	6%	n

Table 15.—Amount of an Annuity of 1

$$s_{\overline{n}|} \text{ at } i = \frac{(1+i)^n - 1}{i}$$

n	½%	¾%	1%	1¼%	1½%	1¾%	n
1	1.0000	1.0000	1.0000	1.0000	1.0000	1.0000	1
2	2.0050	2.0075	2.0100	2.0125	2.0150	2.0175	2
3	3.0150	3.0226	3.0301	3.0377	3.0452	3.0528	3
4	4.0301	4.0452	4.0604	4.0756	4.0909	4.1062	4
5	5.0503	5.0756	5.1010	5.1266	5.1523	5.1781	5
6	6.0755	6.1136	6.1520	6.1907	6.2296	6.2687	6
7	7.1059	7.1595	7.2135	7.2680	7.3230	7.3784	7
8	8.1414	8.2132	8.2857	8.3589	8.4328	8.5075	8
9	9.1821	9.2748	9.3685	9.4634	9.5593	9.6564	9
10	10.2280	10.3443	10.4622	10.5817	10.7027	10.8254	10
11	11.2792	11.4219	11.5668	11.7139	11.8633	12.0148	11
12	12.3356	12.5076	12.6825	12.8604	13.0412	13.2251	12
13	13.3972	13.6014	13.8093	14.0211	14.2368	14.4565	13
14	14.4642	14.7034	14.9474	15.1964	15.4504	15.7095	14
15	15.5365	15.8137	16.0969	16.3863	16.6821	16.9844	15
16	16.6142	16.9323	17.2579	17.5912	17.9324	18.2817	16
17	17.6973	18.0593	18.4304	18.8111	19.2014	19.6016	17
18	18.7858	19.1947	19.6147	20.0462	20.4894	20.9446	18
19	19.8797	20.3387	20.8109	21.2968	21.7967	22.3112	19
20	20.9791	21.4912	22.0190	22.5630	23.1237	23.7016	20
22	23.1944	23.8223	24.4716	25.1431	25.8376	26.5559	22
24	25.4320	26.1885	26.9735	27.7881	28.6335	29.5110	24
26	27.6919	28.5903	29.5256	30.4996	31.5140	32.5704	26
28	29.9745	31.0282	32.1291	33.2794	34.4815	35.7379	28
30	32.2800	33.5029	34.7849	36.1291	37.5387	39.0172	30
32	34.6086	36.0148	37.4941	39.0504	40.6883	42.4122	32
34	36.9606	38.5646	40.2577	42.0453	43.9331	45.9271	34
36	39.3361	41.1527	43.0769	45.1155	47.2760	49.5661	36
38	41.7354	43.7798	45.9527	48.2629	50.7199	53.3336	38
40	44.1588	46.4465	48.8864	51.4896	54.2679	57.2341	40
44	49.0788	51.9009	54.9318	58.1883	61.6889	65.4532	44
48	54.0978	57.5207	61.2226	65.2284	69.5652	74.2628	48
52	59.2180	63.3111	67.7689	72.6271	77.9249	83.7055	52
56	64.4414	69.2771	74.5810	80.4027	86.7975	93.8267	56
60	69.7700	75.4241	81.6697	88.5745	96.2147	104.6752	60
64	75.2060	81.7577	89.0462	97.1626	106.2096	116.3033	64
68	80.7516	88.2834	96.7222	106.1882	116.8179	128.7670	68
72	86.4089	95.0070	104.7099	115.6736	128.0772	142.1263	72
76	92.1801	101.9347	113.0220	125.6423	140.0274	156.4456	76
80	98.0677	109.0725	121.6715	136.1188	152.7109	171.7938	80
84	104.0739	116.4269	130.6723	147.1290	166.1726	188.2450	84
88	110.2012	124.0045	140.0385	158.7002	180.4605	205.8783	88
92	116.4519	131.8119	149.7850	170.8609	195.6251	224.7788	92
96	122.8285	139.8562	159.9273	183.6411	211.7202	245.0374	96
100	129.3337	148.1445	170.4814	197.0723	228.8030	266.7518	100
n	½%	¾%	1%	1¼%	1½%	1¾%	n

TABLE 15.—AMOUNT OF AN ANNUITY OF 1.—*(Continued)*

$$s_{\overline{n}|} \text{ at } i = \frac{(1+i)^n - 1}{i}$$

n	2%	2½%	3%	4%	5%	6%	n
1	1.0000	1.0000	1.0000	1.0000	1.0000	1.0000	1
2	2.0200	2.0250	2.0300	2.0400	2.0500	2.0600	2
3	3.0604	3.0756	3.0909	3.1216	3.1525	3.1836	3
4	4.1216	4.1525	4.1836	4.2465	4.3101	4.3746	4
5	5.2040	5.2563	5.3091	5.4163	5.5256	5.6371	5
6	6.3081	6.3877	6.4684	6.6330	6.8019	6.9753	6
7	7.4343	7.5474	7.6625	7.8983	8.1420	8.3938	7
8	8.5830	8.7361	8.8923	9.2142	9.5491	9.8975	8
9	9.7546	9.9545	10.1591	10.5828	11.0266	11.4913	9
10	10.9497	11.2034	11.4639	12.0061	12.5779	13.1808	10
11	12.1687	12.4835	12.8078	13.4864	14.2068	14.9716	11
12	13.4121	13.7956	14.1920	15.0258	15.9171	16.8699	12
13	14.6803	15.1404	15.6178	16.6268	17.7130	18.8821	13
14	15.9739	16.5190	17.0863	18.2919	19.5986	21.0151	14
15	17.2934	17.9319	18.5989	20.0236	21.5786	23.2760	15
16	18.6393	19.3802	20.1569	21.8245	23.6575	25.6725	16
17	20.0121	20.8647	21.7616	23.6975	25.8404	28.2129	17
18	21.4123	22.3863	23.4144	25.6454	28.1324	30.9057	18
19	22.8406	23.9460	25.1169	27.6712	30.5390	33.7600	19
20	24.2974	25.5447	26.8704	29.7781	33.0660	36.7856	20
21	25.7833	27.1833	28.6765	31.9692	35.7193	39.9927	21
22	27.2990	28.8629	30.5368	34.2480	38.5052	43.3923	22
23	28.8450	30.5844	32.4529	36.6179	41.4305	46.9958	23
24	30.4219	32.3490	34.4265	39.0826	44.5020	50.8156	24
25	32.0303	34.1578	36.4593	41.6459	47.7271	54.8645	25
26	33.6709	36.0117	38.5530	44.3117	51.1135	59.1564	26
27	35.3443	37.9120	40.7096	47.0842	54.6691	63.7058	27
28	37.0512	39.8598	42.9309	49.9676	58.4026	68.5281	28
29	38.7922	41.8563	45.2189	52.9663	62.3227	73.6398	29
30	40.5681	43.9027	47.5754	56.0849	66.4388	79.0582	30
31	42.3794	46.0003	50.0027	59.3283	70.7608	84.8017	31
32	44.2270	48.1503	52.5028	62.7015	75.2988	90.8898	32
33	46.1116	50.3540	55.0778	66.2095	80.0638	97.3432	33
34	48.0338	52.6129	57.7302	69.8579	85.0670	104.1838	34
35	49.9945	54.9282	60.4621	73.6522	90.3203	111.4348	35
36	51.9944	57.3014	63.2759	77.5983	95.8363	119.1209	36
37	54.0343	59.7339	66.1742	81.7022	101.6281	127.2681	37
38	56.1149	62.2273	69.1594	85.9703	107.7095	135.9042	38
39	58.2372	64.7830	72.2342	90.4091	114.0950	145.0585	39
40	60.4020	67.4026	75.4013	95.0255	120.7998	154.7620	40
42	64.8622	72.8398	82.0232	104.8196	135.2318	175.9505	42
44	69.5027	78.5523	89.0484	115.4129	151.1430	199.7580	44
46	74.3306	84.5540	96.5015	126.8706	168.6852	226.5081	46
48	79.3535	90.8596	104.4084	139.2632	188.0254	256.5645	48
50	84.5794	97.4843	112.7969	152.6671	209.3480	290.3359	50
n	2%	2½%	3%	4%	5%	6%	n

TABLE 16.—PRESENT VALUE OF AN ANNUITY OF 1

$$a_{\overline{n}|} \text{ at } i = \frac{1 - (1 + i)^{-n}}{i}$$

n	½%	¾%	1%	1¼%	1½%	1¾%	n
1	0.9950	0.9926	0.9901	0.9877	0.9852	0.9828	1
2	1.9851	1.9777	1.9704	1.9631	1.9559	1.9487	2
3	2.9702	2.9556	2.9410	2.9265	2.9122	2.8980	3
4	3.9505	3.9261	3.9020	3.8781	3.8544	3.8309	4
5	4.9259	4.8894	4.8534	4.8178	4.7826	4.7479	5
6	5.8964	5.8456	5.7955	5.7460	5.6972	5.6490	6
7	6.8621	6.7946	6.7282	6.6627	6.5982	6.5346	7
8	7.8230	7.7366	7.6517	7.5681	7.4859	7.4051	8
9	8.7791	8.6716	8.5660	8.4623	8.3605	8.2605	9
10	9.7304	9.5996	9.4713	9.3455	9.2222	9.1012	10
11	10.6770	10.5207	10.3676	10.2178	10.0711	9.9275	11
12	11.6189	11.4349	11.2551	11.0793	10.9075	10.7395	12
13	12.5562	12.3423	12.1337	11.9302	11.7315	11.5376	13
14	13.4887	13.2430	13.0037	12.7706	12.5434	12.3220	14
15	14.4166	14.1370	13.8651	13.6005	13.3432	13.0929	15
16	15.3399	15.0243	14.7179	14.4203	14.1313	13.8505	16
17	16.2586	15.9050	15.5623	15.2299	14.9076	14.5951	17
18	17.1728	16.7792	16.3983	16.0295	15.6726	15.3269	18
19	18.0824	17.6468	17.2260	16.8193	16.4262	16.0461	19
20	18.9874	18.5080	18.0456	17.5993	17.1686	16.7529	20
22	20.7841	20.2112	19.6604	19.1306	18.6208	18.1303	22
24	22.5629	21.8891	21.2434	20.6242	20.0304	19.4607	24
26	24.3240	23.5422	22.7952	22.0813	21.3986	20.7457	26
28	26.0677	25.1707	24.3164	23.5025	22.7267	21.9870	28
30	27.7941	26.7751	25.8077	24.8889	24.0158	23.1858	30
32	29.5033	28.3557	27.2696	26.2413	25.2671	24.3439	32
34	31.1955	29.9128	28.7027	27.5605	26.4817	25.4624	34
36	32.8710	31.4468	30.1075	28.8473	27.6607	26.5428	36
38	34.5299	32.9581	31.4847	30.1025	28.8051	27.5863	38
40	36.1722	34.4469	32.8347	31.3269	29.9158	28.5942	40
44	39.4082	37.3587	35.4555	33.6864	32.0406	30.5082	44
48	42.5803	40.1848	37.9740	35.9315	34.0426	32.2938	48
52	45.6897	42.9276	40.3942	38.0677	35.9287	33.9597	52
56	48.7378	45.5897	42.7200	40.1004	37.7059	35.5140	56
60	51.7256	48.1734	44.9550	42.0346	39.3803	36.9640	60
64	54.6543	50.6810	47.1029	43.8750	40.9579	38.3168	64
68	57.5253	53.1147	49.1669	45.6262	42.4442	39.5789	68
72	60.3395	55.4768	51.1504	47.2925	43.8447	40.7564	72
76	63.0982	57.7694	53.0565	48.8780	45.1641	41.8550	76
80	65.8023	59.9944	54.8882	50.3867	46.4073	42.8799	80
84	68.4530	62.1540	56.6485	51.8222	47.5786	43.8361	84
88	71.0514	64.2499	58.3400	53.1881	48.6822	44.7282	88
92	73.5985	66.2841	59.9656	54.4879	49.7220	45.5605	92
96	76.0952	68.2584	61.5277	55.7246	50.7017	46.3370	96
100	78.5426	70.1746	63.0289	56.9013	51.6247	47.0615	100
n	½%	¾%	1%	1¼%	1½%	1¾%	n

TABLE 16.—PRESENT VALUE OF AN ANNUITY OF 1.—(*Continued*)

$$a_{\overline{n}|} \text{ at } i = \frac{1 - (1 + i)^{-n}}{i}$$

n	2%	2½%	3%	4%	5%	6%	n
1	0.9804	0.9756	0.9709	0.9615	0.9524	0.9434	1
2	1.9416	1.9274	1.9135	1.8861	1.8594	1.8334	2
3	2.8839	2.8560	2.8286	2.7751	2.7232	2.6730	3
4	3.8077	3.7620	3.7171	3.6299	3.5460	3.4651	4
5	4.7135	4.6458	4.5797	4.4518	4.3295	4.2124	5
6	5.6014	5.5081	5.4172	5.2421	5.0757	4.9173	6
7	6.4720	6.3494	6.2303	6.0021	5.7864	5.5824	7
8	7.3255	7.1701	7.0197	6.7327	6.4632	6.2098	8
9	8.1622	7.9709	7.7861	7.4353	7.1078	6.8017	9
10	8.9826	8.7521	8.5302	8.1109	7.7217	7.3601	10
11	9.7868	9.5142	9.2526	8.7605	8.3064	7.8869	11
12	10.5753	10.2578	9.9540	9.3851	8.8633	8.3838	12
13	11.3484	10.9832	10.6350	9.9856	9.3936	8.8527	13
14	12.1062	11.6909	11.2961	10.5631	9.8986	9.2950	14
15	12.8493	12.3814	11.9379	11.1184	10.3797	9.7122	15
16	13.5777	13.0550	12.5611	11.6523	10.8378	10.1059	16
17	14.2919	13.7122	13.1661	12.1657	11.2741	10.4773	17
18	14.9920	14.3534	13.7535	12.6593	11.6896	10.8276	18
19	15.6785	14.9789	14.3238	13.1339	12.0853	11.1581	19
20	16.3514	15.5892	14.8775	13.5903	12.4622	11.4699	20
21	17.0112	16.1845	15.4150	14.0292	12.8212	11.7641	21
22	17.6580	16.7654	15.9369	14.4511	13.1630	12.0416	22
23	18.2922	17.3321	16.4436	14.8568	13.4886	12.3034	23
24	18.9139	17.8850	16.9355	15.2470	13.7986	12.5504	24
25	19.5235	18.4244	17.4131	15.6221	14.0939	12.7834	25
26	20.1210	18.9506	17.8768	15.9828	14.3752	13.0032	26
27	20.7069	19.4640	18.3270	16.3296	14.6430	13.2105	27
28	21.2813	19.9649	18.7641	16.6631	14.8981	13.4062	28
29	21.8444	20.4535	19.1885	16.9837	15.1411	13.5907	29
30	22.3965	20.9303	19.6004	17.2920	15.3725	13.7648	30
31	22.9377	21.3954	20.0004	17.5885	15.5928	13.9291	31
32	23.4683	21.8492	20.3888	17.8736	15.8027	14.0840	32
33	23.9886	22.2919	20.7658	18.1476	16.0025	14.2302	33
34	24.4986	22.7238	21.1318	18.4112	16.1929	14.3681	34
35	24.9986	23.1452	21.4872	18.6646	16.3742	14.4982	35
36	25.4888	23.5563	21.8323	18.9083	16.5469	14.6210	36
37	25.9695	23.9573	22.1672	19.1426	16.7113	14.7368	37
38	26.4406	24.3486	22.4925	19.3679	16.8679	14.8460	38
39	26.9026	24.7303	22.8082	19.5845	17.0170	14.9491	39
40	27.3555	25.1028	23.1148	19.7928	17.1591	15.0463	40
42	28.2348	25.8206	23.7014	20.1856	17.4232	15.2245	42
44	29.0800	26.5038	24.2543	20.5488	17.6628	15.3832	44
46	29.8923	27.1542	24.7754	20.8847	17.8801	15.5244	46
48	30.6731	27.7732	25.2667	21.1951	18.0772	15.6500	48
50	31.4236	28.3623	25.7298	21.4822	18.2559	15.7619	50
n	2%	2½%	3%	4%	5%	6%	n

TABLE 17.—THE ANNUITY THAT 1 WILL PURCHASE*

$$\frac{1}{a_{\overline{n}|}} \text{ at } i = \frac{i}{1 - v^n}$$

n	½%	¾%	1%	1¼%	1½%	1¾%	n
1	1.00500	1.00750	1.01000	1.01250	1.01500	1.01750	1
2	0.50375	0.50563	0.50751	0.50939	0.51128	0.51316	2
3	33667	33835	34002	34170	34338	34507	3
4	25313	25471	25628	25786	25944	26103	4
5	0.20301	0.20452	0.20604	0.20756	0.20909	0.21062	5
6	16960	17107	17255	17403	17553	17702	6
7	14573	14717	14863	15009	15156	15303	7
8	12783	12926	13069	13213	13358	13504	8
9	11391	11532	11674	11817	11961	12106	9
10	0.10277	0.10417	0.10558	0.10700	0.10843	0.10988	10
11	09366	09505	09645	09787	09929	10073	11
12	08607	08745	08885	09026	09168	09311	12
13	07964	08102	08241	08382	08524	08667	13
14	07414	07551	07690	07831	07972	08116	14
15	0.06936	0.07074	0.07212	0.07353	0.07494	0.07638	15
16	06519	06656	06794	06935	07077	07220	16
17	06151	06287	06426	06566	06708	06852	17
18	05823	05960	06098	06238	06381	06524	18
19	05530	05667	05805	05946	06088	06232	19
20	0.05267	0.05403	0.05542	0.05682	0.05825	0.05969	20
22	04811	04948	05086	05227	05370	05516	22
24	04432	04568	04707	04849	04992	05139	24
26	04111	04248	04387	04529	04673	04820	26
28	03836	03973	04112	04255	04400	04548	28
30	0.03598	0.03735	0.03875	0.04018	0.04164	0.04313	30
32	03389	03527	03667	03811	03958	04108	32
34	03206	03343	03484	03628	03776	03927	34
36	03042	03180	03321	03467	03615	03768	36
38	02896	03034	03176	03322	03472	03625	38
40	0.02765	0.02903	0.03046	0.03192	0.03343	0.03497	40
44	02538	02677	02820	02969	03121	03278	44
48	02349	02489	02633	02783	02938	03097	48
52	02189	02330	02476	02627	02783	02945	52
56	02052	02193	02341	02494	02652	02816	56
60	0.01933	0.02076	0.02224	0.02379	0.02539	0.02705	60
64	01830	01973	02123	02279	02442	02610	64
68	01738	01883	02034	02192	02356	02527	68
72	01657	01803	01955	02115	02281	02454	72
76	01585	01731	01885	02046	02214	02389	76
80	0.01520	0.01667	0.01822	0.01985	0.02155	0.02332	80
84	01461	01609	01765	01930	02102	02281	84
88	01407	01556	01714	01880	02054	02236	88
92	01359	01509	01668	01835	02011	02195	92
96	01314	01465	01625	01795	01972	02158	96
100	0.01273	0.01425	0.01587	0.01757	0.01937	0.02125	100
n	½%	¾%	1%	1¼%	1½%	1¾%	n

* The annuity that will amount to 1 is equal to

$$\frac{1}{s_{\overline{n}|}} = \frac{1}{a_{\overline{n}|}} - i.$$

TABLE 17.—THE ANNUITY THAT 1 WILL PURCHASE*.—(*Continued*)

$$\frac{1}{a_{\overline{n}|}} \text{ at } i = \frac{i}{1 - v^n}$$

n	2%	2½%	3%	4%	5%	6%	n
1	1.02000	1.02500	1.03000	1.04000	1.05000	1.06000	1
2	0.51505	0.51883	0.52261	0.53020	0.53780	0.54544	2
3	34675	35014	35353	36035	36721	37411	3
4	26262	26582	26903	27549	28201	28859	4
5	0.21216	0.21525	0.21835	0.22463	0.23097	0.23740	5
6	17853	18155	18460	19076	19702	20336	6
7	15451	15750	16051	16661	17282	17914	7
8	13651	13947	14246	14853	15472	16104	8
9	12252	12546	12843	13449	14069	14702	9
10	0.11133	0.11426	0.11723	0.12329	0.12950	0.13587	10
11	10218	10511	10808	11415	12039	12679	11
12	09456	09749	10046	10655	11283	11928	12
13	08812	09105	09403	10014	10646	11296	13
14	08260	08554	08853	09467	10102	10758	14
15	0.07783	.08077	0.08377	0.08994	0.09634	0.10296	15
16	07365	07660	07961	08582	09227	09895	16
17	06997	07293	07595	08220	08870	09544	17
18	06670	06967	07271	07899	08555	09236	18
19	06378	06676	06981	07614	08275	08962	19
20	0.06116	0.06415	0.06722	0.07358	0.08024	0.08718	20
21	05878	06179	06487	07128	07800	08500	21
22	05663	05965	06275	06920	07597	08305	22
23	05467	05770	06081	06731	07414	08128	23
24	05287	05591	05905	06559	07247	07968	24
25	0.05122	0.05428	0.05743	0.06401	0.07095	0.07823	25
26	04970	05277	05594	06257	06956	07690	26
27	04829	05138	05456	06124	06829	07570	27
28	04699	05009	05329	06001	06712	07459	28
29	04578	04889	05211	05888	06605	07358	29
30	0.04465	0.04778	0.05102	0.05783	0.06505	0.07265	30
31	04360	04674	05000	05686	06413	07179	31
32	04261	04577	04905	05595	06328	07100	32
33	04169	04486	04816	05510	06249	07027	33
34	04082	04401	04732	05431	06176	06960	34
35	0.04000	0.04321	0.04654	0.05358	0.06107	0.06897	35
36	03923	04245	04580	05289	06043	06839	36
37	03851	04174	04511	05224	05984	06786	37
38	03782	04107	04446	05163	05928	06736	38
39	03717	04044	04384	05106	05876	06689	39
40	0.03656	0.03984	0.04326	0.05052	0.05828	0.06646	40
42	03542	03873	04219	04954	05739	06568	42
44	03439	03773	04123	04866	05662	06501	44
46	03345	03683	04036	04788	05593	06441	46
48	03260	03601	03958	04718	05532	06390	48
50	0.03182	0.03526	0.03887	0.04655	0.05478	0.06344	50
n	2%	2½%	3%	4%	5%	6%	n

* The annuity that will amount to 1 is equal to

$$\frac{1}{s_{\overline{n}|}} = \frac{1}{a_{\overline{n}|}} - i.$$

TABLE 18a.—COMPOUND AMOUNT FOR TIMES LESS THAN A YEAR

$$s = (1 + i)^{\frac{1}{p}}$$

p	½%	¾%	1%	1¼%	1½%	1¾%	p
12	1.0004	1.0006	1.0008	1.0010	1.0012	1.0014	12
4	1.0012	1.0019	1.0025	1.0031	1.0037	1.0043	4
2	1.0025	1.0037	1.0050	1.0062	1.0075	1.0087	2

p	2%	2½%	3%	4%	5%	6%	p
12	1.0017	1.0021	1.0025	1.0033	1.0041	1.0049	12
4	1.0050	1.0062	1.0074	1.0099	1.0123	1.0147	4
2	1.0100	1.0124	1.0149	1.0198	1.0247	1.0296	2

TABLE 18b.—THE VALUE OF $j_{(p)} = p[(1 + i)^{\frac{1}{p}} - 1]$

p	½%	¾%	1%	1¼%	1½%	1¾%	p
12	0.00499	0.00747	0.00995	0.01243	0.01490	0.01736	12
4	00499	00748	00996	01244	01492	01739	4
2	00499	00749	00998	01246	01494	01742	2

p	2%	2½%	3%	4%	5%	6%	p
12	0.01982	0.02472	0.02960	0.03928	0.04889	0.05841	12
4	01985	02477	02967	03941	04909	05870	4
2	01990	02485	02978	03961	04939	05913	2

TABLE 18c.—THE VALUE OF $\dfrac{i}{j_{(p)}} = \dfrac{i}{p[(1 + i)^{\frac{1}{p}} - 1]}$

p	½%	¾%	1%	1¼%	1½%	1¾%	p
12	1.0023	1.0034	1.0046	1.0057	1.0069	1.0080	12
4	1.0019	1.0028	1.0037	1.0047	1.0056	1.0065	4
2	1.0012	1.0019	1.0025	1.0031	1.0037	1.0044	2

p	2%	2½%	3%	4%	5%	6%	p
12	1.0091	1.0114	1.0137	1.0182	1.0227	1.0272	12
4	1.0075	1.0093	1.0112	1.0149	1.0186	1.0222	4
2	1.0050	1.0062	1.0074	1.0099	1.0123	1.0148	2

TABLE 19.—AMERICAN EXPERIENCE TABLE OF MORTALITY

Age x	Number living l_x	Number dying d_x	Yearly probability of dying q_x	Yearly probability of living p_x	Age x	Number living l_x	Number dying d_x	Yearly probability of dying q_x	Yearly probability of living p_x
10	100 000	749	0.007 490	0.992 510	53	66 797	1 091	0.016 333	0.983 667
11	99 251	746	007 516	992 484	54	65 706	1 143	017 396	982 604
12	98 505	743	007 543	992 457	55	64 563	1 199	018 571	981 429
13	97 762	740	007 569	992 431	56	63 364	1 260	019 885	980 115
14	97 022	737	007 596	992 404	57	62 104	1 325	021 335	978 665
15	96 285	735	0.007 634	0.992 366	58	60 779	1 394	0.022 936	0.977 064
16	95 550	732	007 661	992 339	59	59 385	1 468	024 720	975 280
17	94 818	729	007 688	992 312	60	57 917	1 546	026 693	973 307
18	94 089	727	007 727	992 273	61	56 371	1 628	028 880	971 120
19	93 362	725	007 765	992 235	62	54 743	1 713	031 292	968 708
20	92 637	723	0.007 805	0.992 195	63	53 030	1 800	0.033 943	0.966 057
21	91 914	722	007 855	992 145	64	51 230	1 889	036 873	963 127
22	91 192	721	007 906	992 094	65	49 341	1 980	040 129	959 871
23	90 471	720	007 958	992 042	66	47 361	2 070	043 707	956 293
24	89 751	719	008 011	991 989	67	45 291	2 158	047 647	952 353
25	89 032	718	0.008 065	0.991 935	68	43 133	2 243	0.052 002	0.947 998
26	88 314	718	008 130	991 870	69	40 890	2 321	056 762	943 238
27	87 596	718	008 197	991 803	70	38 569	2 391	061 993	938 007
28	86 878	718	008 264	991 736	71	36 178	2 448	067 665	932 335
29	86 160	719	008 345	991 655	72	33 730	2 487	073 733	926 267
30	85 441	720	0.008 427	0.991 573	73	31 243	2 505	0.080 178	0.919 822
31	84 721	721	008 510	991 490	74	28 738	2 501	087 028	912 972
32	84 000	723	008 607	991 393	75	26 237	2 476	094 371	905 629
33	83 277	726	008 718	991 282	76	23 761	2 431	102 311	897 689
34	82 551	729	008 831	991 169	77	21 330	2 369	111 064	888 936
35	81 822	732	0.008 946	0.991 054	78	18 961	2 291	0.120 827	0.879 173
36	81 090	737	009 089	990 911	79	16 670	2 196	131 734	868 266
37	80 353	742	009 234	990 766	80	14 474	2 091	144 466	855 534
38	79 611	749	009 408	990 592	81	12 383	1 964	158 605	841 395
39	78 862	756	009 586	990 414	82	10 419	1 816	174 297	825 703
40	78 106	765	0.009 794	0.990 206	83	8 603	1 648	0.191 561	0.808 439
41	77 341	774	010 008	989 992	84	6 955	1 470	211 359	788 641
42	76 567	785	010 252	989 748	85	5 485	1 292	235 562	764 448
43	75 782	797	010 517	989 483	86	4 193	1 114	265 681	734 319
44	74 985	812	010 829	989 171	87	3 079	933	303 020	696 980
45	74 173	828	0.011 163	0.988 837	88	2 146	744	0.346 692	0.653 308
46	73 345	848	011 562	988 438	89	1 402	555	395 863	604 137
47	72 497	870	012 000	988 000	90	847	385	454 545	545 455
48	71 627	896	012 509	987 491	91	462	246	532 468	467 532
49	70 731	927	013 106	986 894	92	216	137	634 259	365 741
50	69 804	962	0.013 781	0.986 219	93	79	58	734 177	0.265 823
51	68 842	1 001	014 541	985 459	94	21	18	0.857 143	142 857
52	67 841	1 044	015 389	984 611	95	3	3	1.000 000	000 000

TABLE 20.—COMPLETE ELLIPTIC INTEGRALS, K AND E

$$K = \int_0^{\frac{\pi}{2}} \frac{dx}{\sqrt{1 - k^2 \sin^2 x}}, \qquad E = \int_0^{\frac{\pi}{2}} \sqrt{1 - k^2 \sin^2 x}\; dx$$

$\sin^{-1} k$	K	E	$\sin^{-1} k$	K	E	$\sin^{-1} k$	K	E
0°	1.5708	1.5708	50°	1.9356	1.3055	81°.0	3.2553	1.0338
1	5709	5707	51	9539	2963	2	2771	0326
2	5713	5703	52	9729	2870	4	2995	0313
3	5719	5697	53	1.9927	2776	6	3223	0302
4	5727	5689	54	2.0133	2681	81.8	3458	0290
5	1.5738	1.5678	55	2.0347	1.2587	82.0	3.3699	1.0278
6	5751	5665	56	0571	2492	2	3946	0267
7	5767	5650	57	0804	2397	4	4199	0256
8	5785	5630	58	1047	2301	6	4460	0245
9	5805	5611	59	1300	2206	82.8	4728	0234
10	1.5828	1.5589	60	2.1565	1.2111	83.0	3.5004	1.0223
11	5854	5564	61	1842	2015	2	5288	0213
12	5882	5537	62	2132	1921	4	5581	0202
13	5913	5507	63	2435	1826	6	5884	0192
14	5946	5476	64	2754	1732	83.8	6196	0182
15	1.5981	1.5442	65	2.3088	1.1638	84.0	3.6519	1.0172
16	6020	5405	65.5	3261	1592	2	6853	0163
17	6061	5367	66.0	3439	1546	4	7198	0153
18	6105	5326	66.5	3621	1499	6	7557	0144
19	6151	5283	67.0	3809	1454	84.8	7930	0135
20	1.6200	1.5238	67.5	2.4001	1.1408	85.0	3.8317	1.0127
21	6252	5191	68.0	4198	1362	2	8721	0118
22	6307	5141	68.5	4401	1317	4	9142	0110
23	6363	5090	69.0	4610	1273	6	3.9583	0102
24	6426	5037	69.5	4825	1228	85.8	4.0044	0094
25	1.6490	1.4981	70.0	2.5046	1.1184	86.0	4.0528	1.0087
26	6557	4924	70.5	5273	1140	2	1037	0079
27	6627	4864	71.0	5507	1096	4	1574	0072
28	6701	4803	71.5	5749	1053	6	2142	0065
29	6777	4740	72.0	5998	1011	86.8	2744	0059
30	1.6858	1.4675	72.5	2.6256	1.0968	87.0	4.3387	1.0053
31	6941	4608	73.0	6521	0927	2	4073	0047
32	7028	4539	73.5	6796	0885	4	4812	0041
33	7119	4469	74.0	7081	0844	6	5619	0036
34	7214	4397	74.5	7375	0804	87.8	6477	0031
35	1.7312	1.4323	75.0	2.7681	1.0764	88.0	4.7427	1.0026
36	7415	4248	75.5	7998	0725	2	8479	0022
37	7522	4171	76.0	8327	0686	4	4.9654	0017
38	7633	4092	76.5	8669	0648	6	5.0988	0014
39	7748	4013	77.0	9026	0611	88.8	2527	0010
40	1.7868	1.3931	77.5	2.9397	1.0574	89.0	5.4349	1.0008
41	7992	3849	78.0	2.9786	0538	1	5402	0006
42	8122	3765	78.5	3.0192	0502	2	6579	0005
43	8256	3680	79.0	0617	0468	3	7914	0005
44	8396	3594	79.5	1064	0434	4	5.9455	0003
45	1.8541	1.3506	80.0	3.1534	1.0401	89.5	6.1278	1.0002
46	8692	3418	2	1729	0388	6	3504	0001
47	8848	3329	4	1928	0375	7	6.6385	0001
48	9011	3238	6	2132	0363	8	7.0440	0000
49	1.9180	1.3147	80.8	3.2340	1.0350	89.9	7.7371	1.0000

TABLE 21.—VALUES OF THE PROBABILITY INTEGRAL $\dfrac{2}{\sqrt{\pi}}\displaystyle\int_0^x e^{-x^2}\,dx$

x		0	1	2	3	4	5	6	7	8	9
0.0	0.0	000	113	226	338	451	564	676	789	901	*013
1	0.1	125	236	348	459	569	680	790	900	*009	*118
2	0.2	227	335	443	550	657	763	869	974	*079	*183
3	0.3	286	389	491	593	694	794	893	992	*090	*187
4	0.4	284	380	475	569	662	755	847	937	*027	*117
0.5	0.5	205	292	379	465	549	633	716	798	879	959
6	0.6	039	117	194	270	346	420	494	566	638	708
7	0.6	778	847	914	981	*047	*112	*175	*238	*300	*361
8	0.7	421	480	538	595	651	707	761	814	867	918
0.9	0.7	969	*019	*068	*116	*163	*209	*254	*299	*342	*385
1.0	0.8	427	468	508	548	586	624	661	698	733	768
1	0.8	802	835	868	900	931	961	991	*020	*048	*076
2	0.9	103	130	155	181	205	229	252	275	297	319
3	0.9	340	361	381	400	419	438	456	473	490	507
4	0.9	523	539	554	569	583	597	611	624	637	649
1.5	0.9	661	673	684	695	706	716	726	736	745	755
6	0.9	763	772	780	788	796	804	811	818	825	832
7	0.9	838	844	850	856	861	867	872	877	882	886
8	0.9	891	895	899	903	907	911	915	918	922	925
1.9	0.9	928	931	934	937	939	942	944	947	949	951
2.0	0.99	532	552	572	591	609	626	642	658	673	688
1	0.99	702	715	728	741	753	764	775	785	795	805
2	0.99	814	822	831	839	846	854	861	867	874	880
3	0.99	886	891	897	902	906	911	915	920	924	928
4	0.99	931	935	938	941	944	947	950	952	955	957
2.5	0.99	959	961	963	965	967	969	971	972	974	975
6	0.99	976	978	979	980	981	982	983	984	985	986
7	0.99	987	987	988	989	989	990	991	991	992	992
8	0.999	925	929	933	937	941	944	948	951	954	956
2.9	0.999	959	961	964	966	968	970	972	973	975	977

SOME CONSTANTS

Name	Value	Log
π	3.14159 26536	0.49714 98727
e	2.71828 18285	0.43429 44819
$M = \log e$	0.43429 44819	9.63778 43113 − 10
$1/M = \ln 10$	2.30258 50930	0.36221 56887

$g_0 = 32.1740$ feet per second per second

$\quad = 980.665$ centimeters per second per second.

$\quad g = 978.0490(1 + 0.0052884 \sin^2 \varphi - 0.0000059 \sin^2 2\varphi)$

centimeters per second per second, where φ is the latitude.

Correction for altitude above sea level:

$\quad$ −0.3 centimeter per second per second for each 1000 meters;

$\quad$ −0.003 feet per second per second for each 1000 feet.

TABLE 22.—CONVERSION TABLES

a. Length equivalents

Centi-meters	Inches	Feet	Meters	Kilometers	Miles
1	0.3937	0.03281	0.01	10^{-5}	6.214×10^{-6}
2.540	1	0.08333	0.0254	2.54×10^{-5}	1.578×10^{-5}
30.48	12	1	0.3048	3.048×10^{-4}	1.894×10^{-4}
100	39.37	3.281	1	0.001	6.214×10^{-4}
100,000	39,370	3,281	1,000	1	0.6214
160,935	63,360	5,280	1,609	1.609	1

b. Area equivalents

Square meters	Square inches	Square feet	Square rods	Acres	Hectares
1	1,550	10.76	0.0395	2.471×10^{-4}	10^{-4}
6.452×10^{-4}	1	0.006944	2.551×10^{-5}	1.594×10^{-7}	6.452×10^{-8}
0.09290	144	1	0.003673	2.296×10^{-5}	9.290×10^{-6}
25.29	39,204	272.25	1	0.00625	2.529×10^{-3}
4,047	6,272,640	43,560	160	1	0.4047
10,000	1.550×10^7	1.076×10^5	395	2.471	1

c. Volume and capacity equivalents

Cubic inches	Cubic feet	U. S. gallons Liquid	U. S. gallons Dry	Bushels	Liters
1	5.787×10^{-4}	4.329×10^{-3}	3.720×10^{-3}	4.650×10^{-4}	0.01639
1728	1	7.481	6.429	0.8036	28.32
231	0.1337	1	0.8594	0.1074	3.785
268.8	0.1556	1.164	1	0.125	4.405
2150	1.244	9.309	8	1	35.24
61.02	0.03531	0.2642	0.2270	0.02838	1

d. Mass equivalents

Kilo-grams	Pounds Troy	Pounds Avoirdupois	Tons Short	Tons Long	Tons Metric
1	2.6792	2.205	1.102×10^{-3}	9.842×10^{-4}	0.001
0.3732	1	0.8229	4.114×10^{-4}	3.673×10^{-4}	3.732×10^{-4}
0.4536	1.215	1	0.0005	4.464×10^{-4}	4.536×10^{-4}
907.2	2,431	2,000	1	0.8929	0.9072
1,016	2,722	2,240	1.12	1	1.016
1,000	2,679	2,205	1.102	0.9842	1

e. Velocity equivalents

Centi-meters per second	Kilo-meters per hour	Feet per second	Feet per minute	Miles per hour	Knots
1	0.036	0.03281	1.9685	0.02237	0.01943
27.78	1	0.9113	54.68	0.6214	0.5396
30.48	1.097	1	60	0.6818	0.5921
0.5080	0.01829	0.01667	1	0.01136	0.00987
44.70	1.609	1.467	88	1	0.8684
51.48	1.8532	1.6889	101.34	1.15155	1

TABLE 22.—CONVERSION TABLES.—(*Continued*)
f. Energy or work equivalents

Joules = 10^7 ergs	Kilogram-meters	Foot-pounds	Kilowatt-hours	Kilogram-calories	British thermal units
1	0.10197	0.7376	2.778×10^{-7}	2.390×10^{-4}	9.486×10^{-4}
9.80665	1	7.233	2.724×10^{-6}	0.002344	0.009302
1.356	0.1383	1	3.766×10^{-7}	3.241×10^{-4}	0.001286
3.6×10^6	3.671×10^5	2.655×10^6	1	860.5	3,415
4,183	426.6	3,086	0.001162	1	3.968
1,054	107.5	777.52	2.928×10^{-4}	0.2520	1

1 horsepower = 550 foot-pounds per second
= 0.7457 kilowatts
= 0.7074 British thermal units per second.

g. Pressure equivalents

Mega-dynes per square centimeter	Kilograms per square centimeter	Pounds per square inch	Atmospheres	Columns of mercury at 0°C.	
				Meters	Inches
1	1.0197	14.50	0.9869	0.7500	29.53
0.9807	1	14.22	0.9678	0.7355	28.96
0.06895	0.07031	1	0.06804	0.05171	2.036
1.0133	1.0333	14.70	1	0.76	29.92
1.3333	1.3596	19.34	1.316	1	39.37
0.03386	0.03453	0.4912	0.03342	0.02540	1

TABLE 23.—BINOMIAL COEFFICIENTS, $(n)_r = {}_nC_r = {}_nC_{n-r}$

r \ n	0	1	2	3	4	5	6	7	8	9	10
1	1	1				Sum of any two adjacent numbers					
2	1	2	1			in the same row is equal to the					
3	1	3	3	1		number just below the right-hand					
4	1	4	6	4	1	one of them.					
5	1	5	10	10	5	1					
6	1	6	15	20	15	6	1				
7	1	7	21	35	35	21	7	1			
8	1	8	28	56	70	56	28	8	1		
9	1	9	36	84	126	126	84	36	9	1	
10	1	10	45	120	210	252	210	120	45	10	1
11	1	11	55	165	330	462	462	330	165	55	11
12	1	12	66	220	495	792	924	792	495	220	66
13	1	13	78	286	715	1287	1716	1716	1287	715	286
14	1	14	91	364	1001	2002	3003	3432	3003	2002	1001
15	1	15	105	455	1365	3003	5005	6435	6435	5005	3003

TABLE OF INTEGRALS

In the following integral formulas, u denotes a function of an independent variable, as x, or may itself be an independent variable. The letters a, b, c, m, n, p, q, r denote constants which may, in general, be assigned any value for which the expressions in which they occur are then defined (do not become infinite). In particular, the exponents m and n may have any positive or negative values, except as noted. An arbitrary constant of integration C is to be added to the value given for each indefinite integral.

See also the brief table of fundamental integrals in Art. 132. The notation (Art. 69) is used to indicate that an equivalent value of the integral, or some part of it, may be found in the article numbered 69. The notation (150) indicates that integral formula 150 is to be used for exceptional cases of the formula with which the notation occurs, or for the purpose of continuing its evaluation.

INTEGRALS INVOLVING $c + bu$

1. $\displaystyle\int (c + bu)^n du = \frac{(c + bu)^{n+1}}{b(n + 1)}, \qquad n \neq -1.$ $\hspace{2cm}$ (2)

2. $\displaystyle\int (c + bu)^{-1} du = \int \frac{du}{c + bu} = \frac{\ln (c + bu)}{b}.$

3. $\displaystyle\int \frac{u\, du}{c + bu} = \frac{u}{b} - \frac{c \ln (c + bu)}{b^2}.$

4. $\displaystyle\int \frac{du}{u(c + bu)^2} = \frac{1}{c(c + bu)} - \frac{1}{c^2} \ln \frac{c + bu}{u}.$

5. $\displaystyle\int \frac{(c + bu) du}{u} = bu + c \ln u.$

6. $\displaystyle\int \frac{u^2\, du}{c + bu} = \frac{u^2}{2b} - \frac{cu}{b^2} + \frac{c^2 \ln (c + bu)}{b^3}.$

7. $\displaystyle\int \frac{u\, du}{(c + bu)^2} = \frac{c}{b^2(c + bu)} + \frac{\ln (c + bu)}{b^2}.$

8. $\displaystyle\int \frac{du}{u(c + bu)} = \frac{1}{c} \ln \frac{u}{c + bu}.$

INTEGRALS INVOLVING $c + bu$ (*Continued*)

9. $\displaystyle\int \frac{du}{(c + u)(q + u)} = \frac{1}{q - c} \ln \frac{c + u}{q + u}, \quad q \neq c;$

$\displaystyle = -\frac{1}{c + u}, \quad q = c.$

10. $\displaystyle\int \frac{du}{(c + bu)(q + pu)} = \frac{\ln (c + bu) - \ln (q + pu)}{bq - pc},$

$cp \neq bq;$

$\displaystyle = \frac{-1}{p(c + bu)} \text{ or } \frac{-1}{b(q + pu)}, \quad cp = bq.$

11. $\displaystyle\int \frac{du}{u\sqrt{c + bu}} = \frac{2}{\sqrt{-c}} \tan^{-1} \sqrt{\frac{c + bu}{-c}}, \quad c < 0;$

$\displaystyle = -\frac{2}{\sqrt{c}} \tanh^{-1} \sqrt{\frac{c + bu}{c}}^{*}, \text{ or}$

$\displaystyle \frac{1}{\sqrt{c}} \ln \frac{\sqrt{c + bu} - \sqrt{c}}{\sqrt{c + bu} + \sqrt{c}}, \quad c > 0.$

12. $\displaystyle\int \frac{du}{u^2\sqrt{c + bu}} = -\frac{\sqrt{c + bu}}{cu} - \frac{b}{2c}\int \frac{du}{u\sqrt{c + bu}}. \quad (11)$

13. $\displaystyle\int \frac{\sqrt{c + bu}\, du}{u} = 2\sqrt{c + bu} + c\int \frac{du}{u\sqrt{c + bu}}. \quad (11)$

14. $\displaystyle\int \frac{du}{\sqrt{u}\sqrt{c + bu}} = \frac{-1}{\sqrt{-b}} \sin^{-1} \frac{2bu + c}{c}, \quad b < 0;$

$\displaystyle = \frac{1}{\sqrt{b}} \cosh^{-1} \frac{2bu + c}{c}, \quad b > 0. \quad \text{(Art. 69)}$

15. $\displaystyle\int \frac{\sqrt{c + bu}\, du}{u^{3/2}} = -2\sqrt{\frac{c + bu}{u}} + 2\sqrt{b} \sinh^{-1} \sqrt{\frac{bu}{c}}.$

16. $\displaystyle\int u(c + bu)^n du = \frac{(c + bu)^{n+2}}{b^2(n + 2)} - \frac{c(c + bu)^{n+1}}{b^2(n + 1)},$

$n \neq -1, -2. \quad (3, 4)$

* When the expression under the radical is greater than unity $\tanh^{-1}$ must be replaced by $\coth^{-1}$.

INTEGRALS INVOLVING $c + bu$ (*Continued*)

17. $\int u^m(c + bu)^n du$

$$= \frac{u^{m+1}(c + bu)^n}{m + n + 1} + \frac{cn}{m + n + 1} \int u^m(c + bu)^{n-1} du, \text{ or}$$

$$= \frac{u^m(c + bu)^{n+1}}{b(m + n + 1)} - \frac{cm}{b(m + n + 1)} \int u^{m-1}(c + bu)^n du,$$

$$m + n \neq -1.$$

18. $\int \frac{u^m\, du}{(c + bu)^n}$

$$= \frac{u^{m+1}}{c(n - 1)(c + bu)^{n-1}} - \frac{m - n + 2}{c(n - 1)} \int \frac{u^m\, du}{(c + bu)^{n-1}},$$

$$n \neq 1; \quad (19)$$

$$= \frac{u^m}{b(m - n + 1)(c + bu)^{n-1}} - \frac{cm}{b(m - n + 1)} \int \frac{u^{m-1}\, du}{(c + bu)^n},$$

$$n - m \neq 1.$$

19. $\int \frac{u^m\, du}{c + bu} = \frac{u^m}{bm} - \frac{c}{b} \int \frac{u^{m-1}\, du}{c + bu}.$

20. $\int \frac{(c + bu)^n du}{u^m}$

$$= -\frac{(c + bu)^{n+1}}{c(m - 1)u^{m-1}} + \frac{b(n - m + 2)}{c(m - 1)} \int \frac{(c + bu)^n du}{u^{m-1}},$$

$$m \neq 1; \quad (21)$$

$$= \frac{(c + bu)^n}{(n - m + 1)u^{m-1}} + \frac{cn}{n - m + 1} \int \frac{(c + bu)^{n-1}\, du}{u^m},$$

$$m - n \neq 1.$$

21. $\int \frac{(c + bu)^n du}{u} = \frac{(c + bu)^n}{n} + c \int \frac{(c + bu)^{n-1} du}{u}.$

22. $\int \frac{du}{u^m(c + bu)^n}$

$$= \frac{-1}{c(m - 1)u^{m-1}(c + bu)^{n-1}}$$

$$- \frac{b(m + n - 2)}{c(m - 1)} \int \frac{du}{u^{m-1}(c + bu)^n}, \quad m \neq 1;$$

$$= \frac{1}{c(n - 1)u^{m-1}(c + bu)^{n-1}}$$

$$+ \frac{m + n - 2}{c(n - 1)} \int \frac{du}{u^m(c + bu)^{n-1}}, \quad n \neq 1.$$

INTEGRALS INVOLVING $c + bu$ (*Continued*)

23. $\displaystyle\int u^m \sqrt{c + bu}\, du = \frac{2u^m(c + bu)^{3/2}}{b(2m + 3)}$

$\displaystyle\qquad - \frac{2cm}{b(2m + 3)}\int u^{m-1}\sqrt{c + bu}\, du, \qquad m \neq -\tfrac{3}{2}. \quad (15)$

24. $\displaystyle\int \frac{u^m\, du}{\sqrt{c + bu}} = \frac{2u^m\sqrt{c + bu}}{b(2m + 1)} - \frac{2cm}{b(2m + 1)}\int \frac{u^{m-1}\, du}{\sqrt{c + bu}},$

$\displaystyle\qquad\qquad\qquad\qquad\qquad\qquad\qquad m \neq -\tfrac{1}{2}. \quad (14)$

25. $\displaystyle\int \frac{du}{u^m\sqrt{c + bu}} = -\frac{\sqrt{c + bu}}{c(m - 1)u^{m-1}}$

$\displaystyle\qquad - \frac{b(2m - 3)}{2c(m - 1)}\int \frac{du}{u^{m-1}\sqrt{c + bu}}, \qquad m \neq 1. \quad (11)$

26. $\displaystyle\int \frac{\sqrt{c + bu}\, du}{u^m} = -\frac{(c + bu)^{3/2}}{c(m - 1)u^{m-1}}$

$\displaystyle\qquad - \frac{b(2m - 5)}{2c(m - 1)}\int \frac{\sqrt{c + bu}\, du}{u^{m-1}}, \qquad m \neq 1. \quad (13)$

27. $\displaystyle\int (c + bu)^m (q + pu)^n\, du$

$\displaystyle\quad = \frac{(c + bu)^{m+1}(q + pu)^n}{b(m + n + 1)}$

$\displaystyle\qquad + \frac{n(bq - cp)}{b(m + n + 1)}\int (c + bu)^m (q + pu)^{n-1}\, du,$

$\displaystyle\qquad\qquad\qquad\qquad\qquad\qquad m + n \neq -1.$

28. $\displaystyle\int \frac{du}{(c + bu)^m (q + pu)^n}$

$\displaystyle\quad = \frac{1}{(n - 1)(bq - cp)(c + bu)^{m-1}(q + pu)^{n-1}}$

$\displaystyle\qquad + \frac{b(m + n - 2)}{(n - 1)(bq - cp)}\int \frac{du}{(c + bu)^m (q + pu)^{n-1}},$

$\displaystyle\qquad\qquad\qquad\qquad\qquad\qquad n \neq 1,\ bq \neq cp;$

$\displaystyle\quad = \frac{-1}{p(m + n - 1)(c + bu)^m (q + pu)^{n-1}},$

$\displaystyle\qquad\qquad\qquad\qquad\qquad\qquad m + n \neq 1,\ bq = cp.$

INTEGRALS INVOLVING $c + bu$ (*Continued*)

29. $\displaystyle\int \frac{(c + bu)^m du}{(q + pu)^n}$

$$= \frac{(c + bu)^{m+1}}{(n - 1)(bq - cp)(q + pu)^{n-1}}$$

$$- \frac{b(m - n + 2)}{(n - 1)(bq - cp)} \int \frac{(c + bu)^m du}{(q + pu)^{n-1}}, \qquad n \neq 1, bq \neq cp;$$

$$= \frac{(c + bu)^m}{p(m - n + 1)(q + pu)^{n-1}}, \qquad n - m \neq 1, bq = cp.$$

30. $\displaystyle\int \frac{(c + bu)^m du}{(q + pu)^n}$

$$= \frac{(c + bu)^m}{p(m - n + 1)(q + pu)^{n-1}}$$

$$- \frac{m(bq - cp)}{p(m - n + 1)} \int \frac{(c + bu)^{m-1} du}{(q + pu)^n}, \qquad n - m \neq 1.$$

$$= \frac{-(c + bu)^m}{p(n - 1)(q + pu)^{n-1}} + \frac{bm}{p(n - 1)} \int \frac{(c + bu)^{m-1} du}{(q + pu)^{n-1}},$$
$$n \neq 1.$$

INTEGRALS INVOLVING $c + au^2$

31. $\displaystyle\int \sqrt{c + au^2}\, u\, du = \frac{1}{3a}(c + au^2)^{3/2}.$

32. $\displaystyle\int \frac{u\, du}{\sqrt{c + au^2}} = \frac{1}{a}\sqrt{c + au^2}.$

33. $\displaystyle\int \frac{u\, du}{c + au^2} = \frac{1}{2a} \ln (c + au^2).$

34. $\displaystyle\int (c + au^2)^n u\, du = \frac{(c + au^2)^{n+1}}{2a(n + 1)}, \qquad n \neq -1.$ 　　(33)

35. $\displaystyle\int \frac{du}{\sqrt{c + au^2}} = \frac{1}{\sqrt{-a}} \sin^{-1} u\sqrt{\frac{-a}{c}}, \qquad a < 0, c > 0;$

$$= \frac{1}{\sqrt{a}} \sinh^{-1} u\sqrt{\frac{a}{c}}, \qquad a > 0, c > 0;$$

$$= \frac{1}{\sqrt{a}} \cosh^{-1} u\sqrt{\frac{a}{-c}}, \qquad a > 0, c < 0;$$

$$= \frac{1}{\sqrt{a}} \ln (u\sqrt{a} + \sqrt{c + au^2}), \qquad a > 0.$$

36. $\displaystyle\int \sqrt{c + au^2}\, du = \frac{u}{2}\sqrt{c + au^2} + \frac{c}{2} \int \frac{du}{\sqrt{c + au^2}}.$ 　(35)

INTEGRALS INVOLVING $c + au^2$ (*Continued*)

37. $\displaystyle\int \frac{\sqrt{c + au^2}\, du}{u^2} = -\frac{\sqrt{c + au^2}}{u} + a\int \frac{du}{\sqrt{c + au^2}}.$ (35)

38. $\displaystyle\int \frac{du}{u^2\sqrt{c + au^2}} = -\frac{\sqrt{c + au^2}}{cu}.$

39. $\displaystyle\int \frac{u^2\, du}{\sqrt{c + au^2}} = \frac{u\sqrt{c + au^2}}{2a} - \frac{c}{2a}\int \frac{du}{\sqrt{c + au^2}}.$ (35)

40. $\displaystyle\int \frac{u^2\, du}{(c + au^2)^{3/2}} = -\frac{u}{a\sqrt{c + au^2}} + \frac{1}{a}\int \frac{du}{\sqrt{c + au^2}}.$ (35)

41. $\displaystyle\int (c + au^2)^{3/2}du = \frac{u}{8}(2au^2 + 5c)\sqrt{c + au^2}$
$$+ \frac{3c^2}{8}\int \frac{du}{\sqrt{c + au^2}}.$$ (35)

42. $\displaystyle\int \frac{du}{c + au^2} = \frac{1}{\sqrt{ac}}\,\tan^{-1} u\frac{\sqrt{ac}}{c},$ a and c like signs;

$$= \frac{1}{\sqrt{-ac}}\,\tanh^{-1} u\frac{\sqrt{-ac}}{c},$$ a and c unlike signs;

$$= \frac{1}{2\sqrt{-ac}}\ln \frac{\sqrt{c} + u\sqrt{-a}}{\sqrt{c} - u\sqrt{-a}},$$ $a < 0, c > 0;$

$$= \frac{1}{2\sqrt{-ac}}\ln \frac{u\sqrt{a} - \sqrt{-c}}{u\sqrt{a} + \sqrt{-c}},$$ $a > 0, c < 0.$

43. $\displaystyle\int \frac{u^2\, du}{c + au^2} = \frac{u}{a} - \frac{c}{a}\int \frac{du}{c + au^2}.$ (42)

44. $\displaystyle\int \frac{du}{u(c + au^2)} = \frac{1}{2c}\ln \frac{u^2}{c + au^2}.$

45. $\displaystyle\int \frac{du}{(c + au^2)^{3/2}} = \frac{u}{c\sqrt{c + au^2}}.$

46. $\displaystyle\int \frac{du}{u^2(c + au^2)} = -\frac{1}{cu} - \frac{a}{c}\int \frac{du}{c + au^2}.$ (42)

* When the argument is numerically greater than unity $\tanh^{-1}$ must be replaced by $\coth^{-1}$.

INTEGRALS INVOLVING $c + au^2$ (*Continued*)

47. $\displaystyle\int \frac{du}{u\sqrt{c + au^2}} = \frac{1}{\sqrt{-c}} \sec^{-1} u\sqrt{\frac{a}{-c}},\qquad a > 0,\ c < 0;$

$\displaystyle\phantom{\int \frac{du}{u\sqrt{c + au^2}}} = -\frac{1}{\sqrt{c}} \operatorname{csch}^{-1} u\sqrt{\frac{a}{c}},\qquad a > 0,\ c > 0;$

$\displaystyle\phantom{\int \frac{du}{u\sqrt{c + au^2}}} = -\frac{1}{\sqrt{c}} \operatorname{sech}^{-1} u\sqrt{\frac{-a}{c}},\qquad a < 0,\ c > 0;$

$\displaystyle\phantom{\int \frac{du}{u\sqrt{c + au^2}}} = \frac{1}{\sqrt{c}} \ln \frac{\sqrt{c + au^2} - \sqrt{c}}{u},\qquad c > 0.$

48. $\displaystyle\int \frac{\sqrt{c + au^2}\,du}{u} = \sqrt{c + au^2} + c\int \frac{du}{u\sqrt{c + au^2}}.\qquad (47)$

49. $\displaystyle\int (c + au^2)^n du = \frac{u(c + au^2)^n}{2n + 1} + \frac{2cn}{2n + 1}\int (c + au^2)^{n-1}du,$

$$n \neq -\tfrac{1}{2}.\qquad (35)$$

50. $\displaystyle\int \frac{du}{(c + au^2)^n} = \frac{u}{2c(n - 1)(c + au^2)^{n-1}}$

$$+ \frac{2n - 3}{2c(n - 1)}\int \frac{du}{(c + au^2)^{n-1}},\qquad n \neq 1.\quad (42)$$

51. $\displaystyle\int \frac{u\,du}{(c + au^2)^n} = -\frac{1}{2a(n - 1)(c + au^2)^{n-1}},\qquad n \neq 1.\quad (33)$

52. $\displaystyle\int \frac{u^2\,du}{(c + au^2)^n} = \frac{-u}{2a(n - 1)(c + au^2)^{n-1}}$

$$+ \frac{1}{2a(n - 1)}\int \frac{du}{(c + au^2)^{n-1}},\qquad n \neq 1.\quad (43)$$

53. $\displaystyle\int \frac{du}{u(c + au^2)^n} = \frac{1}{2c(n - 1)(c + au^2)^{n-1}}$

$$+ \frac{1}{c}\int \frac{du}{u(c + au^2)^{n-1}},\qquad n \neq 1.\quad (44)$$

54. $\displaystyle\int \frac{du}{u^2(c + au^2)^n} = \frac{1}{2c(n - 1)u(c + au^2)^{n-1}}$

$$+ \frac{2n - 1}{2c(n - 1)}\int \frac{du}{u^2(c + au^2)^{n-1}},\qquad n \neq 1.\quad (46)$$

55. $\displaystyle\int \frac{(c + au^2)^n\,du}{u} = \frac{(c + au^2)^n}{2n} + c\int \frac{(c + au^2)^{n-1}du}{u}.$

INTEGRALS INVOLVING $c + au^2$ (*Continued*)

56. $\displaystyle\int \frac{du}{u^n(c + au^2)} = -\frac{1}{c(n-1)u^{n-1}}$

$$-\frac{a}{c}\int \frac{du}{u^{n-2}(c + au^2)}, \qquad n \neq 1. \quad (44)$$

57. $\displaystyle\int u^n\sqrt{c + au^2}\,du = \frac{u^{n-1}(c + au^2)^{3/2}}{a(n+2)}$

$$-\frac{c(n-1)}{a(n+2)}\int u^{n-2}\sqrt{c + au^2}\,du, \qquad n \neq -2. \quad (37)$$

58. $\displaystyle\int \frac{u^n\,du}{\sqrt{c + au^2}} = \frac{u^{n-1}\sqrt{c + au^2}}{na} - \frac{c(n-1)}{na}\int \frac{u^{n-2}\,du}{\sqrt{c + au^2}},$

$$n \neq 0. \quad (35)$$

59. $\displaystyle\int \frac{\sqrt{c + au^2}\,du}{u^n} = -\frac{(c + au^2)^{3/2}}{c(n-1)u^{n-1}}$

$$-\frac{a(n-4)}{c(n-1)}\int \frac{\sqrt{c + au^2}\,du}{u^{n-2}}, \qquad n \neq 1. \quad (48)$$

60. $\displaystyle\int \frac{du}{u^n\sqrt{(c + au^2)}} = -\frac{\sqrt{c + au^2}}{c(n-1)u^{n-1}}$

$$-\frac{a(n-2)}{c(n-1)}\int \frac{du}{u^{n-2}\sqrt{c + au^2}}, \qquad n \neq -1. \quad (32)$$

61. $\displaystyle\int u^m(c + au^2)^n du = \frac{u^{m-1}(c + au^2)^{n+1}}{a(m + 2n + 1)}$

$$-\frac{c(m-1)}{a(m + 2n + 1)}\int u^{m-2}(c + au^2)^n du, \qquad m + 2n \neq -1;$$

$$= \frac{u^{m+1}(c + au^2)^n}{m + 2n + 1} + \frac{2cn}{m + 2n + 1}\int u^m(c + au^2)^{n-1} du,$$

$$m + 2n \neq -1.$$

62. $\displaystyle\int \frac{u^m\,du}{(c + au^2)^n} = \frac{u^{m+1}}{2c(n-1)(c + au^2)^{n-1}}$

$$+ \frac{2n - m - 3}{2c(n-1)}\int \frac{u^m\,du}{(c + au^2)^{n-1}}, \qquad n \neq 1;$$

$$= \frac{u^{m-1}}{a(m - 2n + 1)(c + au^2)^{n-1}}$$

$$-\frac{c(m-1)}{a(m - 2n + 1)}\int \frac{u^{m-2}\,du}{(c + au^2)^n}, \qquad 2n - m \neq 1.$$

INTEGRALS INVOLVING $c + au^2$ (Continued)

63. $\displaystyle\int \frac{(c + au^2)^n du}{u^m}$

$$= -\frac{(c + au^2)^{n+1}}{c(m - 1)u^{m-1}} - \frac{a(m - 2n - 3)}{c(m - 1)}\int \frac{(c + au^2)^n du}{u^{m-2}},$$

$$m \neq 1;$$

$$= \frac{(c + au^2)^n}{(2n - m + 1)u^{m-1}} + \frac{2cn}{2n - m + 1}\int \frac{(c + au^2)^{n-1}du}{u^m},$$

$$m \neq 2n + 1.$$

64. $\displaystyle\int \frac{du}{u^m(c + au^2)^n}$

$$= \frac{-1}{c(m - 1)u^{m-1}(c + au^2)^{n-1}}$$

$$- \frac{a(m + 2n - 3)}{c(m - 1)}\int \frac{du}{u^{m-2}(c + au^2)^n}, \quad m \neq 1;$$

$$= \frac{1}{2c(n - 1)u^{m-1}(c + au^2)^{n-1}}$$

$$+ \frac{m + 2n - 3}{2c(n - 1)}\int \frac{du}{u^m(c + au^2)^{n-1}}, \quad n \neq 1.$$

INTEGRALS INVOLVING $\sqrt{a^2 - u^2}$

65. $\displaystyle\int \sqrt{a^2 - u^2}\, du = \frac{u}{2}\sqrt{a^2 - u^2} + \frac{a^2}{2}\sin^{-1}\frac{u}{a}.$

66. $\displaystyle\int u\sqrt{a^2 - u^2}\, du = -\frac{1}{3}(a^2 - u^2)^{3/2}.$

67. $\displaystyle\int u^2\sqrt{a^2 - u^2}\, du = -\frac{u}{4}(a^2 - u^2)^{3/2} + \frac{a^2}{4}\int \sqrt{a^2 - u^2}\, du.$

(65)

68. $\displaystyle\int (a^2 - u^2)^{3/2}du = \frac{u}{4}(a^2 - u^2)^{3/2} + \frac{3a^2}{4}\int \sqrt{a^2 - u^2}\, du.$

(65)

69. $\displaystyle\int (a^2 - u^2)^{3/2}u\, du = -\frac{1}{5}(a^2 - u^2)^{5/2}.$

70. $\displaystyle\int (a^2 - u^2)^{3/2}u^2\, du = -\frac{u}{6}(a^2 - u^2)^{5/2} + \frac{a^2u}{24}(a^2 - u^2)^{3/2}$

$$+ \frac{a^4}{8}\int \sqrt{a^2 - u^2}\, du.$$

INTEGRALS INVOLVING $\sqrt{a^2 - u^2}$ (*Continued*)

71. $\displaystyle\int \frac{\sqrt{a^2 - u^2}\, du}{u} = \sqrt{a^2 - u^2} - a \operatorname{sech}^{-1} \frac{u}{a}.$ (76)

72. $\displaystyle\int \frac{\sqrt{a^2 - u^2}\, du}{u^2} = -\frac{\sqrt{a^2 - u^2}}{u} - \sin^{-1} \frac{u}{a}.$

73. $\displaystyle\int \frac{(a^2 - u^2)^{3/2} du}{u} = \frac{4a^2 - u^2}{3}\sqrt{a^2 - u^2} - a^3 \operatorname{sech}^{-1} \frac{u}{a}.$

74. $\displaystyle\int \frac{(a^2 - u^2)^{3/2} du}{u^2} = -\frac{u^2 + 2a^2}{2u}\sqrt{a^2 - u^2} - \frac{3a^2}{2} \sin^{-1} \frac{u}{a}.$

75. $\displaystyle\int \frac{du}{\sqrt{a^2 - u^2}} = \sin^{-1} \frac{u}{a}, \quad \text{or} \quad -\cos^{-1} \frac{u}{a}.$

76. $\displaystyle\int \frac{du}{u\sqrt{a^2 - u^2}} = -\frac{1}{a} \operatorname{sech}^{-1} \frac{u}{a}, \quad \text{or} \quad -\frac{1}{a} \ln \frac{\sqrt{a^2 - u^2} + a}{u}.$

77. $\displaystyle\int \frac{du}{u^2\sqrt{a^2 - u^2}} = -\frac{\sqrt{a^2 - u^2}}{a^2 u}.$

78. $\displaystyle\int \frac{du}{(a^2 - u^2)^{3/2}} = \frac{u}{a^2\sqrt{a^2 - u^2}}.$

79. $\displaystyle\int \frac{du}{u(a^2 - u^2)^{3/2}} = \frac{1}{a^2\sqrt{a^2 - u^2}} - \frac{1}{a^3} \operatorname{sech}^{-1} \frac{u}{a}.$ (76)

80. $\displaystyle\int \frac{du}{u^2(a^2 - u^2)^{3/2}} = \frac{2u^2 - a^2}{a^4 u\sqrt{a^2 - u^2}}.$

81. $\displaystyle\int \frac{u\, du}{\sqrt{a^2 - u^2}} = -\sqrt{a^2 - u^2}.$

82. $\displaystyle\int \frac{u^2\, du}{\sqrt{a^2 - u^2}} = -\frac{u}{2}\sqrt{a^2 - u^2} + \frac{a^2}{2} \sin^{-1} \frac{u}{a}.$

83. $\displaystyle\int \frac{u\, du}{(a^2 - u^2)^{3/2}} = \frac{1}{\sqrt{a^2 - u^2}}.$

84. $\displaystyle\int \frac{u^2\, du}{(a^2 - u^2)^{3/2}} = \frac{u}{\sqrt{a^2 - u^2}} - \sin^{-1} \frac{u}{a}.$

INTEGRALS INVOLVING $\sqrt{u^2 - a^2}$

85. $\displaystyle\int \sqrt{u^2 - a^2}\, du = \frac{u}{2}\sqrt{u^2 - a^2} - \frac{a^2}{2} \cosh^{-1} \frac{u}{a}.$ (Art. 69)

86. $\displaystyle\int u\sqrt{u^2 - a^2}\, du = \tfrac{1}{3}(u^2 - a^2)^{3/2}.$

87. $\displaystyle\int u^2\sqrt{u^2 - a^2}\, du = \frac{u}{4}(u^2 - a^2)^{3/2} + \frac{a^2}{4}\int \sqrt{u^2 - a^2}\, du.$ (85)

INTEGRALS INVOLVING $\sqrt{u^2 - a^2}$ (*Continued*)

88. $\displaystyle\int (u^2 - a^2)^{3/2}du = \frac{u}{4}(u^2 - a^2)^{3/2} - \frac{3a^2}{4}\int \sqrt{u^2 - a^2}\, du.$ (85)

89. $\displaystyle\int (u^2 - a^2)^{3/2}u\, du = \tfrac{1}{5}(u^2 - a^2)^{5/2}.$

90. $\displaystyle\int (u^2 - a^2)^{3/2}u^2\, du = \frac{u(u^2 - a^2)^{3/2}}{24}(4u^2 - 3a^2)$
$$- \frac{a^4}{8}\int \sqrt{u^2 - a^2}\, du. \quad (85)$$

91. $\displaystyle\int \frac{\sqrt{u^2 - a^2}\, du}{u} = \sqrt{u^2 - a^2} - a\, \sec^{-1}\frac{u}{a}.$

92. $\displaystyle\int \frac{\sqrt{u^2 - a^2}\, du}{u^2} = -\frac{\sqrt{u^2 - a^2}}{u} + \cosh^{-1}\frac{u}{a}.$ (Art. 69)

93. $\displaystyle\int \frac{(u^2 - a^2)^{3/2}du}{u} = \frac{(u^2 - 4a^2)}{3}\sqrt{u^2 - a^2} + a^3 \sec^{-1}\frac{u}{a}.$

94. $\displaystyle\int \frac{(u^2 - a^2)^{3/2}du}{u^2} = \frac{u^2 + 2a^2}{2u}\sqrt{u^2 - a^2} - \frac{3a^2}{2}\cosh^{-1}\frac{u}{a}.$

95. $\displaystyle\int \frac{du}{\sqrt{u^2 - a^2}} = \cosh^{-1}\frac{u}{a}, \text{ or } \ln\left(\sqrt{u^2 - a^2} + u\right).$

96. $\displaystyle\int \frac{du}{u\sqrt{u^2 - a^2}} = \frac{1}{a}\sec^{-1}\frac{u}{a}, \text{ or } \frac{1}{a}\cos^{-1}\frac{a}{u}.$

97. $\displaystyle\int \frac{du}{u^2\sqrt{u^2 - a^2}} = \frac{\sqrt{u^2 - a^2}}{a^2 u}.$

98. $\displaystyle\int \frac{du}{(u^2 - a^2)^{3/2}} = -\frac{u}{a^2\sqrt{u^2 - a^2}}.$

99. $\displaystyle\int \frac{du}{u(u^2 - a^2)^{3/2}} = -\frac{1}{a^2\sqrt{u^2 - a^2}} - \frac{1}{a^3}\sec^{-1}\frac{u}{a}.$

100. $\displaystyle\int \frac{du}{u^2(u^2 - a^2)^{3/2}} = \frac{a^2 - 2u^2}{a^4 u\sqrt{u^2 - a^2}}.$

101. $\displaystyle\int \frac{u\, du}{\sqrt{u^2 - a^2}} = \sqrt{u^2 - a^2}.$

102. $\displaystyle\int \frac{u^2\, du}{\sqrt{u^2 - a^2}} = \frac{u}{2}\sqrt{u^2 - a^2} + \frac{a^2}{2}\cosh^{-1}\frac{u}{a}.$ (Art. 69)

103. $\displaystyle\int \frac{u\, du}{(u^2 - a^2)^{3/2}} = -\frac{1}{\sqrt{u^2 - a^2}}.$

104. $\displaystyle\int \frac{u^2\, du}{(u^2 - a^2)^{3/2}} = -\frac{u}{\sqrt{u^2 - a^2}} + \cosh^{-1}\frac{u}{a}.$ (Art. 69)

INTEGRALS INVOLVING $\sqrt{u^2 + a^2}$

105. $\displaystyle\int \sqrt{u^2 + a^2}\, du = \frac{u}{2}\sqrt{u^2 + a^2} + \frac{a^2}{2} \sinh^{-1} \frac{u}{a}.$ (Art. 69)

106. $\displaystyle\int u\sqrt{u^2 + a^2}\, du = \frac{1}{3}(u^2 + a^2)^{3/2}.$

107. $\displaystyle\int u^2\sqrt{u^2 + a^2}\, du = \frac{u}{4}(u^2 + a^2)^{3/2} - \frac{a^2}{4}\int \sqrt{u^2 + a^2}\, du.$

(105)

108. $\displaystyle\int (u^2 + a^2)^{3/2}du = \frac{u}{4}(u^2 + a^2)^{3/2} + \frac{3a^2}{4}\int \sqrt{u^2 + a^2}\, du.$

(105)

109. $\displaystyle\int (u^2 + a^2)^{3/2}u\, du = \frac{1}{5}(u^2 + a^2)^{5/2}.$

110. $\displaystyle\int (u^2 + a^2)^{3/2}u^2\, du = \frac{u}{24}(u^2 + a^2)^{3/2}(4u^2 + 3a^2)$

$$- \frac{a^4}{8}\int \sqrt{u^2 + a^2}\, du. \quad (105)$$

111. $\displaystyle\int \frac{\sqrt{u^2 + a^2}\, du}{u} = \sqrt{u^2 + a^2} - a\, \operatorname{csch}^{-1} \frac{u}{a}.$ (Art. 69)

112. $\displaystyle\int \frac{\sqrt{u^2 + a^2}\, du}{u^2} = -\frac{\sqrt{u^2 + a^2}}{u} + \sinh^{-1} \frac{u}{a}.$ (Art. 69)

113. $\displaystyle\int \frac{(u^2 + a^2)^{3/2}du}{u} = \frac{\sqrt{u^2 + a^2}}{3}(u^2 + 4a^2) - a^3\, \operatorname{csch}^{-1} \frac{u}{a}.$

(Art. 69)

114. $\displaystyle\int \frac{(u^2 + a^2)^{3/2}du}{u^2} = \frac{\sqrt{u^2 + a^2}}{2u}(u^2 - 2a^2) + \frac{3a^2}{2} \sinh^{-1} \frac{u}{a}.$

(Art. 69)

115. $\displaystyle\int \frac{du}{\sqrt{u^2 + a^2}} = \sinh^{-1} \frac{u}{a}, \quad \text{or} \quad \ln\left(\sqrt{u^2 + a^2} + u\right).$

116. $\displaystyle\int \frac{du}{u\sqrt{u^2 + a^2}} = -\frac{1}{a} \operatorname{csch}^{-1} \frac{u}{a}, \quad -\frac{1}{a} \sinh^{-1} \frac{a}{u}, \quad \text{or}$

$$-\frac{1}{a} \ln \frac{\sqrt{u^2 + a^2} + a}{u}.$$

117. $\displaystyle\int \frac{du}{u^2\sqrt{u^2 + a^2}} = -\frac{\sqrt{u^2 + a^2}}{a^2 u}.$

118. $\displaystyle\int \frac{du}{(u^2 + a^2)^{3/2}} = \frac{u}{a^2\sqrt{u^2 + a^2}}.$

INTEGRALS INVOLVING $\sqrt{u^2 + a^2}$ (*Continued*)

119. $\displaystyle\int \frac{du}{u(u^2 + a^2)^{3/2}} = \frac{1}{a^2\sqrt{u^2 + a^2}} - \frac{1}{a^3}\operatorname{csch}^{-1}\frac{u}{a}.$ (Art. 69)

120. $\displaystyle\int \frac{du}{u^2(u^2 + a^2)^{3/2}} = -\frac{2u^2 + a^2}{a^4 u\sqrt{u^2 + a^2}}.$

121. $\displaystyle\int \frac{u\,du}{\sqrt{u^2 + a^2}} = \sqrt{u^2 + a^2}.$

122. $\displaystyle\int \frac{u^2\,du}{\sqrt{u^2 + a^2}} = \frac{u}{2}\sqrt{u^2 + a^2} - \frac{a^2}{2}\sinh^{-1}\frac{u}{a}.$ (Art. 69)

123. $\displaystyle\int \frac{u\,du}{(u^2 + a^2)^{3/2}} = -\frac{1}{\sqrt{u^2 + a^2}}.$

124. $\displaystyle\int \frac{u^2\,du}{(u^2 + a^2)^{3/2}} = -\frac{u}{\sqrt{u^2 + a^2}} + \sinh^{-1}\frac{u}{a}.$ (Art. 69)

INTEGRALS INVOLVING $z = q + pu$, $X = c + bu + au^2$

Let $A = 4ac - b^2$, $B = bp - 2aq$, $C = aq^2 - bpq + cp^2$,

$w = \dfrac{1}{q + pu}.$

125. $\displaystyle\int\frac{du}{X} = \frac{2}{\sqrt{A}}\tan^{-1}\frac{2au + b}{\sqrt{A}},$ $A > 0;$

$\qquad = \dfrac{-2}{\sqrt{-A}}\tanh^{-1}\dfrac{2au + b^*}{\sqrt{-A}},$ $A < 0;$

$\qquad = \dfrac{1}{\sqrt{-A}}\ln\dfrac{2au + b - \sqrt{-A}}{2au + b + \sqrt{-A}},$ $A < 0;$

$\qquad = \dfrac{-2}{2au + b},$ $A = 0.$

126. $\displaystyle\int\frac{z\,du}{X} = \frac{p}{2a}\ln X + \frac{2aq - bp}{2a}\int\frac{du}{X}.$ (125)

127. $\displaystyle\int\frac{du}{\sqrt{X}} = \frac{-1}{\sqrt{-a}}\sin^{-1}\frac{b + 2au}{\sqrt{-A}},$ $a < 0, A < 0;$

$\qquad = \dfrac{1}{\sqrt{a}}\sinh^{-1}\dfrac{b + 2au}{\sqrt{A}},$ $a > 0, A > 0;$

$\qquad = \dfrac{1}{\sqrt{a}}\cosh^{-1}\dfrac{b + 2au}{\sqrt{-A}},$ $a > 0, A < 0;$

$\qquad = \dfrac{1}{\sqrt{a}}\ln(b + 2au + 2\sqrt{aX}),$ $a > 0.$

* When the argument is numerically greater than unity $\tanh^{-1}$ must be replaced by $\coth^{-1}$.

INTEGRALS INVOLVING $z = q + pu$, $X = c + bu + au^2$
(Continued)

128. $\displaystyle\int \frac{du}{X^{3/2}} = \frac{2(b + 2au)}{A\sqrt{X}}.$

129. $\displaystyle\int \frac{du}{X^{\frac{n}{2}}} = \frac{2(b + 2au)}{(n-2)AX^{\frac{n}{2}-1}} + \frac{4a(n-3)}{A(n-2)} \int \frac{du}{X^{\frac{n}{2}-1}}.$

130. $\displaystyle\int \frac{z\,du}{\sqrt{X}} = \frac{p}{a}\sqrt{X} + \frac{2aq - bp}{2a} \int \frac{du}{\sqrt{X}}.$ (127)

131. $\displaystyle\int \frac{du}{z\sqrt{X}} = \frac{1}{\sqrt{-C}} \sin^{-1} \frac{2C + Bz}{zp\sqrt{-A}},$ $C < 0, A < 0;$

$\qquad\qquad = -\frac{1}{\sqrt{C}} \sinh^{-1} \frac{2C + Bz}{zp\sqrt{A}},$ $C > 0, A > 0;$

$\qquad\qquad = -\frac{1}{\sqrt{C}} \cosh^{-1} \frac{2C + Bz}{zp\sqrt{-A}},$ $C > 0, A < 0;$

$\qquad\qquad = -\frac{1}{\sqrt{C}} \ln\left[\frac{p\sqrt{X} + \sqrt{C}}{z} + \frac{B}{2\sqrt{C}}\right],$ $C > 0;$

$\qquad\qquad = \frac{-2p\sqrt{X}}{Bz},$ $C = 0, B \neq 0$

$\qquad\qquad = \frac{-1}{\sqrt{a}\,z},$ $C = B = 0, a > 0.$

132. $\displaystyle\int \frac{du}{zX^{\frac{n}{2}}} = \frac{p}{(n-2)CX^{\frac{n}{2}-1}} - \frac{B}{2C}\int \frac{du}{X^{\frac{n}{2}}} + \frac{p^2}{C}\int \frac{du}{zX^{\frac{n}{2}-1}}.$

133. $\displaystyle\int \sqrt{X}\,du = \frac{b + 2au}{4a}\sqrt{X} + \frac{A}{8a}\int \frac{du}{\sqrt{X}}.$ (127)

134. $\displaystyle\int X^{\frac{n}{2}}\,du = \frac{(2au + b)}{2a(n+1)}X^{\frac{n}{2}} + \frac{nA}{4(n+1)a}\int X^{\frac{n}{2}-1}\,du.$

135. $\displaystyle\int z\sqrt{X}\,du = \frac{pX^{3/2}}{3a} - \frac{B}{2a}\int \sqrt{X}\,du.$ (133)

136. $\displaystyle\int zX^{\frac{n}{2}}\,du = \frac{pX^{\frac{n}{2}+1}}{(n+2)a} - \frac{B}{2a}\int X^{\frac{n}{2}}\,du.$ (134)

137. $\displaystyle\int \frac{X^{\frac{n}{2}}\,du}{z} = \frac{X^{\frac{n}{2}}}{np} + \frac{C}{p^2}\int \frac{X^{\frac{n}{2}-1}}{z}du + \frac{B}{2p^2}\int X^{\frac{n}{2}-1}\,du.$

INTEGRALS INVOLVING $z = q + pu$, $X = c + bu + au^2$
(Continued)

138. $\displaystyle \int z^m X^{\frac{n}{2}}\, du = \frac{1}{(m+n+1)a}\Big[pz^{m-1}X^{\frac{n}{2}+1}$

$- (m + \tfrac{1}{2}n)B \int z^{m-1}X^{\frac{n}{2}}\, du - (m-1)C \int z^{m-2}X^{\frac{n}{2}}\, du \Big],$

$$m + n + 1 \neq 0.$$

139. $\displaystyle \int z^m X^{\frac{n}{2}}\, du = -p^m \int \frac{(\sqrt{a + Bw + Cw^2})^n}{w}\, dw,$

$$m + n + 1 = 0. \quad (137)$$

140. $\displaystyle \int \frac{X^{\frac{n}{2}}}{z^m}\, du = \frac{-1}{(m-1)C}\Big[\frac{pX^{\frac{n}{2}+1}}{z^{m-1}} + \Big(m - \frac{n}{2} - 2\Big)B \int \frac{X^{\frac{n}{2}}}{z^{m-1}}\, du$

$+ (m - n - 3)a \int \frac{X^{\frac{n}{2}}}{z^{m-2}}\, du \Big], \qquad (m-1)\, C \neq 0.$

141. $\displaystyle \int \frac{X^{\frac{n}{2}}}{z^m}\, du = \frac{2}{(2m-n-2)B}\Big[\frac{pX^{\frac{n}{2}+1}}{z^m} + (m-n-2)a \int \frac{X^{\frac{n}{2}}}{z^{m-1}}\, du \Big],$

$$C = 0,\ B(2m - n - 2) \neq 0.$$

142. $\displaystyle \int \frac{X^{\frac{n}{2}}}{z^m}\, du = \frac{-1}{p^{n+1}} \int \frac{(a + Bw)^{\frac{n}{2}}}{w^{n-m+2}}\, dw, \qquad C = 0.$

143. $\displaystyle \int \frac{z}{X^n}\, du = \frac{-p}{2a(n-1)X^{n-1}} + \frac{2aq - bp}{2a} \int \frac{du}{X^n}.$

144. $\displaystyle \int \frac{u^m}{\sqrt{X}}\, du = \frac{u^{m-1}\sqrt{X}}{am} - \frac{c(m-1)}{am} \int \frac{u^{m-2}}{\sqrt{X}}\, du$

$- \frac{b(2m-1)}{2am} \int \frac{u^{m-1}}{\sqrt{X}}\, du.$

INTEGRALS INVOLVING $c + bu^r$

145. $\displaystyle \int \frac{u^{r-1}\, du}{c + bu^r} = \frac{1}{br} \ln (c + bu^r).$

146. $\displaystyle \int (c + bu^r)^n u^{r-1}\, du = \frac{(c + bu^r)^{n+1}}{br(n+1)}, \qquad n \neq -1. \quad (145)$

147. $\displaystyle \int \frac{du}{u(c + bu^r)} = \frac{1}{cr} \ln \frac{u^r}{c + bu^r}.$

148. $\displaystyle\int u^m(c + bu^r)^n du$

$$= \frac{u^{m+1}(c + bu^r)^n}{m + nr + 1} + \frac{crn}{m + nr + 1}\int u^m(c + bu^r)^{n-1}du,$$
$$m + nr \neq -1$$

$$= \frac{u^{m-r+1}(c + bu^r)^{n+1}}{b(m + nr + 1)}$$
$$- \frac{c(m - r + 1)}{b(m + nr + 1)}\int u^{m-r}(c + bu^r)^n du, \qquad m + nr \neq -1$$

149. $\displaystyle\int \frac{(c + bu^r)^n du}{u^m}$

$$= -\frac{(c + bu^r)^{n+1}}{c(m - 1)u^{m-1}} + \frac{b(nr - m + r + 1)}{c(m - 1)}\int \frac{(c + bu^r)^n}{u^{m-r}}du,$$
$$m \neq 1 \quad (148)$$

150. $\displaystyle\int \frac{u^m\,du}{(c + bu^r)^n} = \frac{u^{m-r+1}}{b(m - rn + 1)(c + bu^r)^{n-1}}$

$$- \frac{c(m - r + 1)}{b(m - rn + 1)}\int \frac{u^{m-r}\,du}{(c + bu^r)^n},$$
$$rn - m \neq 1$$

$$= \frac{u^{m+1}}{cr(n - 1)(c + bu^r)^{n-1}}$$
$$- \frac{(m - rn + r + 1)}{cr(n - 1)}\int \frac{u^m\,du}{(c + bu^r)^{n-1}},$$
$$n \neq 1 \quad (148)$$

151. $\displaystyle\int \frac{du}{u^m(c + bu^r)^n}$

$$= \frac{1}{cr(n - 1)u^{m-1}(c + bu^r)^{n-1}}$$
$$+ \frac{m + nr - r - 1}{cr(n - 1)}\int \frac{du}{u^m(c + bu^r)^{n-1}}, \qquad n \neq 1$$

$$= -\frac{1}{c(m - 1)u^{m-1}(c + bu^r)^{n-1}}$$
$$- \frac{b(m + nr - r - 1)}{c(m - 1)}\int \frac{du}{u^{m-r}(c + bu^r)^n}, \qquad m \neq 1$$

Direct Trigonometric Functions

152. $\displaystyle\int \sin u \, du = -\cos u.$

153. $\displaystyle\int \cos u \, du = \sin u.$

154. $\displaystyle\int \tan u \, du = -\ln \cos u,$ or $\ln \sec u.$

155. $\displaystyle\int \cot u \, du = \ln \sin u,$ or $-\ln \csc u.$

156. $\displaystyle\int \sec u \, du = \ln \tan \left(\frac{u}{2} + \frac{\pi}{4}\right) = \mathrm{gd}^{-1}u,$ or

$$\ln (\sec u + \tan u).$$

157. $\displaystyle\int \csc u \, du = \ln \tan \frac{u}{2} = \mathrm{gd}^{-1}\left(u - \frac{\pi}{2}\right),$ or

$$\ln (\csc u - \cot u).$$

158. $\displaystyle\int \sin^2 u \, du = \tfrac{1}{2}u - \tfrac{1}{4} \sin 2u.$

159. $\displaystyle\int \cos^2 u \, du = \tfrac{1}{2}u + \tfrac{1}{4} \sin 2u.$

160. $\displaystyle\int \tan^2 u \, du = \tan u - u.$

161. $\displaystyle\int \cot^2 u \, du = -\cot u - u.$

162. $\displaystyle\int \sec^2 u \, du = \tan u.$

163. $\displaystyle\int \csc^2 u \, du = -\cot u.$

164. $\displaystyle\int \sec^3 u \, du = \tfrac{1}{2} \sec u \tan u + \tfrac{1}{2} \ln \tan \left(\frac{u}{2} + \frac{\pi}{4}\right).$ (156)

165. $\displaystyle\int \csc^3 u \, du = -\tfrac{1}{2} \csc u \cot u + \tfrac{1}{2} \ln \tan \frac{u}{2}.$ (157)

166. $\displaystyle\int \sin^n u \, du = -\frac{\sin^{n-1} u \cos u}{n} + \frac{n-1}{n} \int \sin^{n-2} u \, du.$

167. $\displaystyle\int \cos^n u \, du = \frac{\cos^{n-1} u \sin u}{n} + \frac{n-1}{n} \int \cos^{n-2} u \, du.$

168. $\displaystyle\int \tan^n u \, du = \frac{\tan^{n-1} u}{n-1} - \int \tan^{n-2} u \, du.$ (154)

Direct Trigonometric Functions.—(*Continued*)

169. $\displaystyle\int \cot^n u \, du = -\frac{\cot^{n-1} u}{n-1} - \int \cot^{n-2} u \, du.$ \qquad (155)

170. $\displaystyle\int \sec^n u \, du = \frac{\tan u \, \sec^{n-2} u}{n-1} + \frac{n-2}{n-1}\int \sec^{n-2} u \, du.$

171. $\displaystyle\int \csc^n u \, du = -\frac{\cot u \, \csc^{n-2} u}{n-1} + \frac{n-2}{n-1}\int \csc^{n-2} u \, du.$

172. $\displaystyle\int \sin u \cos u \, du = \tfrac{1}{2} \sin^2 u, \quad -\tfrac{1}{2}\cos^2 u, \quad \text{or} \quad -\tfrac{1}{4}\cos 2u.$

173. $\displaystyle\int \sin u \tan u \, du = \ln \tan\left(\frac{u}{2} + \frac{\pi}{4}\right) - \sin u = \mathrm{gd}^{-1}u - \sin u.$

174. $\displaystyle\int \cos u \cot u \, du = \ln \tan\frac{u}{2} + \cos u = \mathrm{gd}^{-1}\left(u - \frac{\pi}{2}\right) + \cos u.$

175. $\displaystyle\int \tan u \sec u \, du = \sec u.$

176. $\displaystyle\int \cot u \csc u \, du = -\csc u.$

177. $\displaystyle\int \sec u \csc u \, du = \ln \tan u.$

178. $\displaystyle\int \sin au \sin bu \, du = -\frac{\sin (a+b)u}{2(a+b)} + \frac{\sin (a-b)u}{2(a-b)},$
$$a^2 \neq b^2 \quad (158)$$

179. $\displaystyle\int \sin au \cos bu \, du = -\frac{\cos (a+b)u}{2(a+b)} - \frac{\cos (a-b)u}{2(a-b)},$
$$a^2 \neq b^2 \quad (172)$$

180. $\displaystyle\int \cos au \cos bu \, du = \frac{\sin (a+b)u}{2(a+b)} + \frac{\sin (a-b)u}{2(a-b)},$
$$a^2 \neq b^2 \quad (159)$$

181. $\displaystyle\int \sec u \tan^2 u \, du = \tfrac{1}{2} \sec u \tan u - \tfrac{1}{2} \ln \tan\left(\frac{u}{2} + \frac{\pi}{4}\right).$
$$(156)$$

182. $\displaystyle\int \sec u \csc^2 u \, du = -\csc u + \ln \tan\left(\frac{u}{2} + \frac{\pi}{4}\right). \quad (156)$

183. $\displaystyle\int \csc u \cot^2 u \, du = -\tfrac{1}{2} \csc u \cot u - \tfrac{1}{2} \ln \tan \frac{u}{2}. \quad (157)$

184. $\displaystyle\int \csc u \sec^2 u \, du = \sec u + \ln \tan \frac{u}{2}. \qquad (157)$

DIRECT TRIGONOMETRIC FUNCTIONS.—(*Continued*)

185. $\displaystyle\int \sin^2 u \cos^2 u \, du = \frac{u}{8} - \frac{\sin 4u}{32}.$

186. $\displaystyle\int \sec^2 u \csc^2 u \, du = \sec u \csc u - 2 \cot u = -2 \cot 2u.$

187. $\displaystyle\int \sin^n u \cos u \, du = \frac{\sin^{n+1} u}{n+1},$ $\qquad n \neq -1.$ (155)

188. $\displaystyle\int \cos^n u \sin u \, du = -\frac{\cos^{n+1} u}{n+1},$ $\qquad n \neq -1.$ (154)

189. $\displaystyle\int \sec^n u \tan u \, du = \frac{\sec^n u}{n}.$

190. $\displaystyle\int \csc^n u \cot u \, du = -\frac{\csc^n u}{n}.$

191. $\displaystyle\int \tan^n u \sec^2 u \, du = \frac{\tan^{n+1} u}{n+1},$ $\qquad n \neq -1.$ (177)

192. $\displaystyle\int \cot^n u \csc^2 u \, du = -\frac{\cot^{n+1} u}{n+1},$ $\qquad n \neq -1.$ (177)

193. $\displaystyle\int \sin^m u \cos^n u \, du$

$$= \frac{\sin^{m+1} u \cos^{n-1} u}{m+n} + \frac{n-1}{m+n}\int \sin^m u \cos^{n-2} u \, du,$$
$$m \neq -n;$$

$$= -\frac{\sin^{m-1} u \cos^{n+1} u}{m+n} + \frac{m-1}{m+n}\int \sin^{m-2} u \cos^n u \, du,$$
$$m \neq -n;$$

$$= -\frac{\sin^{m+1} u \cos^{n+1} u}{n+1} + \frac{m+n+2}{n+1}\int \sin^m u \cos^{n+2} u \, du,$$
$$n \neq -1;$$

$$= \frac{\sin^{m+1} u \cos^{n+1} u}{m+1} + \frac{m+n+2}{m+1}\int \sin^{m+2} u \cos^n u \, du,$$
$$m \neq -1.$$

194. $\displaystyle\int \sec^m u \tan^n u \, du$

$$= \frac{\sec^{m-2} u \tan^{n+1} u}{m+n-1} + \frac{m-2}{m+n-1}\int \sec^{m-2} u \tan^n u \, du, \text{ or}$$

$$= \frac{\sec^m u \tan^{n-1} u}{m+n-1} - \frac{n-1}{m+n-1}\int \sec^m u \tan^{n-2} u \, du,$$
$$m+n \neq 1.$$

Direct Trigonometric Functions.—(*Continued*)

195. $\displaystyle\int \csc^m u \cot^n u \, du$

$$= -\frac{\csc^{m-2} u \cot^{n+1} u}{m+n-1} + \frac{m-2}{m+n-1}\int \csc^{m-2} u \cot^n u \, du,$$

or

$$= -\frac{\csc^m u \cot^{n-1} u}{m+n-1} - \frac{n-1}{m+n-1}\int \csc^m u \cot^{n-2} u \, du,$$

$$m+n \neq 1.$$

196. $\displaystyle\int \sec^m u \cot^n u \, du$

$$= -\frac{\sec^{m-2} u \cot^{n-1} u}{n-1} + \frac{m-2}{n-1}\int \sec^{m-2} u \cot^{n-2} u \, du,$$

$$n \neq 1.$$

197. $\displaystyle\int \frac{du}{1 - \sin u} = \tan\left(\frac{u}{2} + \frac{\pi}{4}\right)$

198. $\displaystyle\int \frac{du}{1 + \sin u} = \tan\left(\frac{u}{2} - \frac{\pi}{4}\right).$

199. $\displaystyle\int \frac{du}{1 - \cos u} = -\cot\frac{u}{2}.$

200. $\displaystyle\int \frac{du}{1 + \cos u} = \tan\frac{u}{2}.$

201. $\displaystyle\int \sqrt{1 - \cos u}\, du = -2\sqrt{2}\cos\frac{u}{2}.$

202. $\displaystyle\int \sqrt{1 + \cos u}\, du = 2\sqrt{2}\sin\frac{u}{2}.$

203. $\displaystyle\int \sqrt{1 - \sin u}\, du = 2\sqrt{2}\sin\left(\frac{u}{2} + \frac{\pi}{4}\right) = 2\left(\sin\frac{u}{2} + \cos\frac{u}{2}\right)$

204. $\displaystyle\int \sqrt{1 + \sin u}\, du = 2\sqrt{2}\sin\left(\frac{u}{2} - \frac{\pi}{4}\right) = 2\left(\sin\frac{u}{2} - \cos\frac{u}{2}\right)$

205. $\displaystyle\int \frac{du}{a + b\sin u} = \frac{2}{\sqrt{a^2 - b^2}}\tan^{-1}\left[\frac{\sqrt{a^2 - b^2}}{a + b}\tan\left(\frac{u}{2} - \frac{\pi}{4}\right)\right],$

$$a^2 > b^2;$$

$$= \frac{2}{\sqrt{b^2 - a^2}}\tanh^{-1}\left[\frac{\sqrt{b^2 - a^2}}{b + a}\tan\left(\frac{u}{2} - \frac{\pi}{4}\right)\right],^*$$

$$b^2 > a^2. \quad (197, 198)$$

* When the expression within the bracket is numerically greater than unity, $\tanh^{-1}$ must be replaced by $\coth^{-1}$.

Direct Trigonometric Functions.—*(Continued)*

206. $\int \dfrac{du}{a + b \cos u} = \dfrac{2}{\sqrt{a^2 - b^2}} \tan^{-1}\left[\dfrac{\sqrt{a^2 - b^2}}{a + b} \tan \dfrac{u}{2}\right],\ a^2 > b^2;$

$$= \dfrac{2}{\sqrt{b^2 - a^2}} \tanh^{-1}\left[\dfrac{\sqrt{b^2 - a^2}}{b + a} \tan \dfrac{u}{2}\right]^{*},$$

$$b^2 > a^2.\quad (199,\ 200)$$

207. $\int \dfrac{du}{a + b \tan u} = \dfrac{1}{a^2 + b^2}[b \ln (a \cos u + b \sin u) + au].$

208. $\int \dfrac{du}{a + b \cot u} = \dfrac{u}{a} - \dfrac{b}{a}\int \dfrac{du}{a \tan u + b};\quad \dfrac{1}{b} \ln \sec u,$ if $a = 0.$

209. $\int \dfrac{du}{a + b \sec u} = \dfrac{u}{a} - \dfrac{b}{a}\int \dfrac{du}{a \cos u + b};\quad \dfrac{1}{b} \sin u,$ if $a = 0.$

210. $\int \dfrac{du}{a + b \csc u} = \dfrac{u}{a} - \dfrac{b}{a}\int \dfrac{du}{a \sin u + b};\quad -\dfrac{1}{b} \cos u,$ if $a = 0.$

Direct Hyperbolic Functions

211. $\int \sinh u\ du = \cosh u.$ **212.** $\int \cosh u\ du = \sinh u.$

213. $\int \tanh u\ du = \ln \cosh u.$ **214.** $\int \coth u\ du = \ln \sinh u.$

215. $\int \operatorname{sech} u\ du = \operatorname{gd} u = \tan^{-1} \sinh u,\quad$ or $\quad 2 \tan^{-1} e^u.$

216. $\int \operatorname{csch} u\ du = -\coth^{-1} \cosh u,\ -2 \tanh^{-1} e^u,$ or $\ln \tanh \dfrac{u}{2}.$

217. $\int \sinh^2 u\ du = -\dfrac{u}{2} + \dfrac{\sinh 2u}{4}.$

218. $\int \cosh^2 u\ du = \dfrac{u}{2} + \dfrac{\sinh 2u}{4}.$

219. $\int \tanh^2 u\ du = u - \tanh u.$

220. $\int \coth^2 u\ du = u - \coth u.$

221. $\int \operatorname{sech}^2 u\ du = \tanh u.$ **222.** $\int \operatorname{csch}^2 u\ du = -\coth u.$

---

* When the expression within the bracket is numerically greater than unity $\tanh^{-1}$ must be replaced by $\coth^{-1}$.

DIRECT HYPERBOLIC FUNCTIONS.—(*Continued*)

223. $\int \text{sech}^3 u \, du = \frac{1}{2} \tanh u \, \text{sech} \, u + \frac{1}{2} \, \text{gd} \, u.$ (215)

224. $\int \text{csch}^3 u \, du = -\frac{1}{2} \coth u \, \text{csch} \, u - \frac{1}{2} \ln \tanh \frac{u}{2}.$

225. $\int \sinh^n u \, du = \dfrac{\sinh^{n-1} u \cosh u}{n} - \dfrac{n-1}{n} \int \sinh^{n-2} u \, du.$

226. $\int \cosh^n u \, du = \dfrac{\sinh u \cosh^{n-1} u}{n} + \dfrac{n-1}{n} \int \cosh^{n-2} u \, du.$

227. $\int \tanh^n u \, du = -\dfrac{1}{n-1} \tanh^{n-1} u + \int \tanh^{n-2} u \, du,$

$$n \neq 1. \quad (213)$$

228. $\int \coth^n u \, du = -\dfrac{1}{n-1} \coth^{n-1} u + \int \coth^{n-2} u \, du,$

$$n \neq 1. \quad (214)$$

229. $\int \text{sech}^n u \, du = \dfrac{1}{n-1} \sinh u \, \text{sech}^{n-1} u$

$$+ \dfrac{n-2}{n-1} \int \text{sech}^{n-2} u \, du, \qquad n \neq 1. \quad (215)$$

230. $\int \text{csch}^n u \, du = -\dfrac{1}{n-1} \cosh u \, \text{csch}^{n-1} u$

$$- \dfrac{n-2}{n-1} \int \text{csch}^{n-2} u \, du, \qquad n \neq 1. \quad (216)$$

231. $\int \sinh u \cosh u \, du = \frac{1}{2} \sinh^2 u, \text{ or } \frac{1}{2} \cosh^2 u, \text{ or } \frac{1}{4} \cosh 2u.$

232. $\int \sinh u \tanh u \, du = \sinh u - \text{gd} \, u.$ (215)

233. $\int \cosh u \coth u \, du = \ln \tanh \frac{u}{2} + \cosh u.$

234. $\int \cosh u \, \text{csch} \, u \, du = \ln \sinh u, \text{ or } -\ln \text{csch} \, u.$

235. $\int \tanh u \, \text{sech} \, u \, du = -\text{sech} \, u.$

236. $\int \coth u \, \text{csch} \, u \, du = -\text{csch} \, u.$

Direct Hyperbolic Functions.—(*Continued*)

237. $\int \operatorname{sech} u \operatorname{csch} u \, du = \ln \tanh u.$

238. $\int \sinh au \sinh bu \, du = \dfrac{\sinh (a + b)u}{2(a + b)} - \dfrac{\sinh (a - b)u}{2(a - b)},$
$$a^2 \neq b^2. \quad (217)$$

239. $\int \sinh au \cosh bu \, du = \dfrac{\cosh (a + b)u}{2(a + b)} + \dfrac{\cosh (a - b)u}{2(a - b)},$
$$a^2 \neq b^2. \quad (231)$$

240. $\int \cosh au \cosh bu \, du = \dfrac{\sinh (a + b)u}{2(a + b)} + \dfrac{\sinh (a - b)u}{2(a - b)},$
$$a^2 \neq b^2. \quad (218)$$

241. $\int \operatorname{sech} u \tanh^2 u \, du = -\tfrac{1}{2} \operatorname{sech} u \tanh u + \tfrac{1}{2} \operatorname{gd} u.$

242. $\int \operatorname{sech} u \operatorname{csch}^2 u \, du = -\operatorname{csch} u - \operatorname{gd} u.$ \qquad (215)

243. $\int \operatorname{csch} u \coth^2 u \, du = -\tfrac{1}{2} \operatorname{csch} u \coth u + \tfrac{1}{2} \ln \tanh \dfrac{u}{2}.$

244. $\int \operatorname{csch} u \operatorname{sech}^2 u \, du = \operatorname{sech} + \ln \tanh \dfrac{u}{2}.$

245. $\int \sinh^2 u \cosh^2 u \, du = -\dfrac{u}{8} + \dfrac{\sinh 4u}{32}.$

246. $\int \operatorname{sech}^2 u \operatorname{csch}^2 u \, du = -2 \coth 2u.$

247. $\int \sinh^n u \cosh u \, du = \dfrac{\sinh^{n+1} u}{n + 1},$ \qquad $n \neq -1.$ \quad (214)

248. $\int \cosh^n u \sinh u \, du = \dfrac{\cosh^{n+1} u}{n + 1},$ \qquad $n \neq -1.$ \quad (213)

249. $\int \operatorname{sech}^n u \tanh u \, du = -\dfrac{\operatorname{sech}^n u}{n}.$

250. $\int \operatorname{csch}^n u \coth u \, du = -\dfrac{\operatorname{csch}^n u}{n}.$

251. $\int \tanh^n u \operatorname{sech}^2 u \, du = \dfrac{\tanh^{n+1} u}{n + 1},$ \qquad $n \neq -1.$ \quad (237)

252. $\int \coth^n u \operatorname{csch}^2 u \, du = -\dfrac{\coth^{n+1} u}{n + 1},$ \qquad $n \neq -1.$ \quad (237)

DIRECT HYPERBOLIC FUNCTIONS.—(*Continued*)

253. $\int \sinh^m u \cosh^n u \, du$

$$= \frac{\sinh^{m+1} u \cosh^{n-1} u}{m+n}$$
$$+ \frac{n-1}{m+n} \int \sinh^m u \cosh^{n-2} u \, du, \qquad m \neq -n;$$

$$= \frac{\sinh^{m-1} u \cosh^{n+1} u}{m+n}$$
$$- \frac{m-1}{m+n} \int \sinh^{m-2} u \cosh^n u \, du, \qquad m \neq -n;$$

$$= -\frac{\sinh^{m+1} u \cosh^{n+1} u}{n+1}$$
$$+ \frac{m+n+2}{n+1} \int \sinh^m u \cosh^{n+2} u \, du, \qquad n \neq -1;$$

$$= \frac{\sinh^{m+1} u \cosh^{n+1} u}{m+1}$$
$$- \frac{m+n+2}{m+1} \int \sinh^{m+2} u \cosh^n u \, du, \qquad m \neq -1.$$

254. $\int \operatorname{sech}^m u \tanh^n u \, du$

$$= \frac{\operatorname{sech}^{m-2} u \tanh^{n+1} u}{m+n-1}$$
$$+ \frac{m-2}{m+n-1} \int \operatorname{sech}^{m-2} u \tanh^n u \, du, \qquad m+n \neq 1;$$

$$= -\frac{\operatorname{sech}^m u \tanh^{n-1} u}{m+n-1}$$
$$+ \frac{n-1}{m+n-1} \int \operatorname{sech}^m u \tanh^{n-2} u \, du, \qquad m+n \neq 1.$$

255. $\int \operatorname{csch}^m u \coth^n u \, du$

$$= -\frac{\operatorname{csch}^{m-2} u \coth^{n+1} u}{m+n-1}$$
$$- \frac{m-2}{m+n-1} \int \operatorname{csch}^{m-2} u \coth^n u \, du, \qquad m+n \neq 1;$$

$$= -\frac{\operatorname{csch}^m u \coth^{n-1} u}{m+n-1}$$
$$+ \frac{n-1}{m+n-1} \int \operatorname{csch}^m u \coth^{n-2} u \, du, \qquad m+n \neq 1.$$

DIRECT HYPERBOLIC FUNCTIONS.—(*Continued*)

256. $\displaystyle\int \mathrm{sech}^m u \coth^n u \, du$

$$= -\frac{\mathrm{sech}^{m-2} u \coth^{n-1} u}{n-1} - \frac{m-2}{n-1}\int \mathrm{sech}^{m-2} u \coth^{n-2} u \, du.$$

257. $\displaystyle\int \sqrt{\cosh u - 1}\, du = 2\sqrt{2}\,\cosh\frac{u}{2}.$

258. $\displaystyle\int \sqrt{\cosh u + 1}\, du = 2\sqrt{2}\,\sinh\frac{u}{2}.$

259. $\displaystyle\int \frac{du}{\cosh u - 1} = -\coth\frac{u}{2}.$ **260.** $\displaystyle\int \frac{du}{\cosh u + 1} = \tanh\frac{u}{2}.$

261. $\displaystyle\int \frac{du}{a + b\cosh u} = \frac{2}{\sqrt{a^2 - b^2}}\tanh^{-1}\left[\frac{\sqrt{a^2 - b^2}}{a + b}\tanh\frac{u}{2}\right],$ *

$$a^2 \neq b^2. \quad (259, 260)$$

262. $\displaystyle\int \frac{du}{a + b\sinh u} = \frac{2}{\sqrt{a^2 + b^2}}\tanh^{-1}\left[\frac{a\tanh\dfrac{u}{2} - b}{\sqrt{a^2 + b^2}}\right].$ *

263. $\displaystyle\int \frac{du}{a + b\tanh u} = \frac{1}{a^2 - b^2}[au - b\ln(a\cosh u + b\sinh u)],$

$$a^2 \neq b^2;$$

$$= \frac{1}{4a}(2u - e^{-2u}),\ \text{if}\ a = b;\ \frac{1}{4a}(2u + e^{2u}),$$

$$\text{if}\ a = -b.$$

264. $\displaystyle\int \frac{du}{a + b\coth u} = \frac{u}{a} - \frac{b}{a}\int \frac{du}{a\tanh u + b};\ \frac{1}{b}\ln\cosh u,$

$$\text{if}\ a = 0.$$

265. $\displaystyle\int \frac{du}{a + b\,\mathrm{sech}\, u} = \frac{u}{a} - \frac{b}{a}\int \frac{du}{a\cosh u + b};\ \frac{1}{b}\sinh u,\ \text{if}\ a = 0.$

266. $\displaystyle\int \frac{du}{a + b\,\mathrm{csch}\, u} = \frac{u}{a} - \frac{b}{a}\int \frac{du}{a\sinh u + b};\ \frac{1}{b}\cosh u,\ \text{if}\ a = 0.$

EXPONENTIAL FUNCTIONS

267. $\displaystyle\int a^u\, du = \frac{a^u}{\ln a}.$ **268.** $\displaystyle\int e^u\, du = e^u.$

* When the expression within the bracket is numerically greater than unity $\tanh^{-1}$ must be replaced by $\coth^{-1}$.

EXPONENTIAL FUNCTIONS.—(*Continued*)

269. $\int e^{-u}\, du = -e^{-u}.$ **270.** $\int ue^{au}\, du = \dfrac{e^{au}}{a^2}(au - 1).$

271. $\int u^n e^{au}\, du = \dfrac{u^n e^{au}}{a} - \dfrac{n}{a}\int u^{n-1} e^{au}\, du;$

$$= \frac{u^n e^{au}}{a}\left[1 - \frac{n}{au} + \frac{n(n-1)}{a^2 u^2} - \cdots \pm \frac{n!}{a^n u^n}\right],$$

n a positive integer.

272. $\int e^{au} P(u)\, du = \dfrac{e^{au}}{a}\left[P - \dfrac{1}{a}\dfrac{dP}{du} + \dfrac{1}{a^2}\dfrac{d^2P}{du^2} - \cdots\right],$

where P is a polynomial in u.

273. $\int \dfrac{e^{au}\, du}{u} = \ln u + \dfrac{(au)}{1} + \dfrac{(au)^2}{2 \times 2!} + \dfrac{(au)^3}{3 \times 3!} + \cdots$

274. $\int \dfrac{e^{au}\, du}{u^n} = -\dfrac{e^{au}}{(n-1)u^{n-1}} + \dfrac{a}{n-1}\int \dfrac{e^{au}\, du}{u^{n-1}}, \qquad n \neq 1.$

(273)

275. $\int e^{au} \ln u\, du = \dfrac{1}{a}e^{au} \ln u - \dfrac{1}{a}\int \dfrac{e^{au}\, du}{u}.$ (273)

276. $\int e^{au} \sin bu\, du = \dfrac{e^{au}}{a^2 + b^2}(a \sin bu - b \cos bu).$

277. $\int e^{au} \cos bu\, du = \dfrac{e^{au}}{a^2 + b^2}(a \cos bu + b \sin bu).$

278. $\int e^{au} \sin^n bu\, du = \dfrac{e^{au} \sin^{n-1} bu}{a^2 + n^2 b^2}(a \sin bu - nb \cos bu)$

$$+ \frac{n(n-1)b^2}{a^2 + n^2 b^2}\int e^{au} \sin^{n-2} bu\, du.$$

279. $\int e^{au} \cos^n bu\, du = \dfrac{e^{au} \cos^{n-1} bu}{a^2 + n^2 b^2}(a \cos bu + nb \sin bu)$

$$+ \frac{n(n-1)b^2}{a^2 + n^2 b^2}\int e^{au} \cos^{n-2} bu\, du.$$

280. $\int e^{au} \sinh bu\, du = \dfrac{e^{au}}{a^2 - b^2}(a \sinh bu - b \cosh bu),\ a^2 \neq b^2;$

$$\frac{1}{4a}(e^{2au} - 2au),\ a = b;\ -\frac{1}{4a}(e^{2au} - 2au),\ a = -b.$$

Exponential Functions.—(*Continued*)

281. $\displaystyle\int e^{au} \cosh bu\, du = \frac{e^{au}}{a^2 - b^2}(a \cosh bu - b \sinh bu),\ a^2 \neq b^2;$

$$\frac{1}{4a}(e^{2au} + 2au),\ a^2 = b^2.$$

282. $\displaystyle\int e^{au} \sinh^n bu\, du* = \frac{e^{au} \sinh^{n-1} bu}{a^2 - n^2 b^2}(a \sinh bu - nb \cosh bu)$

$$+ \frac{n(n-1)b^2}{a^2 - n^2 b^2}\int e^{au} \sinh^{n-2} bu\, du.$$

283. $\displaystyle\int e^{au} \cosh^n bu\, du* = \frac{e^{au} \cosh^{n-1} bu}{a^2 - n^2 b^2}(a \cosh bu - nb \sinh bu)$

$$- \frac{n(n-1)b^2}{a^2 - n^2 b^2}\int e^{au} \cosh^{n-2} bu\, du.$$

Logarithmic Functions†

284. $\displaystyle\int \ln au\, du = u \ln au - u.$

285. $\displaystyle\int (\ln au)^n du = u(\ln au)^n - n\int (\ln au)^{n-1} du.$

286. $\displaystyle\int \frac{\ln au\, du}{u} = \tfrac{1}{2}(\ln au)^2.$

287. $\displaystyle\int \frac{\ln au\, du}{u^n} = -\frac{\ln au}{(n-1)u^{n-1}} - \frac{1}{(n-1)^2 u^{n-1}},\quad n \neq 1.$

$$(286)$$

288. $\displaystyle\int u^n \ln au\, du = \frac{u^{n+1}}{n+1}\ln au - \frac{u^{n+1}}{(n+1)^2},\quad n \neq -1.$

$$(286)$$

289. $\displaystyle\int \frac{du}{u \ln au} = \ln (\ln au).$

290. $\displaystyle\int \frac{(\ln au)^n du}{u} = \frac{(\ln au)^{n+1}}{n+1},\quad n \neq -1. \qquad (289)$

291. $\displaystyle\int \frac{du}{u(\ln au)^n} = -\frac{1}{(n-1)(\ln au)^{n-1}},\quad n \neq 1. \qquad (289)$

---

* If $a^2 = n^2 b^2$, express the integrand as a function of $z = e^{bu}$.

† To secure corresponding integral formulas involving log au, substitute ln au = ln 10 · log au = 2.3026 log au.

292. $\int u^m (\ln au)^n du = \dfrac{u^{m+1}(\ln au)^n}{m+1} - \dfrac{n}{m+1}\int u^m(\ln au)^{n-1}\,du,$

$$m \neq -1. \quad (290)$$

293. $\int \dfrac{u^m\,du}{(\ln au)^n} = -\dfrac{u^{m+1}}{(n-1)(\ln au)^{n-1}} + \dfrac{m+1}{n-1}\int \dfrac{u^m\,du}{(\ln au)^{n-1}},$

$$n \neq 1.$$

294. $\int u^m \ln (c + bu)du$

$$= \dfrac{u^{m+1}}{m+1}\ln (c+bu) - \dfrac{b}{m+1}\int \dfrac{u^{m+1}\,du}{c+bu},\ m \neq -1;$$

$$= \ln c \cdot \ln u + \dfrac{bu}{c} - \dfrac{1}{2^2}\left(\dfrac{bu}{c}\right)^2 + \dfrac{1}{3^2}\left(\dfrac{bu}{c}\right)^3 - \cdots,\ \text{or}$$

$$\tfrac{1}{2}(\ln bu)^2 - \dfrac{c}{bu} + \dfrac{1}{2^2}\left(\dfrac{c}{bu}\right)^2 - \dfrac{1}{3^2}\left(\dfrac{c}{bu}\right)^3 + \cdots,\ m = -1.$$

295. $\int (c + bu)^m \ln u\,du$

$$= \dfrac{(c+bu)^{m+1}}{b(m+1)}\ln u - \dfrac{1}{b(m+1)}\int \dfrac{(c+bu)^{m+1}}{u}du,\ m \neq -1;$$

$$= \dfrac{1}{b}\ln u \cdot \ln (c+bu) - \dfrac{1}{b}\int \dfrac{\ln(c+bu)du}{u},\ m = -1.$$

Inverse Trigonometric Functions

296. $\int \sin^{-1} u\,du = u\sin^{-1} u + \sqrt{1-u^2}.$

297. $\int \cos^{-1} u\,du = u\cos^{-1} u - \sqrt{1-u^2}.$

298. $\int \tan^{-1} u\,du = u\tan^{-1} u - \tfrac{1}{2}\ln (1+u^2).$

299. $\int \cot^{-1} u\,du = u\cot^{-1} u + \tfrac{1}{2}\ln (1+u^2).$

300. $\int \csc^{-1} u\,du = u\csc^{-1} u + \cosh^{-1} u.$ \quad (Art. 69)

301. $\int \sec^{-1} u\,du = u\sec^{-1} u - \cosh^{-1} u.$ \quad (Art. 69)

302. $\int \mathrm{vers}^{-1} u\,du = (u-1)\,\mathrm{vers}^{-1} u + \sqrt{2u-u^2}.$

Inverse Trigonometric Functions.—(*Continued*)

303. $\displaystyle\int \frac{\sin^{-1} u \, du}{u} = u + \frac{u^3}{2 \cdot 3^2} + \frac{1 \cdot 3 \cdot u^5}{2 \cdot 4 \cdot 5^2} + \frac{1 \cdot 3 \cdot 5 \cdot u^7}{2 \cdot 4 \cdot 6 \cdot 7^2} + \cdots,$

$$u^2 < 1.$$

304. $\displaystyle\int u^n \sin^{-1} u \, du = \frac{u^{n+1} \sin^{-1} u}{n+1} - \frac{1}{n+1} \int \frac{u^{n+1} \, du}{\sqrt{1-u^2}},$

$$n \neq -1. \quad (303)$$

305. $\displaystyle\int \frac{\cos^{-1} u \, du}{u} = \frac{\pi}{2} \ln u - \int \frac{\sin^{-1} u \cdot du}{u}. \qquad (303)$

306. $\displaystyle\int u^n \cos^{-1} u \, du = \frac{u^{n+1} \cos^{-1} u}{n+1} + \frac{1}{n+1} \int \frac{u^{n+1} \, du}{\sqrt{1-u^2}},$

$$n \neq -1. \quad (305)$$

307. $\displaystyle\int \frac{\tan^{-1} u}{u} du = u - \frac{u^3}{3^2} + \frac{u^5}{5^2} - \frac{u^7}{7^2} + \cdots, \; u^2 < 1;$

$$= \frac{\pi}{2} \ln u + \frac{1}{u} - \frac{1}{3^2 u^3} + \frac{1}{5^2 u^5} - \frac{1}{7^2 u^7} + \cdots,$$

$$u > 1;$$

$$= -\frac{\pi}{2} \ln |u| + \frac{1}{u} - \frac{1}{3^2 u^3} + \frac{1}{5^2 u^5} - \frac{1}{7^2 u^7} + \cdots,$$

$$u < -1.$$

308. $\displaystyle\int u^n \tan^{-1} u \, du = \frac{u^{n+1} \tan^{-1} u}{n+1} - \frac{1}{n+1} \int \frac{u^{n+1} \, du}{1+u^2},$

$$n \neq -1. \quad (307)$$

309. $\displaystyle\int \frac{\cot^{-1} u \, du}{u} = \frac{\pi}{2} \ln |u| - \int \frac{\tan^{-1} u \, du}{u}. \qquad (307)$

310. $\displaystyle\int u^n \cot^{-1} u \, du = \frac{u^{n+1} \cot^{-1} u}{n+1} + \frac{1}{n+1} \int \frac{u^{n+1} \, du}{1+u^2},$

$$n \neq -1. \quad (309)$$

311. $\displaystyle\int \frac{\csc^{-1} u \, du}{u} = -\frac{1}{u} - \frac{1}{2 \cdot 3^2 \cdot u^3} - \frac{1 \cdot 3}{2 \cdot 4 \cdot 5^2 \cdot u^5}$

$$- \frac{1 \cdot 3 \cdot 5}{2 \cdot 4 \cdot 6 \cdot 7^2 \cdot u^7} - \cdots, \; u^2 > 1.$$

312. $\displaystyle\int u^n \csc^{-1} u \, du = \frac{u^{n+1} \csc^{-1} u}{n+1} + \frac{1}{n+1} \int \frac{u^n \, du}{\sqrt{u^2-1}},$

$$n \neq -1. \quad (311)$$

313. $\displaystyle\int \frac{\sec^{-1} u \, du}{u} = \frac{\pi}{2} \ln u - \int \frac{\csc^{-1} u \, du}{u}. \qquad (311)$

INVERSE TRIGONOMETRIC FUNCTIONS.—(*Continued*)

314. $\int u^n \sec^{-1} u \, du = \dfrac{u^{n+1} \sec^{-1} u}{n+1} - \dfrac{1}{n+1} \int \dfrac{u^n \, du}{\sqrt{u^2-1}},$

$$n \neq -1. \quad (313)$$

INVERSE HYPERBOLIC FUNCTIONS

315. $\int \sinh^{-1} u \, du = u \sinh^{-1} u - \sqrt{u^2+1}.$

316. $\int \cosh^{-1} u \, du = u \cosh^{-1} u - \sqrt{u^2-1}.$

317. $\int \tanh^{-1} u \, du = u \tanh^{-1} u + \frac{1}{2} \ln (1-u^2).$

318. $\int \coth^{-1} u \, du = u \coth^{-1} u + \frac{1}{2} \ln (u^2-1).$

319. $\int \operatorname{csch}^{-1} u \, du = u \operatorname{csch}^{-1} u + \sinh^{-1} u.$

320. $\int \operatorname{sech}^{-1} u \, du = u \operatorname{sech}^{-1} u + \sin^{-1} u, \ \operatorname{sech}^{-1} u > 0;$

$$= u \operatorname{sech}^{-1} u - \sin^{-1} u, \ \operatorname{sech}^{-1} u < 0.$$

321. $\int \dfrac{\sinh^{-1} u \, du}{u} = u - \dfrac{u^3}{2 \cdot 3^2} + \dfrac{1 \cdot 3 \cdot u^5}{2 \cdot 4 \cdot 5^2}$

$$- \dfrac{1 \cdot 3 \cdot 5 \cdot u^7}{2 \cdot 4 \cdot 6 \cdot 7^2} + \cdots, u^2 < 1;$$

$$= \tfrac{1}{2}(\ln 2u)^2 - \dfrac{1}{2^3 \cdot u^2} + \dfrac{1 \cdot 3}{2 \cdot 4^3 \cdot u^4}$$

$$- \dfrac{1 \cdot 3 \cdot 5}{2 \cdot 4 \cdot 6^3 \cdot u^6} + \cdots, u > 1;$$

$$= -\tfrac{1}{2}(\ln |2u|)^2 + \dfrac{1}{2^3 \cdot u^2} - \dfrac{1 \cdot 3}{2 \cdot 4^3 \cdot u^4}$$

$$+ \dfrac{1 \cdot 3 \cdot 5}{2 \cdot 4 \cdot 6^3 \cdot u^6} - \cdots, u < -1.$$

322. $\int u^n \sinh^{-1} u \, du = \dfrac{u^{n+1} \sinh^{-1} u}{n+1} - \dfrac{1}{n+1} \int \dfrac{u^{n+1} \, du}{\sqrt{u^2+1}},$

$$n \neq -1. \quad (321)$$

323. $\int \dfrac{\cosh^{-1} u \, du}{u} = \pm \left[\tfrac{1}{2}(\ln 2u)^2 + \dfrac{1}{2^3 \cdot u^2} + \dfrac{1 \cdot 3}{2 \cdot 4^3 \cdot u^4} \right.$

$$\left. + \dfrac{1 \cdot 3 \cdot 5}{2 \cdot 4 \cdot 6^3 \cdot u^6} + \cdots \right], \quad u > 1.$$

INVERSE HYPERBOLIC FUNCTIONS.—(*Continued*)

324. $\displaystyle\int u^n \cosh^{-1} u \, du = \frac{u^{n+1} \cosh^{-1} u}{n+1} - \frac{1}{n+1}\int \frac{u^{n+1} \, du}{\sqrt{u^2 - 1}},$
$$\cosh^{-1} u > 0, \qquad n \neq -1; \quad (323)$$
$$= \frac{u^{n+1} \cosh^{-1} u}{n+1} + \frac{1}{n+1}\int \frac{u^{n+1} \, du}{\sqrt{u^2 - 1}},$$
$$\cosh^{-1} u < 0, \qquad n \neq -1. \quad (323)$$

325. $\displaystyle\int \frac{\tanh^{-1} u \, du}{u} = u + \frac{u^3}{3^2} + \frac{u^5}{5^2} + \frac{u^7}{7^2} + \cdots, \qquad u^2 < 1.$

326. $\displaystyle\int u^n \tanh^{-1} u \, du = \frac{u^{n+1} \tanh^{-1} u}{n+1} - \frac{1}{n+1}\int \frac{u^{n+1} \, du}{1 - u^2},$
$$n \neq -1. \quad (325)$$

327. $\displaystyle\int \frac{\coth^{-1} u \, du}{u} = -\frac{1}{u} - \frac{1}{3^2 u^3} - \frac{1}{5^2 u^5} - \frac{1}{7^2 u^7} - \cdots u^2 > 1.$

328. $\displaystyle\int u^n \coth^{-1} u \, du = \frac{u^{n+1} \coth^{-1} u}{n+1} - \frac{1}{n+1}\int \frac{u^{n+1} \, du}{1 - u^2},$
$$n \neq -1. \quad (327)$$

329. $\displaystyle\int \frac{\operatorname{csch}^{-1} u \, du}{u} = -\frac{1}{u} + \frac{1}{2 \cdot 3^2 \cdot u^3} - \frac{1 \cdot 3}{2 \cdot 4 \cdot 5^2 \cdot u^5}$
$$+ \frac{1 \cdot 3 \cdot 5}{2 \cdot 4 \cdot 6 \cdot 7^2 \cdot u^7} - \cdots, \qquad u^2 > 1;$$
$$= \tfrac{1}{2}(\ln u) \ln \frac{4}{u} + \frac{u^2}{2^3} - \frac{1 \cdot 3 \cdot u^4}{2 \cdot 4^3}$$
$$+ \frac{1 \cdot 3 \cdot 5 \cdot u^6}{2 \cdot 4 \cdot 6^3} - \cdots, \qquad 0 < u < 1;$$
$$= -\tfrac{1}{2}(\ln |u|) \ln \left|\frac{4}{u}\right| - \frac{u^2}{2^3} + \frac{1 \cdot 3 \cdot u^4}{2 \cdot 4^3}$$
$$- \frac{1 \cdot 3 \cdot 5 \cdot u^6}{2 \cdot 4 \cdot 6^3} + \cdots, \qquad -1 < u < 0.$$

330. $\displaystyle\int u^n \operatorname{csch}^{-1} u \, du = \frac{u^{n+1} \operatorname{csch}^{-1} u}{n+1} + \frac{1}{n+1}\int \frac{u^n \, du}{\sqrt{u^2 + 1}},$
$$n \neq -1. \quad (329)$$

331. $\displaystyle\int \frac{\operatorname{sech}^{-1} u \, du}{u} = \pm\left[\tfrac{1}{2}(\ln u) \ln \frac{4}{u} - \frac{u^2}{2^3} - \frac{1 \cdot 3 \cdot u^4}{2 \cdot 4^3}\right.$
$$\left. - \frac{1 \cdot 3 \cdot 5 \cdot u^6}{2 \cdot 4 \cdot 6^3} - \cdots \right], \qquad 0 < u < 1.$$

INVERSE HYPERBOLIC FUNCTIONS.—(*Continued*)

332. $\int u^n \operatorname{sech}^{-1} u \, du = \dfrac{u^{n+1} \operatorname{sech}^{-1} u}{n+1} + \dfrac{1}{n+1} \int \dfrac{u^n \, du}{\sqrt{1-u^2}},$

$$\operatorname{sech}^{-1} u > 0, \ n \neq -1, \quad (331);$$

$$= \dfrac{u^{n+1} \operatorname{sech}^{-1} u}{n+1} - \dfrac{1}{n+1} \int \dfrac{u^n \, du}{\sqrt{1-u^2}},$$

$$\operatorname{sech}^{-1} u < 0, \ n \neq -1. \quad (331)$$

DEFINITE INTEGRALS

333. $\displaystyle\int_0^a \sqrt{a^2 - u^2} \, du = \int_0^a \sqrt{2au - u^2} \, du = \dfrac{\pi a^2}{4}.$

334. $\displaystyle\int_0^{\frac{\pi}{2}} \dfrac{du}{a^2 \cos^2 u + b^2 \sin^2 u} = \dfrac{\pi}{2ab}.$

335. $\displaystyle\int_0^\infty \dfrac{du}{a^2 + u^2} = \dfrac{\pi}{2a}.$

336. $\displaystyle\int_0^\infty \dfrac{du}{(a^2 + u^2)(b^2 + u^2)} = \dfrac{\pi}{2ab(a+b)}.$

337. $\displaystyle\int_0^{\frac{\pi}{2}} \ln \sin u \, du = \int_0^{\frac{\pi}{2}} \ln \cos u \, du = -\dfrac{\pi}{2} \ln 2.$

338. $\displaystyle\int_0^\infty \dfrac{du}{(1-u)u^n} = -\pi \cot n\pi, \quad 0 < n < 1.$

339. $\displaystyle\int_0^\infty \dfrac{u^{m-1} \, du}{1+u^n} = \dfrac{\pi}{n} \csc \dfrac{m\pi}{n}, \quad 0 < m < n.$

340. $\displaystyle\int_0^\infty \operatorname{sech} nu \, du = \dfrac{\pi}{2n}. \qquad \int_0^\infty u \operatorname{csch} nu \, du = \dfrac{\pi^2}{4n^2}.$

341. $\displaystyle\int_0^\infty u^{n-1} e^{-u} \, du = \Gamma(n) = (n-1)\Gamma(n-1), \quad n > 0;$

$$= (n-1)!, \quad n \text{ a positive integer.}$$

342. $B(m, n) = \displaystyle\int_0^1 u^{m-1}(1-u)^{n-1} \, du = \dfrac{\Gamma(m)\Gamma(n)}{\Gamma(m+n)}, m > 0, n > 0.$

343. $\displaystyle\int_0^\infty e^{-a^2 u^2} \, du = \dfrac{\sqrt{\pi}}{2a} = \dfrac{1}{2a} \Gamma(\tfrac{1}{2}).$

344. $\displaystyle\int_0^\infty u^{2n} e^{-au^2} \, du = \dfrac{1 \cdot 3 \cdot 5 \cdots (2n-1)}{2^{n+1} a^n} \sqrt{\dfrac{\pi}{a}},$

$$n \text{ a positive integer.}$$

Definite Integrals.—(*Continued*)

345. $\displaystyle\int_0^\infty u^n e^{-a^2 u^2}\, du = \frac{\Gamma\left(\dfrac{n+1}{2}\right)}{2a^{n+1}}.$

346. $\displaystyle\int_0^\infty e^{-nu}\sqrt{u}\, du = \frac{1}{2n}\sqrt{\frac{\pi}{n}}.\qquad \int_0^\infty \frac{e^{-nu}\,du}{\sqrt{u}} = \sqrt{\frac{\pi}{n}}.$

347. $\displaystyle K = \int_0^{\frac{\pi}{2}} \frac{du}{\sqrt{1 - k^2 \sin^2 u}} = \frac{\pi}{2}\left[1 + \left(\frac{1}{2}\right)^2 k^2 + \left(\frac{1\cdot 3}{2\cdot 4}\right)^2 k^4\right.$
$$\left. + \left(\frac{1\cdot 3\cdot 5}{2\cdot 4\cdot 6}\right)^2 k^6 + \cdots\right], \qquad k^2 < 1.$$

348. $\displaystyle E = \int_0^{\frac{\pi}{2}} \sqrt{1 - k^2 \sin^2 u}\; du = \frac{\pi}{2}\left[1 - \left(\frac{1}{2}\right)^2 k^2 - \left(\frac{1\cdot 3}{2\cdot 4}\right)^2\frac{k^4}{3}\right.$
$$\left. - \left(\frac{1\cdot 3\cdot 5}{2\cdot 4\cdot 6}\right)^2\frac{k^6}{5} - \cdots\right], \qquad k^2 < 1.$$

349. $\displaystyle\int_0^\infty \frac{\sin au\, du}{u} = \frac{\pi}{2}$ or $-\frac{\pi}{2}$ according to whether a is positive

or negative.

350. $\displaystyle\int_0^\infty \frac{\tan u\, du}{u} = \frac{\pi}{2}.$

351. $\displaystyle\int_0^\infty \frac{\sin^2 u\, du}{u^2} = \frac{\pi}{2}.$

352. $\displaystyle\int_0^\infty \sin u^2\, du = \int_0^\infty \cos u^2\, du = \frac{1}{2}\sqrt{\frac{\pi}{2}}.$

353. $\displaystyle\int_0^{\frac{\pi}{2}} \frac{du}{1 + a\cos u} = \frac{\cos^{-1} a}{\sqrt{1 - a^2}}, \qquad a < 1.$

354. $\displaystyle\int_0^{2\pi} \frac{du}{1 + a\cos u} = \frac{2\pi}{\sqrt{1 - a^2}}, \qquad a^2 < 1.$

355. $\displaystyle\int_0^\infty e^{-au}\sin bu\, du = \frac{b}{a^2 + b^2}, \qquad a > 0.$

356. $\displaystyle\int_0^\infty e^{-au}\cos bu\, du = \frac{a}{a^2 + b^2}, \qquad a > 0.$

357. $\displaystyle\int_0^\pi \sin mu \sin nu\, du = \int_0^\pi \cos mu \cos nu\, du = 0$ or $\frac{\pi}{2}$ according to whether m and n are unequal or equal integers.

Definite Integrals.—(*Continued*)

358. $\displaystyle\int_0^\infty \frac{\sin u\ du}{\sqrt{u}} = \int_0^\infty \frac{\cos u\ du}{\sqrt{u}} = \sqrt{\frac{\pi}{2}}.$

359. $\displaystyle\int_0^{\frac{\pi}{2}} \sin^m u \cos^n u\ du = \tfrac{1}{2}B\!\left(\frac{m+1}{2}, \frac{n+1}{2}\right) =$

$\Gamma\!\left(\dfrac{m+1}{2}\right)\Gamma\!\left(\dfrac{n+1}{2}\right) \div 2\Gamma\!\left(\dfrac{m+n+2}{2}\right),\qquad m > -1, n > -1.$

360. $\displaystyle\int_0^{\frac{\pi}{2}} \sin^n u \cos u\ du = \int_0^{\frac{\pi}{2}} \sin u \cos^n u\ du = \frac{1}{n+1}, n+1 > 0.$

In formulas 361 and 362, m and n are positive integers and each is greater than unity.

361.* $\displaystyle\int_0^{\frac{\pi}{2}} \sin^m u\ du = \int_0^{\frac{\pi}{2}} \cos^m u\ du$

$$= \frac{(m-1)(m-3)\ \cdots\ (2 \text{ or } 1)}{m(m-2)(m-4)\ \cdots\ (2 \text{ or } 1)}k,$$

where $k = \dfrac{\pi}{2}$ if m is even, $k = 1$ if m is odd.

362.* $\displaystyle\int_0^{\frac{\pi}{2}} \sin^m u \cos^n u\ du$

$$= \frac{(m-1)(m-3)\ \cdots\ (2 \text{ or } 1)(n-1)(n-3)\ \cdots\ (2 \text{ or } 1)}{(m+n)(m+n-2)(m+n-4)\ \cdots\ (2 \text{ or } 1)}k,$$

where $k = \dfrac{\pi}{2}$ if both m and n are even, $k = 1$ in all other cases.

INFINITE SERIES

1. Convergence and Divergence. Let

$$S_n = u_1 + u_2 + u_3 + \cdots + u_n$$

be the sum of the first n terms of an infinite series U:

$$U \equiv u_1 + u_2 + u_3 + \cdots + u_n + \cdots .$$

If $\lim\limits_{n\to\infty} S_n$ exists, the series U *converges*, otherwise it *diverges*.

---

* These formulas are good for any *complete* quadrant, but the sign of the result must be that of the quantity $(\sin^m \theta \cos^n \theta)$ in the quadrant covered by the limits.

If some of the terms of U are positive and some negative and if the series of absolute values:

$$|u_1| + |u_2| + |u_3| + \cdots$$

converges, the series U converges *absolutely*.

If a series containing both positive and negative terms converges, but does not converge absolutely, it is said to converge *conditionally*.

A finite number of terms of an infinite series may be discarded without affecting its convergence or divergence.

If a series converges absolutely, its terms may be rearranged in any way.

An infinite series of positive terms converges if S_n is always less than some definite number, however large n may be.

A necessary (but not sufficient) condition for the convergence of an infinite series is that $\lim_{n \to \infty} u_n = 0$.

2. Conditions for the Convergence of an Alternating Series. The alternating series $u_1 - u_2 + u_3 - u_4 + \cdots$, $u_n > 0$, converges if $\lim_{n \to \infty} u_n = 0$, and $u_{n+1} \gtreqless u_n$ from a certain term on.

The error made in using S_n for the value of this series is less than u_{n+1}.

3. Tests for Convergence and Divergence.   Ratio of $(n + 1)$st Term to nth Term a Rational Function of n.   If

$$\frac{u_{n+1}}{u_n} = A \frac{n^k + an^{k-1} + a'n^{k-2} + \cdots}{n^h + bn^{h-1} + b'n^{h-2} + \cdots},$$

h and k being positive integers, the series $u_1 + u_2 + u_3 + \cdots$ converges or diverges as follows:

Converges absolutely	Converges conditionally	Diverges
1. $k < h$	4. $k = h$	5. $k > h$
2. $k = h, \|A\| < 1$	$A = -1$	6. $k = h, \|A\| > 1$
3. $k = h, \|A\| = 1,$	$0 < b - a \gtreqless 1$	7. $k = h, A = 1, b - a \gtreqless 1$
$\quad 1 < b - a$		8. $k = h, A = -1, b - a \gtreqless 0$

4. Test-Ratio Tests for Convergence and Divergence.   Let $u_1 + u_2 + u_3 + \cdots$ be the series to be tested for convergence.

Case I.   If $\lim\limits_{n \to \infty} \left| \dfrac{u_{n+1}}{u_n} \right| = r,$

   when $r < 1$, series converges absolutely;
   when $r > 1$ (or $r = \infty$), series diverges;
   when $r = 1$, test fails.

Case II.*   If test I fails, then

let $\lim\limits_{n \to \infty} \left[n - \dfrac{u_{n+1}}{u_n}(n + 1) \right] = s,$

   when $s > 0$ (or $s = +\infty$), series converges;
   when $s < 0$ (or $s = -\infty$), series diverges;
   when $s = 0$, test fails.

Case III.*   If tests I and II fail, then

let $\lim\limits_{n \to \infty} \left[n \ln n - \dfrac{u_{n+1}}{u_n}(n + 1) \ln (n + 1) \right] = t,$

   when $t > 0$ (or $t = +\infty$), series converges;
   when $t < 0$ (or $t = -\infty$), series diverges;
   when $t = 0$, test fails.

5. Comparison Tests for Convergence and Divergence.
Let $a_1 + a_2 + a_3 + \cdots,$ $a_n > 0$, be convergent;
   $b_1 + b_2 + b_3 + \cdots,$ $b_n > 0$, be divergent;
and $u_1 + u_2 + u_3 + \cdots,$ $u_n > 0$, be a series to be tested
for convergence.

If, from some term on, $u_n \lesseqgtr a_n$, the u-series converges;
if, from some term on, $u_n \gtreqless b_n$, the u-series diverges;

if $\lim\limits_{n \to \infty} \dfrac{u_n}{a_n}$ exists, the u-series converges;

if $\lim\limits_{n \to \infty} \dfrac{u_n}{b_n}$ exists and is not zero, the u-series diverges.

Useful comparison series are:
   the geometric series, $a + ar + ar^2 + ar^3 + \cdots$, which
      converges when $|r| < 1$ and diverges when $|r| \gtreqless 1$;

   the p-series, $\dfrac{1}{1^p} + \dfrac{1}{2^p} + \dfrac{1}{3^p} + \cdots$, which converges when

   $p > 1$ and diverges when $p \lesseqgtr 1$.

If $a_1 + a_2 + a_3 + \cdots$ is an absolutely convergent series
and $r_1, r_2, r_3, \ldots$ is a set of quantities, each numerically less than
a certain constant M, then the series $r_1 a_1 + r_2 a_2 + r_3 a_3 + \cdots$
converges absolutely.

* In Case II and in Case III, it is assumed that all terms of the series are
positive from some term on.

6. Intervals of Convergence and Divergence of a Power Series. A series of the form

$$a_0 + a_1x + a_2x^2 + a_3x^3 + \cdots,$$

in which $a_0, a_1, a_2, \ldots$, are constants, is a power series in x.

If u_n is the nth term of this series and if

$$\lim_{n \to \infty} \left| \frac{u_{n+1}}{u_n} \right| = Lx^r,$$

when $L = 0$, series converges absolutely, $x^2 < \infty$;

when $L \neq 0$, series converges absolutely, $|x| < \sqrt[r]{1/L}$;

when $L \neq 0$, series diverges, $|x| > \sqrt[r]{1/L}$;

If $|x| = \sqrt[r]{1/L}$, apply the tests of the preceding articles to the numerical series obtained by substituting $x = \pm \sqrt[r]{1/L}$ in the power series.

If a power series in x converges for $x = x_1$, it converges for every value of x numerically less than x_1.

Example. Test for convergence the series:

$$1 + \frac{1}{3}x^3 + \frac{1 \cdot 2}{3 \cdot 5}x^6 + \frac{1 \cdot 2 \cdot 3}{3 \cdot 5 \cdot 7}x^9 + \cdots + \frac{1 \cdot 2 \cdots (n-1)}{3 \cdot 5 \cdots (2n-1)}x^{3(n-1)} + \cdots.$$

Solution: $\dfrac{u_{n+1}}{u_n} = \dfrac{n}{2n+1}x^3$ and $\lim\limits_{n \to \infty} \dfrac{u_{n+1}}{u_n} = \dfrac{x^3}{2}$. Since $L = \frac{1}{2}$, the series converges for $|x| < \sqrt[3]{2}$; diverges for $|x| > \sqrt[3]{2}$.

When $x = \sqrt[3]{2}$, then $\dfrac{u_{n+1}}{u_n} = \dfrac{2n}{2n+1} = \dfrac{n}{n+0.5}$. In the notation of Art. 3, $A = 1$, $k = h$, $a = 0$, $b = 0.5$; whence $b - a = 0.5$. Therefore by (7) of Art. 3, the series diverges for $x = \sqrt[3]{2}$.

When $x = -\sqrt[3]{2}$, then $\dfrac{u_{n+1}}{u_n} = \dfrac{-2n}{2n+1} = -\dfrac{n}{n+0.5}$. Here $A = -1$, $k = h$, $a = 0$, $b = 0.5$; whence $b - a = 0.5$. Therefore by (4) of Art. 3 the series converges conditionally for $x = -\sqrt[3]{2}$.

Result: Series converges, $-\sqrt[3]{2} \gtreqless x < \sqrt[3]{2}$.

7. Differentiation and Integration of Power Series. A power series represents a continuous function within its interval of convergence. If

$$f(x) = a_0 + a_1x + a_2x^2 + \cdots + a_nx^n + \cdots$$

converges for $|x| \gtreqless r$, then

$$f'(x) = a_1 + 2a_2x + \cdots + na_nx^{n-1} + \cdots$$

converges for $|x| < r$; and

$$\int f(x)dx = c + a_0 x + \frac{a_1 x^2}{2} + \cdots + \frac{a_n x^{n+1}}{n+1} + \cdots$$

converges for $|x| < r$.

EXPANSION IN SERIES

Each series converges and represents the function for the values of x indicated.

For the derivation of these series, and the accuracy obtained by using a limited number of terms, see Arts. 124 and 125.

8. Binomial Series.

$$(1 + x)^n = 1 + \frac{n}{1!}x + \frac{n(n-1)}{2!}x^2 + \frac{n(n-1)(n-2)}{3!}x^3 + \cdots$$

$$= 1 + (n)_1 x + (n)_2 x^2 + (n)_3 x^3 + \cdots, \qquad x^2 < 1.$$

(Art. 25 and Table 23)

$$(x + y)^n = x^n \left(1 + \frac{y}{x}\right)^n = x^n + (n)_1 x^{n-1} y + (n)_2 x^{n-2} y^2$$

$$+ (n)_3 x^{n-3} y^3 + \cdots, \; y^2 < x^2.$$

9. Exponential Series.

$$e^x = 1 + \frac{x}{1!} + \frac{x^2}{2!} + \frac{x^3}{3!} + \frac{x^4}{4!} + \frac{x^5}{5!} + \cdots, \qquad x^2 < \infty$$

$$a^x = 1 + \frac{\ln a}{1!}x + \frac{(\ln a)^2}{2!}x^2 + \frac{(\ln a)^3}{3!}x^3 + \cdots, \qquad x^2 < \infty$$

10. Logarithmic Series.

$$\ln (1 + x) = x - \frac{x^2}{2} + \frac{x^3}{3} - \frac{x^4}{4} + \cdots, \qquad -1 < x \lessgtr 1$$

$$\ln (1 - x) = -x - \frac{x^2}{2} - \frac{x^3}{3} - \frac{x^4}{4} - \cdots, \qquad -1 \lessgtr x < 1$$

$$\ln \frac{1+x}{1-x} = 2\left(x + \frac{x^3}{3} + \frac{x^5}{5} + \frac{x^7}{7} + \cdots\right), \qquad x^2 < 1$$

$$\ln \frac{x+1}{x-1} = 2\left(\frac{1}{x} + \frac{1}{3x^3} + \frac{1}{5x^5} + \frac{1}{7x^7} + \cdots\right), \qquad x^2 > 1$$

$$\ln (x + 1) =$$

$$\ln x + 2\left[\frac{1}{2x+1} + \frac{1}{3(2x+1)^3} + \frac{1}{5(2x+1)^5} + \cdots\right],$$

$$x > 0$$

11. Trigonometric Series.

$$\sin x = x - \frac{x^3}{3!} + \frac{x^5}{5!} - \frac{x^7}{7!} + \cdots, \qquad\qquad x^2 < \infty$$

$$\cos x = 1 - \frac{x^2}{2!} + \frac{x^4}{4!} - \frac{x^6}{6!} + \cdots, \qquad\qquad x^2 < \infty$$

$$\tan x = x + \frac{x^3}{3} + \frac{2x^5}{15} + \frac{17x^7}{315} + \cdots$$
$$+ \frac{2^{2n}(2^{2n} - 1)B_n^*}{(2n)!}x^{2n-1} + \cdots, \qquad x^2 < \frac{\pi^2}{4}$$

$$\cot x = \frac{1}{x} - \frac{x}{3} - \frac{x^3}{45} - \frac{2x^5}{945} - \cdots - \frac{2^{2n}B_n^*}{(2n)!}x^{2n-1} - \cdots,$$
$$x^2 < \pi^2.$$

$$\csc x = \frac{1}{x} + \frac{x}{6} + \frac{7x^3}{360} + \cdots$$
$$+ \frac{2(2^{2n-1} - 1)B_n^*}{(2n)!}x^{2n-1} + \cdots, \qquad x^2 < \pi^2.$$

$$\sec x = 1 + \frac{x^2}{2} + \frac{5x^4}{24} + \frac{61x^6}{720} + \cdots$$
$$+ \frac{E_n^* x^{2n}}{(2n)!} + \cdots, \qquad x^2 < \frac{\pi^2}{4}$$

$$\sin^{-1} x = x + \frac{1}{2}\frac{x^3}{3} + \frac{1\cdot 3}{2\cdot 4}\frac{x^5}{5} + \frac{1\cdot 3\cdot 5}{2\cdot 4\cdot 6}\frac{x^7}{7} + \cdots, \quad x^2 \gtreqless 1$$

$$\cos^{-1} x = \frac{\pi}{2} - \sin^{-1} x.$$

$$\tan^{-1} x = x - \frac{x^3}{3} + \frac{x^5}{5} - \frac{x^7}{7} + \cdots, \qquad\qquad x^2 \gtreqless 1$$

$$\cot^{-1} x = \frac{\pi}{2} - \tan^{-1} x.$$

$$\csc^{-1} x = \frac{1}{x} + \frac{1}{2\cdot 3\cdot x^3} + \frac{1\cdot 3}{2\cdot 4\cdot 5\cdot x^5} + \frac{1\cdot 3\cdot 5}{2\cdot 4\cdot 6\cdot 7\cdot x^7}$$
$$+ \cdots, \qquad x^2 > 1.$$

$$\sec^{-1} x = \frac{\pi}{2} - \csc^{-1} x.$$

12. Hyperbolic Series.

$$\sinh x = x + \frac{x^3}{3!} + \frac{x^5}{5!} + \frac{x^7}{7!} + \cdots, \qquad\qquad x^2 < \infty$$

$$\cosh x = 1 + \frac{x^2}{2!} + \frac{x^4}{4!} + \frac{x^6}{6!} + \cdots, \qquad\qquad x^2 < \infty$$

* See Art. 13.

$$\tanh x = x - \frac{x^3}{3} + \frac{2x^5}{15} - \frac{17x^7}{315} + \cdots$$

$$+ (-1)^{n-1}\frac{2^{2n}(2^{2n} - 1)B_n^*}{(2n)!}x^{2n-1} + \cdots \qquad x^2 < \frac{\pi^2}{4}$$

$$\coth x = \frac{1}{x} + \frac{x}{3} - \frac{x^3}{45} + \frac{2x^5}{945} - \cdots$$

$$+ (-1)^{n-1}\frac{2^{2n}B_n^*}{(2n)!}x^{2n-1} + \cdots, \qquad x^2 < \pi^2.$$

$$\operatorname{csch} x = \frac{1}{x} - \frac{x}{6} + \frac{7x^3}{360} - \cdots$$

$$+ (-1)^{n}\frac{2(2^{2n-1} - 1)B_n^*}{(2n)!}x^{2n-1} + \cdots, \qquad x^2 < \pi^2.$$

$$\operatorname{sech} x = 1 - \frac{x^2}{2} + \frac{5x^4}{24} - \frac{61x^6}{720} + \cdots$$

$$+ (-1)^{n}\frac{E_n^*}{(2n)!}x^{2n} + \cdots, \qquad x^2 < \frac{\pi^2}{4}.$$

$$\sinh^{-1} x = x - \frac{1}{2}\frac{x^3}{3} + \frac{1\cdot 3}{2\cdot 4}\frac{x^5}{5} - \frac{1\cdot 3\cdot 5}{2\cdot 4\cdot 6}\frac{x^7}{7} + \cdots, \quad x^2 < 1$$

$$\sinh^{-1} x = \ln(2x) + \frac{1}{2\cdot 2\cdot x^2} - \frac{1\cdot 3}{2\cdot 4\cdot 4\cdot x^4} + \frac{1\cdot 3\cdot 5}{2\cdot 4\cdot 6\cdot 6\cdot x^6}$$

$$- \cdots, \quad x > 1.$$

$$\tanh^{-1} x = x + \frac{x^3}{3} + \frac{x^5}{5} + \frac{x^7}{7} + \cdots, \qquad x^2 < 1$$

$$\cosh^{-1} x = \pm\left[\ln(2x) - \frac{1}{2\cdot 2\cdot x^2} - \frac{1\cdot 3}{2\cdot 4\cdot 4\cdot x^4}\right.$$

$$\left. - \frac{1\cdot 3\cdot 5}{2\cdot 4\cdot 6\cdot 6\cdot x^6} - \cdots\right], \qquad x > 1.$$

$$\coth^{-1} x = \frac{1}{x} + \frac{1}{3x^3} + \frac{1}{5x^5} + \frac{1}{7x^7} + \cdots, \qquad x^2 > 1.$$

$$\operatorname{csch}^{-1} x = \frac{1}{x} - \frac{1}{2\cdot 3\cdot x^3} + \frac{1\cdot 3}{2\cdot 4\cdot 5\cdot x^5} - \frac{1\cdot 3\cdot 5}{2\cdot 4\cdot 6\cdot 7\cdot x^7}$$

$$+ \cdots, \qquad x^2 > 1.$$

$$\operatorname{csch}^{-1} x = \ln\frac{2}{x} + \frac{x^2}{2\cdot 2} - \frac{1\cdot 3\cdot x^4}{2\cdot 4\cdot 4} + \frac{1\cdot 3\cdot 5\cdot x^6}{2\cdot 4\cdot 6\cdot 6} - \cdots,$$

$$0 < x < 1.$$

$$\operatorname{sech}^{-1} x = \pm\left[\ln\frac{2}{x} - \frac{x^2}{2\cdot 2} - \frac{1\cdot 3\cdot x^4}{2\cdot 4\cdot 4} - \frac{1\cdot 3\cdot 5\cdot x^6}{2\cdot 4\cdot 6\cdot 6} - \cdots\right],$$

$$0 < x < 1.$$

* See Art. 13.

$$\text{gd } x = x - \frac{x^3}{6} + \frac{x^5}{24} - \cdot \cdot \cdot + \frac{(-1)^n E_n^*}{(2n+1)!} x^{2n+1} + \cdot \cdot \cdot ,$$

$$x^2 < \frac{\pi^2}{4}$$

$$\text{gd}^{-1} x = x + \frac{x^3}{6} + \frac{x^5}{24} + \cdot \cdot \cdot + \frac{E_n^* x^{2n+1}}{(2n+1)!} + \cdot \cdot \cdot ,$$

$$x^2 < \frac{\pi^2}{4}$$

13. Euler's Numbers (E) and Bernoulli's Numbers (B).

$$E_1 = 1, \qquad E_2 = 5, \qquad E_3 = 61, \qquad E_4 = 1{,}385$$
$$E_5 = 50{,}521, \qquad E_6 = 2{,}702{,}765, \qquad E_7 = 199{,}360{,}981,$$
$$E_n = (2n)_2 E_{n-1} - (2n)_4 E_{n-2} + \cdot \cdot \cdot + (-1)^{n-2}(2n)_{2n-2} E_1 + (-1)^{n-1}.$$

$$B_1 = \frac{1}{6}, \qquad B_2 = \frac{1}{30}, \qquad B_3 = \frac{1}{42}, \qquad B_4 = \frac{1}{30}, \qquad B_5 = \frac{5}{66},$$

$$B_6 = \frac{691}{2730}, \qquad B_7 = \frac{7}{6}, \qquad B_8 = \frac{3617}{510},$$

$$B_n = \frac{1}{(2n+2)_2}[(2n+2)_4 B_{n-1} - (2n+2)_6 B_{n-2} + \cdot \cdot \cdot$$

$$+ (-1)^n (2n+2)_{2n} B_1 + (-1)^{n-1} n]$$

See Table 23 for values of binomial coefficients $(n)_r$.

14. Reversion of Series.   If

$$y = ax + bx^2 + cx^3 + dx^4 + ex^5 + \cdot \cdot \cdot , \qquad a \neq 0,$$

then

$$x = \frac{1}{a}y - \frac{b}{a^3}y^2 + \frac{2b^2 - ac}{a^5}y^3 + \frac{5abc - a^2 d - 5b^3}{a^7}y^4$$

$$+ \frac{6a^2 bd + 3a^2 c^2 + 14b^4 - a^3 e - 21ab^2 c}{a^9}y^5 + \cdot \cdot \cdot .$$

15. Fourier's Series.   If $f(x)$ is finite and has a finite number of discontinuities and maxima and minima in an interval, $-c \gtreqless x \gtreqless c$, then within this interval

$$f(x) = \frac{a_0}{2} + a_1 \cos \frac{\pi x}{c} + a_2 \cos \frac{2\pi x}{c} + a_3 \cos \frac{3\pi x}{c} + \cdot \cdot \cdot$$

$$+ b_1 \sin \frac{\pi x}{c} + b_2 \sin \frac{2\pi x}{c} + b_3 \sin \frac{3\pi x}{c} + \cdot \cdot \cdot ,$$

where $a_n = \dfrac{1}{c}\displaystyle\int_{-c}^{c} f(x) \cos \dfrac{n\pi x}{c} dx, \ b_n = \dfrac{1}{c}\displaystyle\int_{-c}^{c} f(x) \sin \dfrac{n\pi x}{c} dx.$

* See Art. 13.

If $f(-x) = f(x)$, the b's are all zero and $f(x)$ is represented by a cosine series; if $f(-x) = -f(x)$, the a's are all zero and $f(x)$ is represented by a sine series.

FORMULAS FROM PLANE GEOMETRY

Notation

h = altitude.	p = perimeter.
A = area.	n = number of sides.
b = lower base.	r = radius of circle (inscribed).
b' = upper base.	R = radius of circle (circumscribed).
a, b, c = sides of triangle.	C = circumference of circle.
$s = \frac{1}{2}(a + b + c)$.	d = diameter of circle.

16. Areas.

Rectangle $\qquad A = bh.$

Square $\qquad A = b^2.$

Parallelogram $\qquad A = bh.$

Triangle $\qquad A = \frac{1}{2}bh = rs =$
$$\sqrt{s(s - a)(s - b)(s - c)} = \frac{abc}{4R}.$$

Right triangle (hypotenuse c) $\qquad A = \frac{1}{2}ab.$

Equilateral triangle $\qquad A = \dfrac{a^2\sqrt{3}}{4}.$

Trapezoid $\qquad A = \frac{1}{2}(b + b')h.$

Regular polygon (side a)
$$A = \frac{1}{2}pr = \frac{na^2}{4} \cot \frac{\pi}{n} = nr^2 \tan \frac{\pi}{n}$$
$$= \frac{nR^2}{2} \sin \frac{2\pi}{n}.$$

Circle $\qquad A = \pi r^2 = \dfrac{\pi}{4}d^2.$

Circular sector $\qquad A = \frac{1}{2}r(\text{arc}) = \dfrac{\theta}{360}\pi r^2$, where θ is the number of degrees in the angle of the sector.

Quadrilateral (sides a, b, c, d) $\qquad A = \sqrt{K - abcd \cos^2 \alpha}$,

where $t = \frac{1}{2}(a + b + c + d)$, $K = (t - a)(t - b)(t - c)(t - d)$, and α is one-half the sum of any two opposite angles.

Quadrilateral inscriba-
ble in a circle $\quad A = \sqrt{(t-a)(t-b)(t-c)(t-d)}.$

Quadrilateral circum-
scribed about a circle $\quad A = \sqrt{abcd}\,\sin\alpha.$

Quadrilateral inscribed
in one circle and
circumscribed about
another circle $\quad A = \sqrt{abcd}.$

17. Line Values.

Triangle

$$\text{Altitude on side } b = \frac{2}{b}\sqrt{s(s-a)(s-b)(s-c)}.$$

$$\text{Median on side } b = \tfrac{1}{2}\sqrt{2(a^2+c^2)-b^2}.$$

$$\text{Bisector of } \angle B = \frac{2}{a+c}\sqrt{acs(s-b)}.$$

$$r = \frac{\sqrt{s(s-a)(s-b)(s-c)}}{s}.$$

$$R = \frac{abc}{4\sqrt{s(s-a)(s-b)(s-c)}}.$$

$a^2 = b^2 + c^2 \pm 2bm,$* where m is the projection of c on b.

Right triangle (hypotenuse c)

$$c^2 = a^2 + b^2, \qquad R = \frac{c}{2}, \qquad r = \frac{ab}{a+b+c}.$$

Circle $\qquad C = 2\pi r = \pi d.$

$$\text{Length of arc} = \frac{\theta}{360}(2\pi r),$$

where θ is the number of degrees in the arc.

FORMULAS FROM SOLID GEOMETRY

Notation.

a, b, c = dimensions of paral-
lelepiped.

h = altitude.

s = slant height.

e = edge or element.

p = perimeter of lower
base.

p' = perimeter of upper
base.

p_r = perimeter of right
section.

m = perimeter of mid-section.

* Use $+$ if A is obtuse, $-$ if A is acute.

c = circumference of lower base.

c' = circumference of upper base.

r = radius of lower base

r' = radius of upper base.

B = area of lower base.

B' = area of upper base.

M = area of mid-section.

R = radius of sphere.

D = diameter of sphere.

S = sum of angles.

n = number of sides.

E = spherical excess
$\quad = S - (n - 2)180$.

θ = number of degrees in angle of lune.

A = lateral area.

T = total area.

V = volume.

18. Areas and Volumes of Solids.

Solid	Area	Volume
Parallelepiped		Bh
Rectangular parallelepiped	$T = 2(ab + ac + bc)$	abc
Prism	$A = p_r e$	Bh
Right prism	$A = pe = ph$	Bh
Pyramid		$\frac{1}{3}Bh$
Regular pyramid	$A = \frac{1}{2}ps$	$\frac{1}{3}Bh$
Frustum of pyramid		$\frac{1}{3}(B + B' + \sqrt{BB'})h$
Frustum of regular pyramid	$A = ms = \frac{1}{2}(p + p')s$	$\frac{1}{3}(B + B' + \sqrt{BB'})h$
Prismatoid*		$\frac{1}{6}(B + B' + 4M)h$
Cylinder		Bh
Right circular cylinder	$A = ch = 2\pi rh.$ $T = 2\pi r(r + h).$	$\pi r^2 h$
Cone		$\frac{1}{3}Bh$
Right circular cone	$A = \frac{1}{2}cs = \pi rs.$ $T = \pi r(r + s).$	$\frac{1}{3}\pi r^2 h$
Frustum of cone		$\frac{1}{3}(B + B' + \sqrt{BB'})h$

* Formula also applicable to frustums of cones and pyramids and segments of spheres.

Solid	Area	Volume

Frustum of right

circular cone $\quad A = ms = \frac{1}{2}(c + c')s.\quad \frac{\pi}{3}(r^2 + r'^2 + rr')h$

$\qquad\qquad\qquad = \pi(r + r')s.$

Sphere $\qquad\qquad T = 4\pi R^2 = \pi D^2.\qquad \frac{4}{3}\pi R^3 = \frac{1}{6}\pi D^3.$

Spherical segment

of one base $\quad A = 2\pi Rh,$ zone. $\qquad \frac{1}{2}\pi r^2 h + \frac{1}{6}\pi h^3 =$

$$\frac{\pi h^2}{3}(3R - h).$$

Spherical segment

of two bases $\quad A = 2\pi Rh,$ zone. $\qquad \frac{1}{2}\pi(r^2 + r'^2)h + \frac{1}{6}\pi h^3$

Spherical

sector $\qquad\quad A = 2\pi Rh,$ zone. $\qquad \frac{1}{3}BR = \frac{2}{3}\pi R^2 h.$

Spherical

wedge $\qquad\quad A = \dfrac{\pi R^2 \theta}{90},$ lune. $\qquad \dfrac{\pi \theta R^3}{270}$

Spherical

pyramid $\qquad\quad A = \dfrac{\pi R^2 E}{180},$ polygon. $\qquad \frac{1}{3}BR = \dfrac{\pi R^3 E}{540}.$

19. Cavalieri's Theorem. Assume two solids to have their bases in the same plane. If the plane section of one solid at every distance x above the base is equal in area to the plane section of the other solid at the same distance x above the base, then the volumes of the two solids will be equal.

ALGEBRA

20. Function. A variable y is said to be a *function* of another variable x if, when x is given, y is determined. The variable x to which arbitrary values may be assigned is called the *independent* variable; the variable y which is then determined is called the *dependent* variable.

The symbols $f(x)$, $F(x)$, $\theta(x)$, . . . represent different functions of x. The symbol $f(a)$ denotes the value of the function $f(x)$ when x is equal to a.

If the functional relation between x (independent variable) and y (dependent variable) is expressed by

$$y = f(x),$$

y is said to be an *explicit* function of x; if expressed by

$$f(x, y) = 0,$$

y is said to be an *implicit* function of x.

21. Laws of Exponents.

$$a^m \cdot a^n = a^{m+n}, \qquad (ab)^m = a^m b^m,$$

$$\frac{a^m}{a^n} = a^{m-n}, \qquad a^{\frac{m}{n}} = (\sqrt[n]{a})^m = \sqrt[n]{a^m},$$

$$(a^m)^n = a^{mn}, \qquad a^{-n} = \frac{1}{a^n},$$

$$a^0 = 1, \text{ if } a \neq 0.$$

22. Logarithms.
If $B^L = N (B > 0, B \neq 1)$, then L is the *logarithm* of N to the base B. This is written

$$L = \log_B N.$$

Properties of logarithms.

$$\log_B (N_1 N_2 \cdots N_n) = \log_B N_1 + \log_B N_2 + \cdots + \log_B N_n.$$

$$\log_B \frac{M}{N} = \log_B M - \log_B N.$$

$$\log_B N^n = n \log_B N.$$

$$\log_B N = \log_A N \log_B A; \qquad \log_B A = \frac{1}{\log_A B}.$$

23. Quadratic Equation.
The roots r_1 and r_2 of the quadratic equation $ax^2 + bx + c = 0$ are

$$r_1 = \frac{-b + \sqrt{b^2 - 4ac}}{2a}, \qquad r_2 = \frac{-b - \sqrt{b^2 - 4ac}}{2a}.$$

If a, b, c are real, then
the roots are real and equal when $b^2 - 4ac$ is zero,
                    real and unequal when $b^2 - 4ac$ is greater than zero,
                    real and rational when $b^2 - 4ac$ is a perfect square,
                    conjugate imaginaries when $b^2 - 4ac$ is less than zero.
The sum and product of the roots are, respectively,

$$r_1 + r_2 = -\frac{b}{a}, \qquad r_1 r_2 = \frac{c}{a}.$$

24. Square of a Polynomial.
The square of a polynomial is equal to the sum of the squares of the separate terms plus twice the product of each term by every following term. For example

$$(a + b + c)^2 = a^2 + b^2 + c^2 + 2ab + 2ac + 2bc.$$

25. Binomial Theorem.

$$(a + b)^n = a^n + \frac{n}{1!}a^{n-1}b + \frac{n(n-1)}{2!}a^{n-2}b^2 + \cdots$$

$$+ \frac{n(n-1)(n-2) \cdots (n-r+1)}{r!}a^{n-r}b^r + \cdots$$

$$= a^n + (n)_1 a^{n-1}b + (n)_2 a^{n-2}b^2 + \cdots +$$

$$(n)_r a^{n-r}b^r + \cdots ,$$

where

$$r! = 1 \cdot 2 \cdot 3 \cdots (r-1)r$$

and

$$(n)_r = \frac{n(n-1)(n-2) \cdots (n-r+1)}{r!}.$$

If n is a positive integer, the expansion has $n + 1$ terms; otherwise the formula gives an infinite series which converges if b is numerically less than a.   For coefficients, see Table 23.

26. Arithmetic Progression.
If a_n is the nth term, d the common difference, and s_n the sum of the first n terms,

$$a_n = a_1 + (n-1)d, \qquad s_n = \frac{n}{2}(a_1 + a_n) = \frac{n}{2}[2a_1 + (n-1)d].$$

The *arithmetic mean* of $a_1, a_2, \cdots, a_n$ is $\dfrac{a_1 + a_2 + \cdots + a_n}{n}$.

27. Geometric Progression.
If a_n is the nth term, r the common ratio, and s_n the sum of the first n terms,

$$a_n = a_1 r^{n-1}, \qquad s_n = \frac{ra_n - a_1}{r-1} = \frac{a_1(1-r^n)}{1-r}.$$

If $r^2 < 1$, then s_n approaches a limit s as n increases without limit:

$$s = \frac{a_1}{1-r}.$$

The *geometric mean* of a and b is $\sqrt{ab}$.

28. Harmonic Progression.
The numbers of a sequence are in *harmonic progression* if their reciprocals are in arithmetic progression.

The harmonic mean of a and b is $\dfrac{2ab}{a+b}$.

29. Ratio and Proportion. If $\dfrac{a}{b} = \dfrac{c}{d}$, then

$$\frac{a}{c} = \frac{b}{d}, \qquad \frac{a+b}{b} = \frac{c+d}{d}, \qquad \frac{a-b}{b} = \frac{c-d}{d}, \qquad \frac{a+b}{a-b} = \frac{c+d}{c-d}.$$

If $\dfrac{a}{b} = \dfrac{c}{d} = \dfrac{e}{f} = \cdots$, then

$$\frac{a}{b} = \frac{c}{d} = \frac{e}{f} = \cdots = \frac{a+c+e+\cdots}{b+d+f+\cdots} = \frac{ra+sc+te+\cdots}{rb+sd+tf+\cdots}.$$

30. Variation. The quantity y varies *directly* as the quantity x when the ratio of y to x is constant for all pairs of corresponding values of x and y. This variation is expressed by

$$y = kx,$$

where k is a constant. If $y = y_1$ when $x = x_1$, then

$$y = kx = \frac{y_1}{x_1}x.$$

y varies *inversely* as x if $y = \dfrac{k}{x} = \dfrac{x_1 y_1}{x}$.

31. Rational Integral Equation.

$$a_0 x^n + a_1 x^{n-1} + a_2 x^{n-2} + \cdots + a_{n-1}x + a_n = 0,$$

where $a_0 \neq 0$ and n is a positive integer, is a rational integral equation of the nth degree.

If $r_1, r_2, r_3, \ldots, r_n$ are the roots of this equation, then

the sum of the roots $\qquad\qquad\qquad\qquad\qquad = -\dfrac{a_1}{a_0}$,

the sum of the products of the roots, two at a time $\quad = \dfrac{a_2}{a_0}$,

the sum of the products of the roots, three at a time $= -\dfrac{a_3}{a_0}$,

. . . .

the product of the roots $\qquad\qquad\qquad\qquad\quad = \dfrac{(-1)^n a_n}{a_0}$.

32. Remainder Theorem and Factor Theorem for a Polynomial.
Let $f(x) \equiv a_0 x^n + a_1 x^{n-1} + a_2 x^{n-2} + \cdots + a_{n-1}x + a_n$, where n is a positive integer. Denote by $Q(x)$ the quotient, and by R

the remainder, in the division of $f(x)$ by $x - r$.   Then

$$f(x) \equiv Q(x)(x - r) + R.$$

Therefore

(a) $f(r) = R$.

(b) If $f(r) = 0$, then $x - r$ is a factor of $f(x)$, and conversely, if $x - r$ is a factor of $f(x)$, then $f(r) = 0$.

If $f(r) = 0$, then r is a root of the equation $f(x) = 0$, and conversely, if r is a root of the equation $f(x) = 0$, then $f(r) = 0$.

33. Synthetic Division.   To divide a polynomial

$$f(x) \equiv a_0 x^n + a_1 x^{n-1} + a_2 x^{n-2} + \cdots + a_n$$

by $x - r$, write the coefficients $a_0, a_1, a_2, \ldots , a_n$ in a horizontal row, supplying a zero for each missing coefficient, and write r at the right.

Bring down a_0, multiply a_0 by r and add the product to a_1; multiply this sum by r and add this product to a_2; continue this process until the last product has been added to a_n.

The last sum is equal to the remainder $R = f(r)$; a_0 and the other sums are the coefficients in the quotient $Q(x)$.

Example: Divide $f(x) = 3x^4 - 10x^3 + 5x - 7$ by $x - 3$.
Solution:

$$
\begin{array}{r}
3 - 10 + 0 + 5 - 7 \underline{|3} \\
+ 9 - 3 - 9 - 12 \\
\hline
3 - 1 - 3 - 4 - 19
\end{array}
$$

The remainder in the division is $R = -19 = f(3)$, and the quotient is $Q(x) = 3x^3 - x^2 - 3x - 4$.

34. Upper and Lower Limits.   Rational Roots.   Let

$$f(x) \equiv a_0 x^n + a_1 x^{n-1} + \cdots + a_{n-1} x + a_n = 0, \qquad a_0 \neq 0,$$

be a rational integral equation in x with real coefficients.

(a) If $f(a)$ and $f(b)$ have opposite signs, $f(x) = 0$ has an odd number of real roots between $x = a$ and $x = b$.   This is also true for any continuous function $f(x)$.

(b) If $r > 0$ and if the numbers in the third row of the synthetic division of $f(x)$ by $x - r$ all have the same sign, $f(x) = 0$ has no real root greater than r.

(c) If $r < 0$ and if the numbers in the third row of the synthetic division of $f(x)$ by $x - r$ alternate in sign, $f(x) = 0$ has no real root less than r.

(d) If $a_0, a_1, \ldots, a_n$ ($a_n \neq 0$) are integers, the numerator of any rational root of $f(x) = 0$ (expressed as a common fraction reduced to lowest terms) is an integral factor of a_n and the denominator is an integral factor of a_0. If a_n is zero, the zero root should be removed before applying this principle.

The rational roots of $f(x) = 0$ can be found by substituting (synthetic division) in the equation all rational numbers which fulfill these two conditions. If the coefficients are rational fractions, the equation can be reduced to one having integral coefficients by clearing of fractions.

35. Approximate Values of the Real Roots of an Equation.
Let $f(x) = 0$, where $f(x)$ is a continuous function of x.

Approximate values of the real roots of $f(x) = 0$ can be found by plotting $y = f(x)$ and reading the abscissas of the points where the curve crosses the x-axis.

It is often more convenient to write $f(x) = 0$ as $f_1(x) = f_2(x)$, plot $y = f_1(x)$ and $y = f_2(x)$ on the same axes, and read the abscissas of the points of intersections of these two curves.

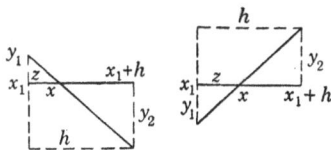

FIG. 1.

Let x_1 be an approximate value of one of the real roots of $f(x) = 0$. By substituting in $f(x) = 0$ values of x near x_1, a number h, numerically small, can be found for which $f(x_1 + h)$ will differ in sign from $f(x_1)$. The root being approximated lies between $x = x_1$ and $x = x_1 + h$ (Art. 34).

For definiteness assume $x_1 < x_1 + h$, and let $y_1 = f(x_1)$, $y_2 = f(x_1 + h)$. Since h is small the graph of $y = f(x)$ between (x_1, y_1) and $(x_1 + h, y_2)$ can be replaced by the straight line joining these two points (Fig. 1). Then $\dfrac{z}{h} = \dfrac{|y_1|}{|y_1| + |y_2|}$ and a second approximate value for this root is

$$x = x_1 + h\frac{|y_1|}{|y_1| + |y_2|}.$$

This process can be repeated until the root is found to the desired accuracy.

If a sufficiently enlarged scale is used in plotting the line, the successive approximate values of the root can be read from the figure.

36. Removal of the Term Containing x^{n-1}. If the substitution $x = y - \dfrac{a_1}{a_0 n}$ is made in the rational integral equation

$$a_0 x^n + a_1 x^{n-1} + a_2 x^{n-2} + \cdots + a_{n-1} x + a_n = 0, \qquad a_0 \neq 0,$$

a rational integral equation of degree n in y is obtained from which the term in y^{n-1} is missing.

37. Graphical Solutions of Quadratic and Cubic Equations. The real roots of $ax^2 + bx + c = 0$ are the abscissas of the points of intersection of $y = x^2$ and $y = -\dfrac{bx + c}{a}$.

A cubic equation can be reduced to the form $x^3 + px + q = 0$ by the use of the substitution given in Art. 36 and division by a_0. The real roots of $x^3 + px + q = 0$ are the abscissas of the points of intersection of $y = x^3$ and $y = -(px + q)$.

38. Roots of Cubic Equation with Real Coefficients. The general cubic equation can be reduced to the form (Art. 36)

$$x^3 + px + q = 0.$$

Let $D = \left(\dfrac{p}{3}\right)^3 + \left(\dfrac{q}{2}\right)^2$.

Case I. $D > 0$.   One real root and two conjugate imaginary roots.

If $p > 0$, let $\sinh \theta = -\dfrac{q}{2}\left(\dfrac{3}{p}\right)^{3/2}$; the roots are

$$2\sqrt{\dfrac{p}{3}} \sinh \dfrac{\theta}{3}, \qquad \sqrt{\dfrac{p}{3}}\left(-\sinh \dfrac{\theta}{3} \pm j\sqrt{3} \cosh \dfrac{\theta}{3}\right).$$

If $p < 0$, $q > 0$, let $\cosh \theta = \dfrac{q}{2}\left(\dfrac{3}{-p}\right)^{3/2}$; the roots are

$$-2\sqrt{-\dfrac{p}{3}} \cosh \dfrac{\theta}{3}, \qquad \sqrt{\dfrac{-p}{3}}\left(\cosh \dfrac{\theta}{3} \pm j\sqrt{3} \sinh \dfrac{\theta}{3}\right).$$

If $p < 0$, $q < 0$, let $\cosh \theta = -\dfrac{q}{2}\left(\dfrac{3}{-p}\right)^{3/2}$; the roots are

$$2\sqrt{\dfrac{-p}{3}} \cosh \dfrac{\theta}{3}, \qquad \sqrt{\dfrac{-p}{3}}\left(-\cosh \dfrac{\theta}{3} \pm j\sqrt{3} \sinh \dfrac{\theta}{3}\right).$$

Case II. $D < 0$.   Three real and distinct roots.

If $\cos \theta = -\dfrac{q}{2}\left(\dfrac{3}{-p}\right)^{\frac{3}{2}}$, the roots are

$$2\sqrt{\frac{-p}{3}}\cos\frac{\theta}{3}, \quad 2\sqrt{\frac{-p}{3}}\cos\left(120° \pm \frac{\theta}{3}\right).$$

Case III. $D = 0$.   Three real roots, two of which are equal. The three roots are

$$-2\sqrt[3]{\frac{q}{2}}, \quad \sqrt[3]{\frac{q}{2}}, \quad \sqrt[3]{\frac{q}{2}}.$$

39. Permutations. Each selection in a definite order, or each different arrangeme t, which can be made from a given number of things, taking all or any part of them at a time, is a permutation. The number of permutations $_nP_r$ of n distinct things taken r at a time is

$$_nP_r = n(n-1)(n-2)\cdots(n-r+1) = \frac{n!}{(n-r)!} = (n)_r r!.$$

The number of permutations P of n things taken all at a time, of which p are alike, q others are alike, and so on, is

$$P = \frac{n!}{p!q!\cdots}.$$

40. Combinations. Each selection or group which can be made from a given number of things, without regard to order or arrangement, is a combination. The number of combinations $_nC_r$ of n things taken r at a time is (see Table 23)

$$_nC_r = \frac{_nP_r}{r!} = \frac{n(n-1)(n-2)\cdots(n-r+1)}{r!} =$$

$$\frac{n!}{r!(n-r)!} = (n)_r = {}_nC_{n-r}.$$

The total number of combinations of n things taken $1, 2, 3, \ldots,$ n at a time is

$$(n)_1 + (n)_2 + (n)_3 + \cdots + (n)_n = 2^n - 1.$$

41. Probability. If a given event may occur in m different ways and may fail to occur in n different ways, and if any one of the $m + n$ ways is as likely to happen as any other, then the probability p that the event will occur on any particular occasion is

$$p = \frac{m}{m + n};$$

and the probability (q) that the event will fail to occur is

$$q = \frac{n}{m + n}.$$

Note: $p + q = 1$.

The probability of the simultaneous occurrence of independent events is the product of their respective probabilities.

The probability of the occurrence of one or the other of two mutually exclusive events is the sum of their probabilities.

The probability of exactly r successes in n trials is $(n)_r p^r q^{n-r}$.

The probability of at least r successes in n trials is

$$p^n + (n)_1 p^{n-1} q + \cdots + (n)_r p^r q^{n-r}.$$

42. Determinants.

(*a*) Determinant of the second order:

$$\begin{vmatrix} a_1 & b_1 \\ a_2 & b_2 \end{vmatrix} = a_1 b_2 - a_2 b_1.$$

(*b*) Determinant of the third order:

$$\begin{vmatrix} a_1 & b_1 & c_1 \\ a_2 & b_2 & c_2 \\ a_3 & b_3 & c_3 \end{vmatrix} = a_1 \begin{vmatrix} b_2 & c_2 \\ b_3 & c_3 \end{vmatrix} - a_2 \begin{vmatrix} b_1 & c_1 \\ b_3 & c_3 \end{vmatrix} + a_3 \begin{vmatrix} b_1 & c_1 \\ b_2 & c_2 \end{vmatrix}$$

$$= a_1(b_2 c_3 - b_3 c_2) - a_2(b_1 c_3 - b_3 c_1) + a_3(b_1 c_2 - b_2 c_1).$$

The quantities a_1, b_1, c_1, etc., are called elements.

The *minor* of any element is the determinant (of next lower order) obtained by deleting the elements of the row and of the column in which it lies.

The value of a determinant of order n is the algebraic sum of n terms, each term consisting of two factors; one factor of each term is an element of a certain row (or column), each element being used only once, and the other is the minor of that element; the element in each factor is used with its sign unchanged, or changed, according as the sum of the number of the row and of the column in which it lies is even or odd.

By the use of (c) of the next article any determinant can be transformed into another of the same order in which all the elements of a designated row (or column) are zeros, except one. The value of the original determinant can then be immediately expressed as a determinant of the next lower order.

43. Properties of Determinants. (a) The columns may be changed to rows and rows to columns without altering the value of the determinant.

$$(b) \quad \begin{vmatrix} a_1 + d_1 & b_1 & c_1 \\ a_2 + d_2 & b_2 & c_2 \\ a_3 + d_3 & b_3 & c_3 \end{vmatrix} = \begin{vmatrix} a_1 & b_1 & c_1 \\ a_2 & b_2 & c_2 \\ a_3 & b_3 & c_3 \end{vmatrix} + \begin{vmatrix} d_1 & b_1 & c_1 \\ d_2 & b_2 & c_2 \\ d_3 & b_3 & c_3 \end{vmatrix}$$

$$(c) \quad \begin{vmatrix} a_1 & b_1 & c_1 \\ a_2 & b_2 & c_2 \\ a_3 & b_3 & c_3 \end{vmatrix} = \begin{vmatrix} a_1 + mb_1 & b_1 & c_1 \\ a_2 + mb_2 & b_2 & c_2 \\ a_3 + mb_3 & b_3 & c_3 \end{vmatrix}$$

(d) Interchanging two adjacent rows (or columns) changes the sign of the result.

(e) If two rows (or columns) are equal, the determinant is zero.

(f) If all the elements of a row (or column) are multiplied by m, the determinant is multiplied by m.

44. Solution of Simultaneous Linear Equations.

If
$$\begin{aligned} a_1x + b_1y + c_1z &= d_1 \\ a_2x + b_2y + c_2z &= d_2 \\ a_3x + b_3y + c_3z &= d_3 \end{aligned} \qquad \text{and if} \qquad D = \begin{vmatrix} a_1 & b_1 & c_1 \\ a_2 & b_2 & c_2 \\ a_3 & b_3 & c_3 \end{vmatrix} \neq 0,$$

then
$$x = \frac{D_1}{D}, \qquad y = \frac{D_2}{D}, \qquad z = \frac{D_3}{D},$$

where $D_1 = \begin{vmatrix} d_1 & b_1 & c_1 \\ d_2 & b_2 & c_2 \\ d_3 & b_3 & c_3 \end{vmatrix}$, $D_2 = \begin{vmatrix} a_1 & d_1 & c_1 \\ a_2 & d_2 & c_2 \\ a_3 & d_3 & c_3 \end{vmatrix}$, $D_3 = \begin{vmatrix} a_1 & b_1 & d_1 \\ a_2 & b_2 & d_2 \\ a_3 & b_3 & d_3 \end{vmatrix}$.

If $d_1 = d_2 = d_3 = 0$, the system of equations will have a solution other than $x = y = z = 0$ if, and only if, $D = 0$.

Similar formulas hold for n simultaneous linear equations in n unknowns, $n = 2, 3, 4, \cdots$.

45. Partial Fractions. Let $\dfrac{G(x)}{f(x)}$ be a fraction in which $G(x)$ and $f(x)$ are polynomials in x with real coefficients. If the degree of $G(x)$ is equal to or greater than that of $f(x)$, then by division write the fraction in the form:

$$\frac{G(x)}{f(x)} = Q(x) + \frac{g(x)}{f(x)},$$

where $g(x)$ is of lower degree than $f(x)$.

Let $f(x)$ be resolved into linear factors $(ax + b)$, and quadratic factors $(lx^2 + mx + n)$ which do not have real linear factors. Equate the proper fraction $\dfrac{g(x)}{f(x)}$ to a sum of fractions determined as follows:

For a factor of $f(x)$ of the form:	Put in the sum a term of the form:
$ax + b$	$\dfrac{A}{ax + b}$
$(ax + b)^k$	$\dfrac{A_1}{(ax + b)} + \dfrac{A_2}{(ax + b)^2} + \cdots + \dfrac{A_k}{(ax + b)^k}$
$lx^2 + mx + n$	$\dfrac{Cx + D}{lx^2 + mx + n}$
$(lx^2 + mx + n)^k$	$\dfrac{C_1x + D_1}{lx^2 + mx + n} + \dfrac{C_2x + D_2}{(lx^2 + mx + n)^2} + \cdots + \dfrac{C_kx + D_k}{(lx^2 + mx + n)^k}$

Assume the identity:

$$\frac{g(x)}{f(x)} = \frac{g(x)}{(ax + b) \cdots (lx^2 + mx + n)^h} = \frac{A}{ax + b} + \cdots$$
$$+ \frac{A_k}{(ax + b)^k} + \frac{Cx + D}{lx^2 + mx + n} + \cdots + \frac{C_hx + D_h}{(lx^2 + mx + n)^h}.$$

Clear of fractions. Secure as many linear equations for the undetermined coefficients $A, \ldots, A_k, C, D, \ldots, C_h, D_h$ as the number of these coefficients: (1) by substituting numerical values of x (using first the zeros of the linear factors) in the cleared equation; or (2) by equating coefficients of like powers of x in the two members of the cleared equation.

Solve these equations for the undetermined coefficients $(A, C, D, \ldots)$, and substitute the values obtained for them in the assumed identity.

INTEREST AND ANNUITIES

46. Rates of Interest. The *effective* rate of interest i is the total amount to be paid for the use of one dollar for one year. If money is worth a *nominal* rate j per annum, to be converted into principal m times a year,

$$i = \left(1 + \frac{j}{m}\right)^m - 1.$$

The rate of interest is to be expressed as a decimal fraction.

47. Amount. Let S be the amount at the end of n years of a principal P invested at an annual effective rate of interest i (annual nominal rate j).

At simple interest $S = P(1 + ni)$,

At interest compounded annually $S = P(1 + i)^n$,

At interest compounded m times a year $S = P\left(1 + \dfrac{j}{m}\right)^{mn}$.

48. Present Value. Let P be the present value of an amount S due at the end of n years, the annual effective rate of interest being i (annual nominal rate j).

At simple interest $P = \dfrac{S}{1 + ni}$,

At interest compounded annually $P = S(1 + i)^{-n}$,

At interest compounded m times a year $P = S\left(1 + \dfrac{j}{m}\right)^{-mn}$.

49. Annuities. An annuity is a series of payments made at regular intervals of time. It is here assumed that the payments are equal in amount and made at the ends of equal intervals of time. The time to elapse between the beginning of the first payment interval and the end of the last is called the *term* of the annuity.

Denote by R the *annual rent* (annual payment) of the annuity; by A its *present value* (the sum of the present values obtained by discounting each annuity payment to the beginning of the term); by K its *amount* (the sum of the amounts obtained by accumulating each annuity payment to the close of the term).

Let $a_{\overline{n}|}$ denote the present value and $s_{\overline{n}|}$ the amount of an annuity of 1 per annum, payable annually for n years, the effective rate of interest being i. . Then

$$A = Ra_{\overline{n}|} = R\frac{1 - (1 + i)^{-n}}{i}; \qquad K = Rs_{\overline{n}|} = R\frac{(1 + i)^n - 1}{i}.$$

If the annuity is payable p times a year for n years, each payment being $\dfrac{R}{p}$, these formulas become

$$A = Ra_{\overline{n}|}^{(p)} = R\frac{1 - (1 + i)^{-n}}{p\left[(1 + i)^{\frac{1}{p}} - 1\right]}, \qquad K = Rs_{\overline{n}|}^{(p)} = R\frac{(1 + i)^n - 1}{p\left[(1 + i)^{\frac{1}{p}} - 1\right]}.$$

If the nominal rate is j, convertible m times a year, replace $1 + i$ by $\left(1 + \dfrac{j}{m}\right)^m$.

If $m = p$, then

$$A = \frac{R}{p}a_{\overline{np}|} \text{ at rate } \frac{j}{p}, \qquad K = \frac{R}{p}s_{\overline{np}|} \text{ at rate } \frac{j}{p}.$$

The annual rent of the annuity that A will purchase is

$$R = A\left(\frac{1}{a_{\overline{n}|}^{(p)}}\right).$$

The annual rent of the annuity that will amount to K in n years is

$$R = K\left(\frac{1}{s_{\overline{n}|}^{(p)}}\right).$$

A perpetuity is an annuity whose payments are to continue forever. The present value of a perpetuity whose payments R are made at intervals of k years is $\left(\dfrac{R}{i}\right)\left(\dfrac{1}{s_{\overline{k}|}}\right)$. If $k = 1$, the present value is $\dfrac{R}{i}$.

To compute $\dfrac{1}{s_{\overline{n}|}}$, $a_{\overline{n}|}^{(p)}$, and $s_{\overline{n}|}^{(p)}$ from Tables 15 to 18, use

$$\frac{1}{s_{\overline{n}|}} = \frac{1}{a_{\overline{n}|}} - i, \qquad a_{\overline{n}|}^{(p)} = a_{\overline{n}|}\frac{i}{j_{(p)}}, \qquad s_{\overline{n}|}^{(p)} = s_{\overline{n}|}\frac{i}{j_{(p)}}.$$

50. Equation of Value. To compare two plans for meeting the same set of obligations, accumulate or discount each to the same date at a specified rate of interest, usually the current rate. To say that the two plans are equivalent means that the amounts thus determined must be equal.

COMPLEX NUMBERS

51. Definition and Representation. If x and y are real numbers and $j = \sqrt{-1}$ ($j^2 = -1$), $z = x + jy$ is a *complex number*. It is represented by the vector OP (Fig. 2).

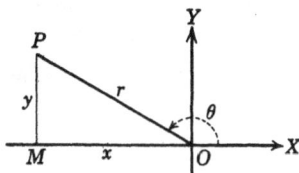

Rectangular form $z = x + jy.$
Polar form
$z = r(cos\ \theta + j\ sin\ \theta) = r\ cis\ \theta.$
Exponential form $z = re^{j\theta}.$

52. Operations with Complex Numbers. Let

FIG. 2.

$$z_1 = x_1 + jy_1, \qquad z_2 = x_2 + jy_2, \cdot\cdot\cdot, \qquad z_n = x_n + jy_n$$

represent complex numbers.

Sum

$$z_1 + z_2 + \cdot\cdot\cdot + z_n = (x_1 + x_2 + \cdot\cdot\cdot + x_n)$$
$$+ j(y_1 + y_2 + \cdot\cdot\cdot + y_n).$$

Difference $z_1 - z_2 = (x_1 - x_2) + j(y_1 - y_2).$

Product $z_1z_2 \cdot\cdot\cdot z_n = r_1r_2 \cdot\cdot\cdot r_n\ cis\ (\theta_1 + \theta_2 + \cdot\cdot\cdot + \theta_n)$
$$= r_1r_2 \cdot\cdot\cdot r_ne^{j(\theta_1 + \cdots + \theta_n)}.$$
$$z_1z_2 = (x_1x_2 - y_1y_2) + j(x_1y_2 + x_2y_1).$$

Quotient $\dfrac{z_1}{z_2} = \dfrac{r_1}{r_2}\ cis\ (\theta_1 - \theta_2) = \dfrac{r_1}{r_2}e^{j(\theta_1 - \theta_2)}$

$$= \frac{x_1x_2 + y_1y_2}{x_2{}^2 + y_2{}^2} + j\frac{x_2y_1 - x_1y_2}{x_2{}^2 + y_2{}^2}.$$

53. Roots and Powers. De Moivre's Theorem:

$$(\cos\ \theta + j\ \sin\ \theta)^n = \cos\ n\theta + j\ \sin\ n\theta.$$
$$z^n = r^n\ cis\ n\theta = r^ne^{jn\theta}, \qquad n \text{ a rational number.}$$
$$\sqrt[n]{z} = \sqrt[n]{r}\ cis\ \frac{\theta + 2k\pi}{n} = \sqrt[n]{r}\ e^{j\frac{\theta + 2k\pi}{n}},$$

where n is an integer and $k = 0, 1, 2, \cdot\cdot\cdot, n - 1.$

54. Properties of Complex Numbers. If

$$x_1 + jy_1 = x_2 + jy_2, \quad \text{then} \quad x_1 = x_2 \quad \text{and} \quad y_1 = y_2.$$
$$e^{j\theta} = \cos\theta + j\sin\theta, \quad e^{-j\theta} = \cos\theta - j\sin\theta.$$

$$\cos\theta = \tfrac{1}{2}(e^{j\theta} + e^{-j\theta}), \quad \sin\theta = \frac{1}{2j}(e^{j\theta} - e^{-j\theta}).$$

$\ln z = \ln r + j\theta + 2k\pi j$, where k is any integer.

$$e^z = e^x(\cos y + j\sin y).$$

$$\sin z = \sin x \cosh y + j\cos x \sinh y, \qquad \sin jy = j\sinh y;$$
$$\cos z = \cos x \cosh y - j\sin x \sinh y, \qquad \cos jy = \cosh y;$$
$$\sinh z = \sinh x \cos y + j\cosh x \sin y, \qquad \sinh jy = j\sin y;$$
$$\cosh z = \cosh x \cos y + j\sinh x \sin y, \qquad \cosh jy = \cos y$$

55. Functions of a Complex Variable. A function

$$w = u(x, y) + jv(x, y)$$

is a function of the complex variable $z = x + jy$ if, and only if,
$\dfrac{\partial w}{\partial y} = j\dfrac{\partial w}{\partial x}.$ If $w = f(z)$, then $u(x, y)$ and $v(x, y)$ satisfy the
relations

$$\frac{\partial^2 u}{\partial x^2} + \frac{\partial^2 u}{\partial y^2} = 0, \qquad \frac{\partial^2 v}{\partial x^2} + \frac{\partial^2 v}{\partial y^2} = 0.$$

The formulas for the differentiation of functions of a complex
variable are the same as in the case of functions of a real variable;
for example, $d(e^z) = e^z\,dz$, $d(\sin z) = \cos z\,dz$.

TRIGONOMETRY

56. Relations between Degrees and Radians.

$$180° = \pi \text{ radians} = 3.1415927 \cdots \text{ radians.}$$

$$1° = \frac{\pi}{180} \text{ radians} = 0.01745329 \cdots \text{ radians.}$$

$$1 \text{ radian} = \frac{180}{\pi} \text{ degrees} = 57.29578° = 57°17'44.8''.$$

57. Trigonometric Functions.* (a) *Definitions* (Fig. 3).

$$\text{sine (sin) } \theta = \frac{y}{r}, \quad \text{cosine (cos) } \theta = \frac{x}{r}, \quad \text{tangent (tan) } \theta = \frac{y}{x},$$

$$\text{cosecant (csc) } \theta = \frac{r}{y}, \text{ secant (sec) } \theta = \frac{r}{x}, \text{ cotangent (cot) } \theta = \frac{x}{y},$$

* See Figs. 36–42.

versine (vers) $\theta = 1 - \cos \theta$,

$$\text{haversine (hav) } \theta = \tfrac{1}{2}(1 - \cos \theta) = \sin^2 \frac{\theta}{2}.$$

(b) *Signs of the trigonometric functions.*

If the angle is in quadrant	I	II	III	IV
the sine and the cosecant are	+	+	−	−
the cosine and the secant are	+	−	−	+
the tangent and the cotangent are	+	−	+	−

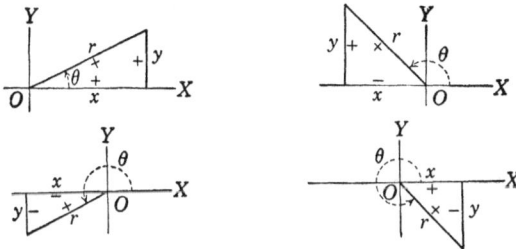

Fig. 3.

(c) *Special values of the trigonometric functions.*

ANGLE	sin	cos	tan	cot	sec	csc
0°	0	1	0	∞	1	∞
30°	$\tfrac{1}{2}$	$\tfrac{1}{2}\sqrt{3}$	$\tfrac{1}{3}\sqrt{3}$	$\sqrt{3}$	$\tfrac{2}{3}\sqrt{3}$	2
45°	$\tfrac{1}{2}\sqrt{2}$	$\tfrac{1}{2}\sqrt{2}$	1	1	$\sqrt{2}$	$\sqrt{2}$
60°	$\tfrac{1}{2}\sqrt{3}$	$\tfrac{1}{2}$	$\sqrt{3}$	$\tfrac{1}{3}\sqrt{3}$	2	$\tfrac{2}{3}\sqrt{3}$
90°	1	0	∞	0	∞	1
180°	0	−1	0	∞	−1	∞
270°	−1	0	∞	0	∞	−1
360°	0	1	0	∞	1	∞

(d) *Functions of angles in any quadrant in terms of angles in first quadrant* (see pages 22 and 23).

58. Trigonometric Identities.

$$\sin x \csc x = 1, \quad \cos x \sec x = 1, \quad \tan x \cot x = 1,$$
$$\sin^2 x + \cos^2 x = 1, \quad 1 + \tan^2 x = \sec^2 x, \quad 1 + \cot^2 x = \csc^2 x,$$
$$\sin (x \pm y) = \sin x \cos y \pm \cos x \sin y,$$
$$\cos (x \pm y) = \cos x \cos y \mp \sin x \sin y,$$
$$\tan (x \pm y) = \frac{\tan x \pm \tan y}{1 \mp \tan x \tan y},$$

$$\sin 2x = 2 \sin x \cos x,$$
$$\cos 2x = \cos^2 x - \sin^2 x = 2 \cos^2 x - 1 = 1 - 2 \sin^2 x,$$
$$\sin 3x = 3 \sin x - 4 \sin^3 x,$$
$$\cos 3x = 4 \cos^3 x - 3 \cos x,$$
$$\sin nx = n \sin x \cos^{n-1} x - (n)_3 \sin^3 x \cos^{n-3} x +$$
$$(n)_5 \sin^5 x \cos^{n-5} x - \cdots,$$
$$\cos nx = \cos^n x - (n)_2 \sin^2 x \cos^{n-2} x + (n)_4 \sin^4 x \cos^{n-4} x$$
$$- \cdots, \quad \text{(Table 23.)}$$

$$\tan\left(\frac{x}{2} + \frac{\pi}{4}\right) = \sec x + \tan x$$

$$\sin^2 x = \tfrac{1}{2}(1 - \cos 2x), \qquad \cos^2 x = \tfrac{1}{2}(1 + \cos 2x),$$

$$1 - \cos x = 2 \sin^2 \frac{x}{2}, \qquad 1 + \cos x = 2 \cos^2 \frac{x}{2},$$

$$\tan \frac{x}{2} = \pm\sqrt{\frac{1 - \cos x}{1 + \cos x}} = \frac{\sin x}{1 + \cos x} = \frac{1 - \cos x}{\sin x},$$

$$\sin x + \sin y = 2 \sin \tfrac{1}{2}(x + y) \cos \tfrac{1}{2}(x - y),$$
$$\sin x - \sin y = 2 \cos \tfrac{1}{2}(x + y) \sin \tfrac{1}{2}(x - y),$$
$$\cos x + \cos y = 2 \cos \tfrac{1}{2}(x + y) \cos \tfrac{1}{2}(x - y),$$
$$\cos x - \cos y = -2 \sin \tfrac{1}{2}(x + y) \sin \tfrac{1}{2}(x - y),$$
$$\sin x \sin y = -\tfrac{1}{2} \cos (x + y) + \tfrac{1}{2} \cos (x - y),$$
$$\cos x \cos y = \tfrac{1}{2} \cos (x + y) + \tfrac{1}{2} \cos (x - y),$$
$$\sin x \cos y = \tfrac{1}{2} \sin (x + y) + \tfrac{1}{2} \sin (x - y),$$
$$\sin x + \sin (x + y) + \sin (x + 2y) + \cdots + \sin [x + (n - 1)y]$$
$$= \frac{\sin [x + \tfrac{1}{2}(n - 1)y] \sin \tfrac{1}{2}ny}{\sin \tfrac{1}{2}y},$$
$$\cos x + \cos (x + y) + \cos (x + 2y) + \cdots + \cos [x + (n - 1)y]$$
$$= \frac{\cos [x + \tfrac{1}{2}(n - 1)y] \sin \tfrac{1}{2}ny}{\sin \tfrac{1}{2}y}.$$

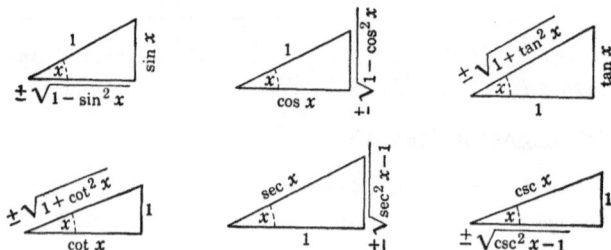

Fig. 4.

59. Functions of an Angle in Terms of Each of the Other Functions.
From the right triangles of Fig. 4 any trigonometric

function of an angle can be expressed readily in terms of any other trigonometric function of that angle. The sign of the radical is determined by the quadrant in which the terminal side of the angle lies.

60. Principal Values of the Inverse Trigonometric Functions. The symbol $\sin^{-1} x$ ($= $ arc sin x) denotes any angle whose sine is x. The inverse trigonometric functions are infinitely many valued. But unless otherwise indicated, the symbol $\sin^{-1} x$, $\cos^{-1} x$, etc., will be used to designate principal values as defined by the following table:

Function	Principal values	Function	Principal values
$\sin^{-1} x$	$-\dfrac{\pi}{2} \gtreqqless \sin^{-1} x \gtreqqless \dfrac{\pi}{2}$	$\cos^{-1} x$	$0 \gtreqqless \cos^{-1} x \gtreqqless \pi$
$\tan^{-1} x$	$-\dfrac{\pi}{2} < \tan^{-1} x < \dfrac{\pi}{2}$	$\cot^{-1} x$	$0 < \cot^{-1} x < \pi$
$\sec^{-1} x$	$0 \gtreqqless \sec^{-1} x < \dfrac{\pi}{2}(x > 0)$	$\csc^{-1} x$	$0 < \csc^{-1} x \gtreqqless \dfrac{\pi}{2}(x > 0)$
$\sec^{-1} x$	$-\pi \gtreqqless \sec^{-1} x < -\dfrac{\pi}{2}(x < 0)$	$\csc^{-1} x$	$-\pi < \csc^{-1} x \gtreqqless -\dfrac{\pi}{2}(x < 0)$

The principal values of these functions are indicated in Figs. 44–49 by the heavier parts of the curves.

61. Solution of Trigonometric Equations. In general, change all the functions into a single function and then solve the equation algebraically for that function.

If $\sin x = a$, $x = (-1)^n \sin^{-1} a + n\pi$;
if $\cos x = a$, $x = \pm\cos^{-1} a + 2n\pi$;
if $\tan x = a$, $x = \tan^{-1} a + n\pi$, where n is an integer.
If $a \sin x + b \cos x = c$, $(c^2 \gtreqqless a^2 + b^2)$, then

$$x = \pm\cos^{-1} \frac{c}{\sqrt{a^2 + b^2}} + \theta + 2n\pi,$$

where $\sin \theta = \dfrac{a}{\sqrt{a^2 + b^2}}, \cos \theta = \dfrac{b}{\sqrt{a^2 + b^2}}.$

62. Plane Triangles.

Notation:

A, B, C = angles. a, b, c = sides opposite A, B, C,
     K = area.   respectively.

r = radius of inscribed circle.

h_b = altitude to the side b.

$s = \frac{1}{2}(a + b + c)$.

R = radius of circumscribed circle.

Formulas:

$A + B + C = 180° = \pi$ radians.

$$\frac{\sin A}{a} = \frac{\sin B}{b} = \frac{\sin C}{c} \left(= \frac{1}{2R} \right), \text{ law of sines.}$$

$$a^2 = b^2 + c^2 - 2bc \cos A,^* \text{ law of cosines.}$$

$$\frac{a - b}{a + b} = \frac{\tan \frac{1}{2}(A - B)}{\tan \frac{1}{2}(A + B)},^* \text{ law of tangents.}$$

$$r = \sqrt{\frac{(s - a)(s - b)(s - c)}{s}} = (s - a) \tan \frac{A}{2}.^*$$

$$\sin \frac{A}{2} = \sqrt{\frac{(s - b)(s - c)}{bc}},^* \qquad \cos \frac{A}{2} = \sqrt{\frac{s(s - a)}{bc}}.^*$$

$$\text{hav } A = \frac{(s - b)(s - c)}{bc},^* \qquad \tan \frac{A}{2} = \frac{r}{s - a}.^*$$

$$K = \frac{1}{2}bh_b^* = \frac{1}{2}ab \sin C^* =$$

$$\frac{a^2 \sin B \sin C^*}{2 \sin (B + C)} = \sqrt{s(s - a)(s - b)(s - c)} = rs = \frac{abc}{4R}.$$

63. Solution of a Right Triangle. Let $C = 90°$.
Then $A + B = 90°$, and $a^2 + b^2 = c^2$.
Case I. Given any side and the angle A.

$$B = 90° - A \qquad \text{and} \qquad \frac{a}{\sin A} = \frac{b}{\cos A} = c.$$

Case II. Given a and b.

$$\tan A = \frac{a}{b} \qquad \text{and} \qquad c = \frac{a}{\sin A}, \qquad B = 90° - A.$$

Case III. Given a and c.

$$\sin A = \frac{a}{c} \text{ and } b = c \cos A = \sqrt{(c + a)(c - a)}, \ B = 90° - A.$$

64. Solution of Plane Triangles. For notation, see Art. 62.
Case I. Given two angles and one side, A, B, a.

---

* Two more formulas are obtained by replacing a by b, b by c, c by a, A by B, B by C, and C by A, as they occur in the formula, and then by repeating this process for the formula thus obtained.

$C = 180° - (A + B), \quad b = a \sin B \csc A, \quad c = a \sin C \csc A.$

Case II.   Given two sides and the included angle, a, b, C.

(a) Draw altitude from A and solve two right triangles.

$\tan B = \dfrac{b \sin C}{a - b \cos C}, \quad A = 180° - (B + C), \quad c = b \sin C \csc B.$

(b) $\frac{1}{2}(A + B) = \frac{1}{2}(180° - C),$

$$\tan \tfrac{1}{2}(A - B) = \frac{a - b}{a + b} \tan \tfrac{1}{2}(A + B),$$

$A = \frac{1}{2}(A + B) + \frac{1}{2}(A - B), \qquad B = \frac{1}{2}(A + B) - \frac{1}{2}(A - B),$
$$c = a \sin C \csc A = b \sin C \csc B.$$

(c) $\qquad\qquad c = \sqrt{a^2 + b^2 - 2ab \cos C},$

$$\sin A = \frac{a \sin C}{c},$$

$$\sin B = \frac{b \sin C}{c},$$

where A, B, C are in same order of magnitude as a, b, c.

Case III.   Given three sides, a, b, c.

(a) Drop a perpendicular from one angle, say C, to the side c. Let m be the segment of c adjacent to a, and n the segment adjacent to b.   Then $n + m = c$ and $n - m = \dfrac{(b - a)(b + a)}{c}.$ Solve these two equations for m and n.   The angles of the original triangle can be found by solving the two right triangles into which it is divided by the perpendicular.

(b) $\tan \dfrac{A}{2} = \dfrac{r}{s - a}, \qquad \tan \dfrac{B}{2} = \dfrac{r}{s - b}, \qquad \tan \dfrac{C}{2} = \dfrac{r}{s - c}.$

(c) Find largest angle, say A, from $\cos A = \dfrac{b^2 + c^2 - a^2}{2bc}.$

$\sin B = \dfrac{b \sin A}{a}, \sin C = \dfrac{c \sin A}{a},$ where $B < 90°$, $C < 90°$.

(d) If only one angle is required, use first formula in (c) or use

$$\operatorname{hav} A = \sin^2 \tfrac{1}{2}A = \frac{(s - b)(s - c)}{bc}, \text{ or}$$

$$\cos \tfrac{1}{2}A = \sqrt{\frac{s(s - a)}{bc}}.$$

Case IV.   Given two sides and angle opposite one of them, a, b, A.   Find B from $\sin B = \dfrac{b \sin A}{a}.$

If $\sin B > 1$ ($\log \sin B > 0$), no solution; if $B = 90°$, one solution; otherwise take $B_1 < 90°$ and $B_2 = 180° - B_1 > 90°$, and proceed as in Case I with B_1 and B_2 separately.

If $A + B_2 \lessgtr 180°$, use B_1 only.

Check formulas: Law of sines and $A + B + C = 180°$ when these have not been used in the solution; $a = b \cos C + c \cos B$;

$$\frac{a - b}{c} = \frac{\sin \frac{1}{2}(A - B)}{\cos \frac{1}{2}C}; \qquad \frac{a + b}{c} = \frac{\cos \frac{1}{2}(A - B)}{\sin \frac{1}{2}C}.$$

SPHERICAL TRIGONOMETRY

65. Spherical Triangle. Let a, b, c denote the sides and A, B, C the opposite angles, respectively, of a spherical triangle; a', b', c' the sides and A', B', C' the opposite angles, respectively, of its polar triangle.

Only triangles in which each angle and each side are less than $180°$ are considered here.

In two polar triangles:

Each angle of the one is the supplement of the opposite side in the other.

In any spherical triangle:

(*a*) Each side is less than the sum of the other two sides.

(*b*) $a + b + c < 360°$.   (*c*) $180° < A + B + C < 540°$.

(*d*) Angles opposite equal sides are equal, and conversely.

(*e*) The greater of two angles is opposite the greater side, and conversely.

66. Right Spherical Triangle. If any two parts (other than $C = 90°$) are given, any other part can be found from one of the formulas:

$$
\begin{array}{ll}
\sin a = \sin A \sin c & \sin a = \tan b \cot B \\
\sin b = \sin B \sin c & \sin b = \tan a \cot A \\
\cos c = \cos a \cos b & \cos c = \cot A \cot B \\
\cos A = \cos a \sin B & \cos A = \tan b \cot c \\
\cos B = \cos b \sin A & \cos B = \tan a \cot c
\end{array}
$$

To resolve the ambiguity concerning any part found from its sine, use one of the rules:

(1) An oblique angle and the side opposite are in the same quadrant.

(2) An odd number of sides cannot be between $90°$ and $180°$.

Always express each unknown part in terms of the two given parts. Check logarithmic computation by using the formula which involves the three computed parts.

Example: Given A and c; find a, b, B.
Solution: Select formulas involving a, A, c; b, A, c; and B, A, c. Solve these for the functions of a, b, B, obtaining:

$\sin a = \sin A \sin c$, $\tan b = \cos A \tan c$, $\cot B = \tan A \cos c$.
Check: $\sin a = \tan b \cot B$.

In a right spherical triangle $(C = 90°)$, $\bar{A} = 90° - A$, $\bar{B} = 90° - B$, $\bar{c} = 90° - c$, a, and b are called *circular parts*. To any circular part (as $\bar{A}$), two of the remaining parts $(b, \bar{c})$ are adjacent and the other two $(a, \bar{B})$ are opposite (Fig. 5).

The 10 equations given for the solution of right triangles can be found from the rule (Napier):

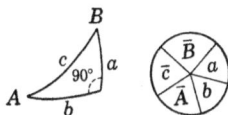

FIG. 5.

The *sine* of any circular part is equal to

(a) the product of the tangents of its adjacent parts;
(b) the product of the cosines of its opposite parts.

67. Oblique Spherical Triangle.*

Law of sines

$$\frac{\sin A}{\sin a} = \frac{\sin B}{\sin b} = \frac{\sin C}{\sin c}.$$

Laws of cosines

$$\cos a = \cos b \cos c + \sin b \sin c \cos A,$$
$$\cos A = -\cos B \cos C + \sin B \sin C \cos a.$$

A spherical triangle is determined by any three of its parts.
Case I.   Given the three sides, a, b, c.

$$(a) \quad s = \frac{a + b + c}{2}, \quad r = \sqrt{\frac{\sin(s - a)\sin(s - b)\sin(s - c)}{\sin s}},$$

$$\tan \tfrac{1}{2}A = \frac{r}{\sin(s - a)}, \quad \tan \tfrac{1}{2}B = \frac{r}{\sin(s - b)},$$

$$\tan \tfrac{1}{2}C = \frac{r}{\sin(s - c)}.$$

Check: Law of sines.
(b) If only one angle is wanted, use

* Additional formulas can be obtained by cyclic change of letters.

$$\text{hav } A = \sin^2 \tfrac{1}{2}A = \frac{\sin (s - b) \sin (s - c)}{\sin b \sin c} \qquad \text{or}$$

$$\cos^2 \tfrac{1}{2}A = \frac{\sin s \sin (s - a)}{\sin b \sin c}.$$

Case II. Given three angles, A, B, C. Reduce to Case I by use of the polar triangle.

Case III. Given two sides and the included angle, a, b, C.
(a) $\tan \tfrac{1}{2}(A - B) = \sin \tfrac{1}{2}(a - b) \csc \tfrac{1}{2}(a + b) \cot \tfrac{1}{2}C$,
$\tan \tfrac{1}{2}(A + B) = \cos \tfrac{1}{2}(a - b) \sec \tfrac{1}{2}(a + b) \cot \tfrac{1}{2}C$,
$\qquad \tan \tfrac{1}{2}c = \tan \tfrac{1}{2}(a - b) \sin \tfrac{1}{2}(A + B) \csc \tfrac{1}{2}(A - B)$,
Check: $\tan \tfrac{1}{2}c = \tan \tfrac{1}{2}(a + b) \cos \tfrac{1}{2}(A + B) \sec \tfrac{1}{2}(A - B)$.
$A = \tfrac{1}{2}(A + B) + \tfrac{1}{2}(A - B), \qquad B = \tfrac{1}{2}(A + B) - \tfrac{1}{2}(A - B)$.
(b) If c only is wanted, use

$$\text{hav } c = \text{hav } (a - b) + \sin a \sin b \text{ hav } C.$$

Case IV. Given two angles and the included side, A, B, c. Reduce to Case III by use of the polar triangle.

Case V. Given two sides and the angle opposite one of them, a, b, A.

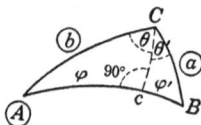

Fig. 6.

(a) Find B from $\sin B = \dfrac{\sin b}{\sin a} \sin A$. If $\log \sin B > 0$, no solution; if $B = 90°$, right triangle; otherwise two values B_1 and $B_2 = 180° - B_1$, subject to the restrictions that $B > A$ when $b > a$ and $B < A$ when $b < a$.

$$\tan \tfrac{1}{2}c = \sin \tfrac{1}{2}(A + B) \csc \tfrac{1}{2}(A - B) \tan \tfrac{1}{2}(a - b).$$
$$\cot \tfrac{1}{2}C = \sin \tfrac{1}{2}(a + b) \csc \tfrac{1}{2}(a - b) \tan \tfrac{1}{2}(A - B).$$

Check: $\cot \tfrac{1}{2}C = \cos \tfrac{1}{2}(a + b) \sec \tfrac{1}{2}(a - b) \tan \tfrac{1}{2}(A + B)$.
(b) Draw altitude from C and solve two right triangles (see Fig. 6).

$$\tan \varphi = \cos A \tan b,$$
$$\cos \varphi' = \cos \varphi \cos a \sec b,$$
$$c = \varphi + \varphi',$$
$$\cot B = \cot A \sin \varphi' \csc \varphi,$$
$$\cot \theta = \cos b \tan A,$$
$$\cot \theta' = \cos a \tan B.$$
$$C = \theta + \theta'.$$

Check: Law of sines.

Angle φ should be chosen in the first quadrant when tan φ is positive, it may be chosen in either the second quadrant (positive) or the fourth quadrant (negative) when tan φ is negative.   Use both positive and negative values of φ', provided

$$0 < \varphi + \varphi' < 180°.$$

θ must be taken in the same quadrant as φ and θ' in the same quadrant as φ'.

Case VI.   Two angles and the side opposite one of them, *A, B, a.*   Reduce to Case V by use of the polar triangle.

HYPERBOLIC FUNCTIONS

68. Definitions.   The hyperbolic functions and the abbreviations used in writing them are

$$\text{Hyperbolic sine of } x = \sinh x = \frac{e^x - e^{-x}}{2},$$

$$\text{Hyperbolic cosine of } x = \cosh x = \frac{e^x + e^{-x}}{2};$$

in terms of these the hyperbolic tangent, cotangent, secant, and cosecant are, respectively,

$$\tanh x = \frac{\sinh x}{\cosh x}, \qquad \coth x = \frac{\cosh x}{\sinh x} = \frac{1}{\tanh x},$$

$$\text{sech } x = \frac{1}{\cosh x}, \qquad \text{csch } x = \frac{1}{\sinh x}.$$

69. Inverse Hyperbolic Functions.   If $x = \sinh y$, then
$$y = \sinh^{-1} x = \text{inverse hyperbolic sine of } x.$$
The inverse hyperbolic functions can be expressed as logarithms:

$$\sinh^{-1}\frac{x}{a} = \ln \frac{x + \sqrt{a^2 + x^2}}{a}; \quad \text{csch}^{-1}\frac{x}{a} = \ln \frac{a + \sqrt{a^2 + x^2}}{x};$$

$$\tanh^{-1}\frac{x}{a} = \tfrac{1}{2} \ln \frac{a + x}{a - x}; \qquad \coth^{-1}\frac{x}{a} = \tfrac{1}{2} \ln \frac{x + a}{x - a};$$

$$\cosh^{-1}\frac{x}{a} = \pm\ln \frac{x + \sqrt{x^2 - a^2}}{a};$$

$$\text{sech}^{-1}\frac{x}{a} = \pm\ln \frac{a + \sqrt{a^2 - x^2}}{x}.$$

70. The Gudermannian.*   The relation

$$x = \ln \tan\left(\frac{\varphi}{2} + \frac{\pi}{4}\right), \text{ or } \varphi = 2\tan^{-1} e^x - \frac{\pi}{2}, \left(-\frac{\pi}{2} < \varphi < \frac{\pi}{2}\right)$$

defines a function φ called the *gudermannian* of x, denoted by gd x; x is the *inverse gudermannian* of φ, denoted by gd^{-1} φ. The gudermannian may also be defined by any of the following

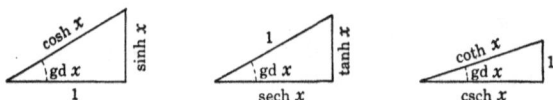

Fig. 7.

consistent relations between the trigonometric functions of $\varphi = $ gd x, and hyperbolic functions of x (see Fig. 7):

$$\sinh x = \tan \text{ gd } x, \quad \cosh x = \sec \text{ gd } x, \quad \tanh x = \sin \text{ gd } x,$$
$$\text{csch } x = \cot \text{ gd } x, \quad \text{sech } x = \cos \text{ gd } x, \quad \coth x = \csc \text{ gd } x.$$
$$\text{gd } x = \sin^{-1} \tanh x = \cos^{-1} \text{ sech } x = \tan^{-1} \sinh x,$$
$$\text{gd}^{-1} \varphi = \sinh^{-1} \tan \varphi = \cosh^{-1} \sec \varphi = \tanh^{-1} \sin \varphi.$$

71. Hyperbolic Identities.

$$\cosh^2 x - \sinh^2 x = 1, \quad \tanh^2 x + \text{sech}^2 x = 1,$$
$$\coth^2 x - \text{csch}^2 x = 1,$$
$$\sinh(x \pm y) = \sinh x \cosh y \pm \cosh x \sinh y,$$
$$\cosh(x \pm y) = \cosh x \cosh y \pm \sinh x \sinh y,$$
$$\tanh(x \pm y) = \frac{\tanh x \pm \tanh y}{1 \pm \tanh x \tanh y},$$
$$\sinh 2x = 2 \sinh x \cosh x, \quad \tanh 2x = \frac{2 \tanh x}{1 + \tanh^2 x},$$
$$\cosh 2x = \cosh^2 x + \sinh^2 x = 2\cosh^2 x - 1 = 1 + 2\sinh^2 x,$$
$$\sinh 3x = 4\sinh^3 x + 3 \sinh x, \quad \cosh 3x = 4\cosh^3 x - 3 \cosh x,$$
$$\sinh nx = n \cosh^{n-1} x \sinh x + (n)_3 \cosh^{n-3} x \sinh^3 x + \cdots$$
$$\cosh nx = \cosh^n x + (n)_2 \cosh^{n-2} x \sinh^2 x + \cdots .$$

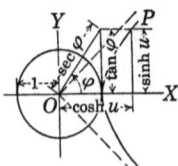

Fig. A.

* Since $\cosh^2 u - \sinh^2 u = 1$, the coordinates of a point P on the equilateral hyperbola $x^2 - y^2 = 1$, in terms of a parameter u, are $x = \cosh u$, $y = \sinh u$. From Fig. A, $\sinh u = \tan \varphi$. Since $\sec^2 \varphi - \tan^2 \varphi = 1$, $\cosh u = \sec \varphi$. Then $\text{csch } u = \cot \varphi$, $\text{sech } u = \cos \varphi$, $\tanh u = \sin \varphi$ and $\coth u = \csc \varphi$, where $\varphi = $ gd u. These relations are shown in Fig. 7, where u has been replaced by x.

$$\sinh x + \sinh y = 2 \sinh \tfrac{1}{2}(x + y) \cosh \tfrac{1}{2}(x - y),$$
$$\sinh x - \sinh y = 2 \cosh \tfrac{1}{2}(x + y) \sinh \tfrac{1}{2}(x - y),$$
$$\cosh x + \cosh y = 2 \cosh \tfrac{1}{2}(x + y) \cosh \tfrac{1}{2}(x - y),$$
$$\cosh x - \cosh y = 2 \sinh \tfrac{1}{2}(x + y) \sinh \tfrac{1}{2}(x - y).$$
$$\cosh x + \sinh x = e^{x}, \qquad \cosh x - \sinh x = e^{-x},$$
$$\sinh^{-1} \operatorname{csch} x = \cosh^{-1} \coth x = \tanh^{-1} \operatorname{sech} x = \coth^{-1} \cosh x$$
$$= \operatorname{sech}^{-1} \tanh x = \operatorname{csch}^{-1} \sinh x = \operatorname{gd}^{-1}\left(\frac{\pi}{2} - \operatorname{gd} x\right).$$
$$\sinh (-x) = -\sinh x, \qquad \cosh (-x) = \cosh x,$$
$$\tanh (-x) = -\tanh x,$$
$$\operatorname{gd} (-x) = -\operatorname{gd} x, \qquad \operatorname{gd}^{-1} (-x) = -\operatorname{gd}^{-1} (x).$$
$$\sinh (x + nj\pi) = (-1)^{n} \sinh x,$$
$$\cosh (x + nj\pi) = (-1)^{n} \cosh x.$$

PLANE ANALYTIC GEOMETRY

72. Coordinates of a Point. The rectangular coordinates of the point P are (x, y); the polar coordinates of P are (r, θ).

From Fig. 8:

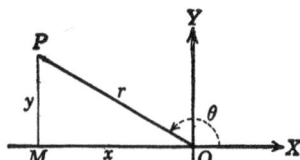

$$x = r \cos \theta, \qquad y = r \sin \theta;$$
$$r = \sqrt{x^2 + y^2}, \qquad \tan \theta = \frac{y}{x},$$
$$\sin \theta = \frac{y}{\sqrt{x^2 + y^2}}, \qquad \cos \theta = \frac{x}{\sqrt{x^2 + y^2}}.$$

Fig. 8.

73. Directed Segments. A directed segment has both magnitude and direction. The directed segment from P_1 to P_2 is denoted by P_1P_2, from P_2 to P_1 by P_2P_1; and $P_2P_1 = -P_1P_2$.

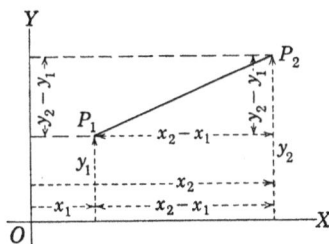

The projection (orthogonal) of P_1P_2 on the x-axis or on any line parallel to it is $x_2 - x_1$; the projection of P_1P_2 on the y-axis or on any line parallel to it is $y_2 - y_1$.

Fig. 9.

74. Projection of a Broken Line. The projection of a segment PQ on a line AB is equal to the algebraic sum of the projections on AB of the segments of any broken line connecting P and Q. See Fig. 10.

If $PP_1P_2P_3Q$ is a broken line connecting P and Q, then
$$MN = MM_1 + M_1M_2 + M_2M_3 + M_3N.$$

75. Distance Formula. Point of Division Formulas. Slope Formula. The distance d between $P_1(x_1, y_1)$ and $P_2(x_2, y_2)$ is

$$d = \sqrt{(x_2 - x_1)^2 + (y_2 - y_1)^2}.$$

The distance d between $P_1(r_1, \theta_1)$ and $P_2(r_2, \theta_2)$ is

$$d = \sqrt{r_1^2 + r_2^2 - 2r_1r_2 \cos (\theta_1 - \theta_2)}.$$

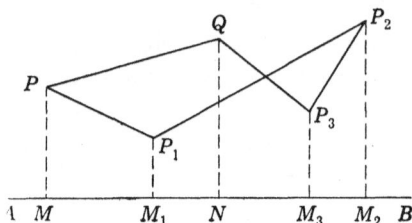

FIG. 10.

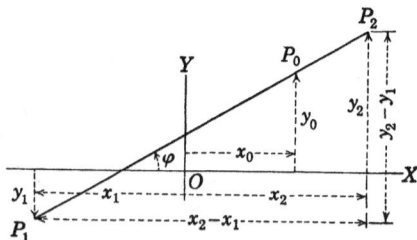

FIG. 11.

The coordinates (x_0, y_0) of the point P_0 which divides the segment from $P_1(x_1, y_1)$ to $P_2(x_2, y_2)$ in the ratio $r_1:r_2$

$$(P_1P_0:P_0P_2 = r_1:r_2)$$

are

$$x_0 = \frac{r_2x_1 + r_1x_2}{r_1 + r_2}, \qquad y_0 = \frac{r_2y_1 + r_1y_2}{r_1 + r_2}.$$

The coordinates (x_0, y_0) of the mid-point of the segment P_1P_2 are:

$$x_0 = \tfrac{1}{2}(x_1 + x_2), \qquad y_0 = \tfrac{1}{2}(y_1 + y_2).$$

The coordinates of a point $\dfrac{1}{n}$th the way from P_1 to P_2 are

$$x_0 = x_1 + \frac{1}{n}(x_2 - x_1), \qquad y_0 = y_1 + \frac{1}{n}(y_2 - y_1).$$

The slope m of the line passing through the points $P_1(x_1, y_1)$ and $P_2(x_2, y_2)$ is the tangent of the inclination (φ) of the line:

$$m = \tan \varphi = \frac{y_2 - y_1}{x_2 - x_1}.$$

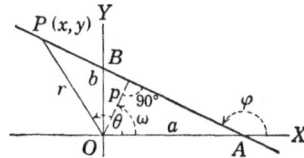

Fig. 12.

76. Equations of a Straight Line.

Notation (Fig. 12)

φ = inclination.
$m = \tan \varphi$ = slope.
a = x-intercept.

b = y-intercept.
p = normal intercept.
ω = inclination of normal axis.

Line	Equation	
Parallel to the x-axis	$y = b$.	(1)
Parallel to the y-axis	$x = a$.	(2)
Intercept form	$\dfrac{x}{a} + \dfrac{y}{b} = 1$.	(3)
Slope-intercept form	$y = mx + b$.	(4)
Line through $(x_1,\ y_1)$ with slope m	$y - y_1 = m(x - x_1)$.	(5)
Line through (x_1,y_1) and (x_2,y_2)	$y - y_1 = \dfrac{y_2 - y_1}{x_2 - x_1}(x - x_1)$.	(6)
Normal form	$x \cos \omega + y \sin \omega = p$.	(7)
General form	$Ax + By + C = 0$.	(8)

Note: $\quad m = -\dfrac{A}{B}, \quad a = -\dfrac{C}{A}, \quad b = -\dfrac{C}{B}, \quad p = \dfrac{-C}{\pm\sqrt{A^2 + B^2}},$

$$\cos \omega = \frac{A}{\pm\sqrt{A^2 + B^2}}, \qquad \sin \omega = \frac{B}{\pm\sqrt{A^2 + B^2}}.$$

Polar equation, line parallel to x-axis	$r \sin \theta = b$.	(9)
Polar equation, line parallel to y-axis	$r \cos \theta = a$.	(10)
Polar equation of a line	$r \cos (\theta - \omega) = p$.	(11)

77. Angle between Two Lines.

The angle (β) which a line having inclination φ_2 and slope m_2 makes with a line having inclination φ_1 and slope m_1 can be found from

$$\beta = \varphi_2 - \varphi_1 \qquad \text{or} \qquad \tan \beta = \frac{m_2 - m_1}{1 + m_1 m_2}.$$

The lines are parallel if $m_1 = m_2$,

$$\text{perpendicular if } m_1 = -\frac{1}{m_2}.$$

The lines $Ax + By + C = 0$ and $A'x + B'y + C' = 0$ are

$$\begin{aligned}
&\text{parallel if} &&A:A' = B:B', \\
&\text{perpendicular if } AA' = -BB', \\
&\text{identical if} &&A:A' = B:B' = C:C'.
\end{aligned}$$

78. Systems of Lines. Let k be any arbitrary constant (parameter).

Equation	Special property
$y - y_1 = k(x - x_1)$	Through (x_1, y_1)
$y = kx + b$	y-intercept b
$x = ky + a$	x-intercept a
$y = mx + k$	Slope m
$Ax + By = k$	$\parallel$ to $Ax + By + C = 0$
$Bx - Ay = k$	$\perp$ to $Ax + By + C = 0$
$Ax + By + C$ $+ k(A'x + B'y + C') = 0$	Through the intersection of $Ax + By + C = 0$ and $A'x + B'y + C' = 0$.

79. Distance from a Line to a Point. The distance d between the line $Ax + By + C = 0$ and the point (x_1, y_1) is

$$d = \left| \frac{Ax_1 + By_1 + C}{\sqrt{A^2 + B^2}} \right|.$$

80. Bisectors of the Angles Formed by Two Lines. The equations of the bisectors of the angles formed by the lines $Ax + By + C = 0$ and $A'x + B'y + C' = 0$ are

$$\frac{Ax + By + C}{\sqrt{A^2 + B^2}} \pm \frac{A'x + B'y + C'}{\sqrt{A'^2 + B'^2}} = 0.$$

81. Transformation of Coordinates. Let (x, y) be the coordinates of a point referred to axes OX, OY and (x', y') be its coordinates referred to new axes, $O'X'$, $O'Y'$.

If the new origin is at the point (h, k), the new axes being parallel to the old, then

$$x = x' + h, \qquad y = y' + k.$$

If O' coincides with O and OX' makes an angle θ with OX, then
$$x = x' \cos \theta - y' \sin \theta, \qquad y = x' \sin \theta + y' \cos \theta,$$

or

$$x = \frac{x' - my'}{\sqrt{1 + m^2}}, \qquad y = \frac{mx' + y'}{\sqrt{1 + m^2}},$$

where $m = \tan \theta$.

THE CIRCLE*

82. Rectangular Equations of a Circle. (Radius R.)

Center at $(0, 0)$	$x^2 + y^2 = R^2.$	(1)
Center at (h, k)	$(x - h)^2 + (y - k)^2 = R^2.$	(2)
General form	$x^2 + y^2 + Dx + Ey + F = 0,$	(3)

in which $h = -\dfrac{D}{2}, \; k = -\dfrac{E}{2}, \; R = \sqrt{\dfrac{D^2}{4} + \dfrac{E^2}{4} - F}.$

The equation of the radical axis of the circles

$$x^2 + y^2 + Dx + Ey + F = 0$$

and

$$x^2 + y^2 + D'x + E'y + F' = 0$$

is

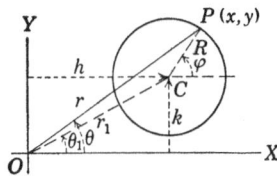

FIG. 13.

$$(D - D')x + (E - E')y + (F - F') = 0.$$

If two circles intersect, their radical axis is their common chord; if they are tangent to each other, their radical axis is their common tangent.

The radical axis is the locus of points from which tangents to the two circles are of equal length.

83. Parametric Equations of a Circle. (Radius R.)

Center at $(0, 0)$	$x = R \cos \phi, \; y = R \sin \phi.$	(1)
Center at (h, k)	$\begin{cases} x = h + R \cos \phi, \\ y = k + R \sin \phi. \end{cases}$	(2)

84. Polar Equations of a Circle. (Radius R.)

Center at $(0, 0)$	$r = R.$	(1)
Center at $(R, 0)$	$r = 2R \cos \theta.$	(2)

* See Figs. 28, 70, 71, 72.

Center at $\left(R, \dfrac{\pi}{2}\right)$ $\qquad\qquad$ $r = 2R\sin\theta.$ $\qquad$ (3)

Center at (r_1, θ_1) $\qquad$ $r^2 + r_1^2 - 2rr_1\cos(\theta - \theta_1)$
$$= R^2. \qquad (4)$$

THE PARABOLA*

85. Definitions. The *parabola* is the locus of a point which moves so that its distance from a fixed point F (the focus) is always equal to its distance from a fixed line (the directrix).

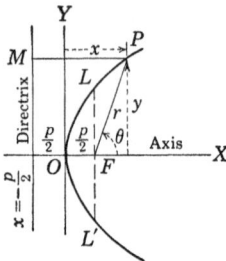

The point O halfway from the focus to the directrix is the *vertex*. The distance from the focus to the directrix is denoted by p.

The line through the focus perpendicular to the directrix is the *axis* of the parabola.

The *latus rectum* is the breadth of the parabola at the focus:

FIG. 14.

$$\textit{Latus rectum} = LL' = 2p.$$

The *eccentricity* e of the parabola is the ratio $FP{:}PM = 1$.

86. Rectangular Equations of a Parabola.

Axis $y = k$, vertex (h, k),
focus $(h \pm \tfrac{1}{2}p, k)$ $\qquad$ $(y - k)^2 = \pm 2p(x - h).$ $\qquad$ (1)
Axis $x = h$, vertex (h, k),
focus $(h, k \pm \tfrac{1}{2}p)$ $\qquad$ $(x - h)^2 = \pm 2p(y - k).$ $\qquad$ (2)
Axis $\|OX$ $\qquad\qquad\qquad$ $cy^2 + dx + ey + f = 0.$ $\qquad$ (3)
Axis $\|OY$ $\qquad\qquad\qquad$ $ax^2 + dx + ey + f = 0.$ $\qquad$ (4)

Reduce (3) to (1) and (4) to (2) by completing the square.

To sketch a parabola: draw the axis; locate vertex, focus, and points L, L' (Fig. 14) distant p from the focus on perpendicular to the axis. Draw curve through L, L', and vertex.

87. Polar Equations of a Parabola.

Axis $\theta = 0$, vertex $\left(\pm\dfrac{p}{2}, 0\right)$, focus $(0, 0)$ $\quad r = \dfrac{p}{1 \pm \cos\theta}.$ $\quad$ (1)

Axis $\theta = \dfrac{\pi}{2}$, vertex $\left(\pm\dfrac{p}{2}, \dfrac{\pi}{2}\right)$, focus $(0, 0)$ $\quad r = \dfrac{p}{1 \pm \sin\theta}.$ $\quad$ (2)

* See Figs. 30, 32, 92, 93.

The Ellipse*

88. Definitions. The *ellipse* is the locus of a point which moves so that the sum of its distances from two fixed points (foci) is a constant (2a). The ellipse may also be defined as the locus of a point which moves so that its distance from a fixed point

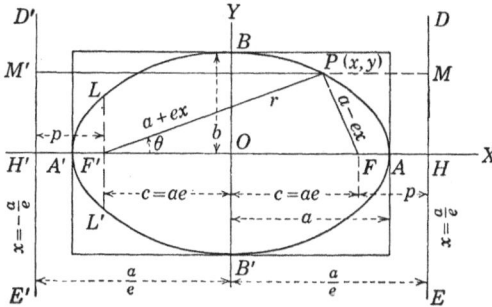

FIG. 15.

(focus) divided by its distance from a fixed line (directrix) is a constant (e) less than unity.

F and F' are foci, ED and $E'D'$ are the corresponding directrices.

Major axis $= A'A = 2a$; minor axis $= B'B = 2b$.

Distance between the foci $= F'F = 2c$.

Eccentricity $= \dfrac{FP}{DM} = \dfrac{a - ex}{\dfrac{a}{e} - x} = e = \dfrac{F'P}{PM'}$.

p = distance from a focus to the corresponding directrix $= FH$

$$= \frac{a}{e}(1 - e^2) = F'H'.$$

Latus rectum = double ordinate through a focus $= LL' = \dfrac{2b^2}{a}$.

$b^2 = a^2 - c^2 = a^2(1 - e^2); e = \dfrac{c}{a}$.

89. Rectangular Equations of an Ellipse. $(a > b)$.

Center $(0, 0)$, major axis along x-axis	$\dfrac{x^2}{a^2} + \dfrac{y^2}{b^2} = 1.$	(1)
Center $(0, 0)$, major axis along y-axis	$\dfrac{x^2}{b^2} + \dfrac{y^2}{a^2} = 1.$	(2)
Center (h, k), major axis along $y = k$	$\dfrac{(x - h)^2}{a^2} + \dfrac{(y - k)^2}{b^2} = 1.$	(3)

* See Figs. 29, 90.

Center (h, k), major axis along $x = h$

$$\frac{(x - h)^2}{b^2} + \frac{(y - k)^2}{a^2} = 1. \quad (4)$$

Axes of ellipse parallel to coordinate axes

$$Ax^2 + Cy^2 + Dx + Ey + F = 0, \quad (5)$$

A and C having like signs. Reduce to (3), or (4) by completing squares.

To sketch the ellipse $\dfrac{(x - h)^2}{a^2} + \dfrac{(y - k)^2}{b^2} = 1$; locate vertices on the axes, a to the right and left of the center, b above and below the center. Through vertices, draw lines parallel to the axes, forming a rectangle. The ellipse lies inside this rectangle, tangent to its sides at the vertices of the ellipse.

90. Parametric Equations of an Ellipse.

Center $(0, 0)$

$$x = a \cos \phi, \; y = b \sin \phi. \quad (1)$$

Center (h, k), major axis $y = k$

$$\begin{cases} x = h + a \cos \phi, \\ y = k + b \sin \phi. \end{cases} \quad (2)$$

91. Polar Equations of an Ellipse.

Center $(\pm c, 0)$, major axis along $\theta = 0$

$$r = \frac{ep}{1 \mp e \cos \theta}. \quad (1)$$

Center $\left(\pm c, \dfrac{\pi}{2} \right)$, major axis along $\theta = \dfrac{\pi}{2}$

$$r = \frac{ep}{1 \mp e \sin \theta}. \quad (2)$$

$$a = \frac{ep}{1 - e^2}, \qquad b = \frac{ep}{\sqrt{1 - e^2}}, \qquad c = ae.$$

THE HYPERBOLA*

92. Definitions. The *hyperbola* is the locus of a point which moves so that the difference of its distances from two fixed points (foci) is a constant $(2a)$. The hyperbola may also be defined as the locus of a point which moves so that its distance from a fixed point (focus) divided by its distance from a fixed line (directrix) is a constant (e) greater than unity.

F and F' are foci, HD and $H'D'$ are the corresponding directrices. Transverse axis $= A'A = 2a$; conjugate axis $= B'B = 2b$.

* See Figs. 31, 33, 91.

Distance between the foci $= F'F = 2c$.

Eccentricity $= \dfrac{FP}{PM} = \dfrac{ex - a}{x - \dfrac{a}{e}} = e = \dfrac{F'P}{PM'}$.

$p =$ distance from a focus to the corresponding directrix

$$= FH = \frac{a}{e}(e^2 - 1) = F'H'.$$

Latus rectum $=$ double ordinate through a focus $= L'L = \dfrac{2b^2}{a}$.

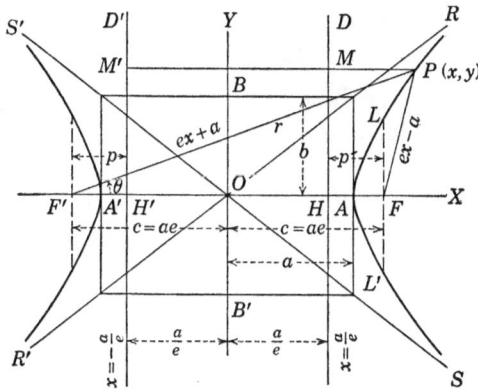

Fig. 16.

The lines $R'R$ and $S'S$ are asymptotes of the hyperbola. The slope of $R'R$ is $\dfrac{b}{a}$ and that of $S'S$ is $-\dfrac{b}{a}$.

$$b^2 = c^2 - a^2 = a^2(e^2 - 1), \quad e = \frac{c}{a}.$$

93. Rectangular Equations of a Hyperbola.

Center $(0, 0)$, transverse axis along x-axis, asymptotes

$$y = \pm\frac{b}{a}x \qquad\qquad \frac{x^2}{a^2} - \frac{y^2}{b^2} = 1. \qquad (1)$$

Center $(0, 0)$, transverse axis along y-axis, asymptotes

$$y = \pm\frac{a}{b}x \qquad\qquad \frac{y^2}{a^2} - \frac{x^2}{b^2} = 1. \qquad (2)$$

Center (h, k), transverse axis
$y = k$, asymptotes

$$y - k = \pm\frac{b}{a}(x - h) \qquad\qquad \frac{(x - h)^2}{a^2} - \frac{(y - k)^2}{b^2} = 1. \quad (3)$$

Center (h, k), transverse axis
$x = h$, asymptotes

$$y - k = \pm\frac{a}{b}(x - h) \qquad\qquad \frac{(y - k)^2}{a^2} - \frac{(x - h)^2}{b^2} = 1. \quad (4)$$

Axes of the hyperbola parallel to coordinate axes

$$Ax^2 + Cy^2 + Dx + Ey + F = 0, \quad (5)$$

A and C having unlike signs.
Reduce to (3) or (4) by completing squares.
Asymptotes $ax + by + c = 0$ and $a'x + b'y + c' = 0$:

$$(ax + by + c)(a'x + b'y + c') = k. \quad (6)$$

To sketch $\dfrac{(x - h)^2}{a^2} - \dfrac{(y - k)^2}{b^2} = 1$: draw the axes of the curve
and locate on them points a to the right and left of the center,
b above and below the center. Through these four points draw
lines parallel to the axes, forming a rectangle. Draw the
diagonals of this rectangle (the asymptotes). The hyperbola
lies outside this rectangle and between its diagonals (extended),
tangent to its sides at the points (the vertices) where they are cut
by the transverse axis.

The two hyperbolas $\dfrac{x^2}{a^2} - \dfrac{y^2}{b^2} = 1$ and $\dfrac{x^2}{a^2} - \dfrac{y^2}{b^2} = -1$ are conjugate hyperbolas. The transverse and conjugate axes of one are,
respectively, the conjugate and transverse axes of the other.
Conjugate hyperbolas have the same asymptotes.

94. Parametric Equations of a Hyperbola.

Center $(0, 0)$, transverse axis x-axis:

$$x = a \sec \phi, \ y = b \tan \phi; \ \text{or} \ x = a \cosh u, \ y = b \sinh u. \quad (1)$$

Center (h, k), transverse axis $y = k$:

$$\begin{cases} x = h + a \sec \phi, \\ y = k + b \tan \phi. \end{cases} \quad \text{or} \quad \begin{cases} x = h + a \cosh u, \\ y = k + b \sinh u. \end{cases} \quad (2)$$

95. Polar Equations of a Hyperbola.

Center $(\pm c,\ 0)$, transverse axis $\theta = 0$

$$r = \frac{ep}{1 \pm e\cos\theta}. \tag{1}$$

Center $\left(\pm c,\ \dfrac{\pi}{2}\right)$, transverse axis $\theta = \dfrac{\pi}{2}$

$$r = \frac{ep}{1 \pm e\sin\theta}. \tag{2}$$

$$a = \frac{ep}{e^2 - 1}, \qquad b = \frac{ep}{\sqrt{e^2 - 1}}, \qquad c = ae.$$

96. Diameters.
A diameter of a circle, parabola, ellipse, or hyperbola is the locus of the mid-points of a system of parallel chords. Let the chords have slope m.

Curve	Diameter
$(y - k)^2 = 2p(x - h)$	$y - k = \dfrac{p}{m}$
$\dfrac{(x-h)^2}{a^2} + \dfrac{(y-k)^2}{b^2} = 1$	$y - k = -\dfrac{b^2}{a^2 m}(x - h)$
$\dfrac{(x-h)^2}{a^2} - \dfrac{(y-k)^2}{b^2} = 1$	$y - k = \dfrac{b^2}{a^2 m}(x - h)$

97. Tangent Line to a Conic.

Equation of curve	Tangent at point (x_1, y_1)	Tangent having slope m
$x^2 + y^2 = a^2$	$xx_1 + yy_1 = a^2$	$y = mx \pm a\sqrt{1 + m^2}$
$y^2 = 2px$	$yy_1 = p(x + x_1)$	$y = mx + \dfrac{p}{2m}$
$x^2 = 2py$	$xx_1 = p(y + y_1)$	$y = mx - \dfrac{pm^2}{2}$
$\dfrac{x^2}{a^2} + \dfrac{y^2}{b^2} = 1$	$\dfrac{xx_1}{a^2} + \dfrac{yy_1}{b^2} = 1$	$y = mx \pm \sqrt{a^2 m^2 + b^2}$
$\dfrac{x^2}{a^2} - \dfrac{y^2}{b^2} = 1$	$\dfrac{xx_1}{a^2} - \dfrac{yy_1}{b^2} = 1$	$y = mx \pm \sqrt{a^2 m^2 - b^2}$
$\dfrac{y^2}{a^2} - \dfrac{x^2}{b^2} = 1$	$\dfrac{yy_1}{a^2} - \dfrac{xx_1}{b^2} = 1$	$y = mx \pm \sqrt{a^2 - b^2 m^2}$

The equation of the tangent to

(a) $Ax^2 + Bxy + Cy^2 + Dx + Ey + F = 0$ at (x_1, y_1) is

$$Axx_1 + \frac{B}{2}(xy_1 + x_1 y) + Cyy_1 + \frac{D}{2}(x + x_1) + \frac{E}{2}(y + y_1) + F = 0.$$

To find the equation of the tangent having slope m, replace y, in the equation of the curve, by $mx + k$, and determine k so that the resulting quadratic equation in x shall have a double root.

If $C = A$, $B = 0$, the curve (a) is a circle and the length (t) of the segment of the tangent from an external point (x_1, y_1) to the point of contact is

$$t = \sqrt{x_1^2 + y_1^2 + \frac{D}{A}x_1 + \frac{E}{A}y_1 + \frac{F}{A}}.$$

98. General Equation of the Second Degree. The locus of

$$ax^2 + bxy + cy^2 + dx + ey + f = 0$$

is a conic section, classified as follows:

$b^2 - 4ac < 0$, ellipse (circle, imaginary lines);
$b^2 - 4ac = 0$, parabola (parallel lines);
$b^2 - 4ac > 0$, hyperbola (intersecting lines).

The slopes of the asymptotes are $\dfrac{1}{2c}[-b \pm \sqrt{b^2 - 4ac}]$.

To reduce the equation to standard form, proceed as follows:
(a) *Central conics.* The center is at (h, k), obtained by solving

$$2ah + bk + d = 0, \qquad bh + 2ck + e = 0.$$

Translating the axes to (h, k) as a new origin, the equation becomes:

$$ax'^2 + bx'y' + cy'^2 + F = 0, \qquad \text{where} \qquad F = f + \tfrac{1}{2}(dh + ek).$$

Rotating the axes about the new origin through an angle θ, $\left(\text{where } \tan 2\theta = \dfrac{b}{a - c}\right)$, the equation becomes:

$$Ax''^2 + Cy''^2 + F = 0$$

where $A + C = a + c$,

$$A - C = \pm\sqrt{(a - c)^2 + b^2} \left[= \frac{b}{\sin 2\theta} \right].$$

$$\tan \theta = \frac{(A - C) - (a - c)}{b} = \frac{b}{(A - C) + (a - c)}.$$

If the sign before $\sqrt{(a - c)^2 + b^2}$ is taken as minus for the ellipse, and opposite to the sign of F for the hyperbola, the foci of the curve will fall on the x''-axis.

(b) *The parabola.* Write the equation in the form

$$(y - mx)^2 + dx + ey + f = 0.$$

Rotate the axes through $\tan^{-1} m$, obtaining

$$(1 + m^2)y'^2 + \frac{1}{\sqrt{1 + m^2}}[(d + me)x' + (e - md)y'] + f = 0.$$

By completing the square and grouping, reduce to the form

$$(y' - k)^2 = C(x' - h).$$

SOLID ANALYTIC GEOMETRY

99. Coordinates of a Point. Coordinates of the point P are

Rectangular	(x, y, z);
Cylindrical	(r', θ, z); $(r' = OB)$
Spherical	(r, θ, φ);
Polar	$(r; \alpha, \beta, \gamma)$,

where $\cos^2 \alpha + \cos^2 \beta + \cos^2 \gamma = 1$.

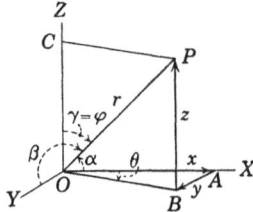

FIG. 17.

100. Relations among Coordinates.

Rectangular and cylindrical coordinates:

$$x = r' \cos \theta, \qquad y = r' \sin \theta, \qquad z = z; \qquad (1)$$

Rectangular and spherical coordinates:

$$x = r \sin \varphi \cos \theta, \qquad y = r \sin \varphi \sin \theta, \qquad z = r \cos \varphi; \qquad (2)$$

Rectangular and polar coordinates:

$$x = r \cos \alpha, \qquad y = r \cos \beta, \qquad z = r \cos \gamma. \qquad (3)$$

101. Distance Formula. Point of Division Formula. Direction Cosines. See Fig. 18.

The projection of P_1P_2 on x-axis $= x_2 - x_1$;

on y-axis $= y_2 - y_1$;

on z-axis $= z_2 - z_1$.

The distance d between P_1 and P_2 is

$$d = \sqrt{(x_2 - x_1)^2 + (y_2 - y_1)^2 + (z_2 - z_1)^2}.$$

The coordinates (x_0, y_0, z_0) of P_0 which divides the segment P_1P_2 in the ratio $r_1 : r_2$ ($P_1P_0 : P_0P_2 = r_1 : r_2$) are

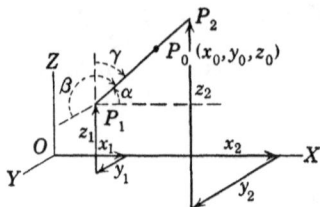

FIG. 18.

$$x_0 = \frac{r_2 x_1 + r_1 x_2}{r_2 + r_1},$$

$$y_0 = \frac{r_2 y_1 + r_1 y_2}{r_2 + r_1},$$

$$z_0 = \frac{r_2 z_1 + r_1 z_2}{r_2 + r_1}.$$

The coordinates (x_0, y_0, z_0) of the mid-point of P_1P_2 are:

$$x_0 = \tfrac{1}{2}(x_1 + x_2), \qquad y_0 = \tfrac{1}{2}(y_1 + y_2), \qquad z_0 = \tfrac{1}{2}(z_1 + z_2).$$

The direction cosines of P_1P_2 are

$$\cos \alpha = \frac{x_2 - x_1}{d}, \qquad \cos \beta = \frac{y_2 - y_1}{d}, \qquad \cos \gamma = \frac{z_2 - z_1}{d}.$$

If $\cos \alpha : \cos \beta : \cos \gamma = l : m : n$, then

$$\cos \alpha = \frac{l}{\sqrt{l^2 + m^2 + n^2}}, \qquad \cos \beta = \frac{m}{\sqrt{l^2 + m^2 + n^2}},$$

$$\cos \gamma = \frac{n}{\sqrt{l^2 + m^2 + n^2}}.$$

Here l, m, n are called *direction numbers*.

THE PLANE

102. Equation of a Plane.   See Fig. 19.

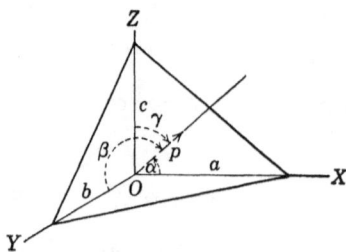

FIG. 19.

Normal form: $\quad x \cos \alpha + y \cos \beta + z \cos \gamma = p$,   (1) where p is the length and α, β, γ are the direction angles of the perpendicular from the origin to the plane.

Intercept form $\dfrac{x}{a} + \dfrac{y}{b} + \dfrac{z}{c} = 1$.   (2)

General equation $Ax + By + Cz + D = 0$.   (3)

To reduce the general equation to the normal form divide by $\pm\sqrt{A^2 + B^2 + C^2}$ with sign opposite to the sign of D.

103. Distance from a Point to a Plane. The distance d from (x_1, y_1, z_1) to $Ax + By + Cz + D = 0$ is

$$d = \left| \frac{Ax_1 + By_1 + Cz_1 + D}{\sqrt{A^2 + B^2 + C^2}} \right|.$$

104. Angle between Two Planes. The angle (θ) between the planes $Ax + By + Cz + D = 0$ and $A'x + B'y + C'z + D' = 0$ is given by the formula

$$\cos \theta = \frac{\pm (AA' + BB' + CC')}{\sqrt{A^2 + B^2 + C^2} \sqrt{A'^2 + B'^2 + C'^2}}.$$

These two planes are

$$\text{parallel if } \frac{A}{A'} = \frac{B}{B'} = \frac{C}{C'};$$
$$\text{perpendicular if } AA' + BB' + CC' = 0.$$

THE STRAIGHT LINE

105. Equations of a Straight Line. Through (x_1, y_1, z_1), direction numbers l, m, n:

$$\frac{x - x_1}{l} = \frac{y - y_1}{m} = \frac{z - z_1}{n} \qquad \text{(symmetric form).} \qquad (1)$$

$x = x_1 + lp, \quad y = y_1 + mp, \quad z = z_1 + np$ (parametric form).

Through (x_1, y_1, z_1) and (x_2, y_2, z_2):

$$\frac{x - x_1}{x_2 - x_1} = \frac{y - y_1}{y_2 - y_1} = \frac{z - z_1}{z_2 - z_1} \qquad \text{(two-point form).} \qquad (2)$$

General form (intersection of two planes):

$$Ax + By + Cz + D = 0, \qquad A'x + B'y + C'z + D' = 0. \qquad (3)$$

If (x_1, y_1, z_1) is any point on this line, then

$$\frac{x - x_1}{BC' - B'C} = \frac{y - y_1}{CA' - C'A} = \frac{z - z_1}{AB' - A'B},$$

provided no denominator is zero. If any denominator is zero, the corresponding numerator is also zero.

Through (x_1, y_1, z_1), perpendicular to $Ax + By + Cz + D = 0$:

$$\frac{x - x_1}{A} = \frac{y - y_1}{B} = \frac{z - z_1}{C}. \qquad (4)$$

106. Angle between Two Lines. The angle (θ) between two lines having direction angles α_1, β_1, γ_1 and α_2, β_2, γ_2 is given by the formula:

$$\cos \theta = \cos \alpha_1 \cos \alpha_2 + \cos \beta_1 \cos \beta_2 + \cos \gamma_1 \cos \gamma_2.$$

These lines are

parallel if $\qquad \cos \alpha_1 : \cos \beta_1 : \cos \gamma_1 = \cos \alpha_2 : \cos \beta_2 : \cos \gamma_2;$

perpendicular if $\cos \alpha_1 \cos \alpha_2 + \cos \beta_1 \cos \beta_2 + \cos \gamma_1 \cos \gamma_2 = 0.$

107. Distance d from a Point to a Line. Given the line

$$\frac{x - x_1}{l} = \frac{y - y_1}{m} = \frac{z - z_1}{n}$$

(where l, m, n are direction cosines) and a point $P_2(x_2, y_2, z_2)$. Then

$$d^2 = (x_2 - x_1)^2 + (y_2 - y_1)^2 + (z_2 - z_1)^2$$
$$- [l(x_2 - x_1) + m(y_2 - y_1) + n(z_2 - z_1)]^2.$$

108. Distance d between Two Nonintersecting Lines. Given the lines

$$\frac{x - x_1}{l_1} = \frac{y - y_1}{m_1} = \frac{z - z_1}{n_1}$$

and

$$\frac{x - x_2}{l_2} = \frac{y - y_2}{m_2} = \frac{z - z_2}{n_2}$$

(where l_1, m_1, n_1 and l_2, m_2, n_2 are direction cosines). Then

$$d = \begin{vmatrix} x_1 - x_2 & l_1 & l_2 \\ y_1 - y_2 & m_1 & m_2 \\ z_1 - z_2 & n_1 & n_2 \end{vmatrix} \times \frac{1}{\sin \theta},$$

where θ is the angle between the given lines.

109. Special Surfaces. The equation of the surface obtained by rotating the curve $f(x, y) = 0$, $z = 0$

about the x-axis is $f(x, \sqrt{y^2 + z^2}) = 0,$
about the y-axis is $f(\sqrt{z^2 + x^2}, y) = 0.$

The locus, in space of three dimensions, of the equation $f(x, y) = 0$ is a cylinder whose directrix is the curve $f(x, y) = 0$, $z = 0$, and whose elements are parallel to the z-axis.

Two other formulas may be obtained in each case by advancing x, y, z in cyclic order.

The locus of a homogeneous equation of the second degree in x, y, z, is a cone with vertex at the origin.

110. General Equation of the Second Degree. The locus defined by the equation

$$Ax^2 + By^2 + Cz^2 + Eyz + Fzx + Gxy + Lx + My + Nz + D = 0$$

is a quadric surface.

This equation either has no locus or represents a cylinder, a cone, an ellipsoid, a hyperboloid, a paraboloid, a sphere, a pair of planes, or a point.

SPECIAL QUADRIC SURFACES

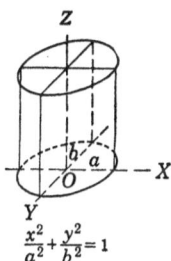

Elliptic Cylinder

$$\frac{x^2}{a^2} + \frac{y^2}{b^2} = 1$$

FIG. 20.

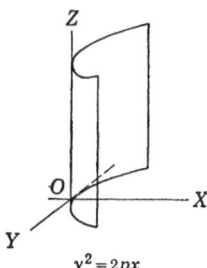

Parabolic Cylinder

$$y^2 = 2px$$

FIG. 21.

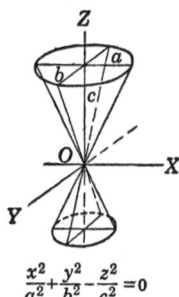

Elliptic Cone

$$\frac{x^2}{a^2} + \frac{y^2}{b^2} - \frac{z^2}{c^2} = 0$$

FIG. 22.

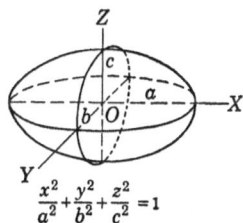

Ellipsoid

$$\frac{x^2}{a^2} + \frac{y^2}{b^2} + \frac{z^2}{c^2} = 1$$

FIG. 23.

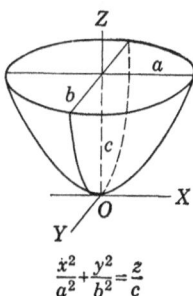

Elliptic Paraboloid

$$\frac{x^2}{a^2} + \frac{y^2}{b^2} = \frac{z}{c}$$

FIG 24.

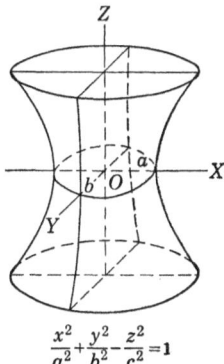

Hyperboloid of One Sheet

$$\frac{x^2}{a^2} + \frac{y^2}{b^2} - \frac{z^2}{c^2} = 1$$

FIG. 25.

SPECIAL QUADRIC SURFACES.—(*Continued*)

Hyperboloid of Two Sheets

Hyperbolic Paraboloid

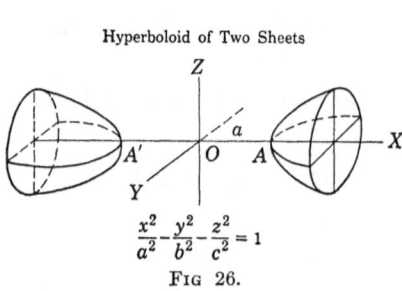

$$\frac{x^2}{a^2} - \frac{y^2}{b^2} - \frac{z^2}{c^2} = 1$$

FIG 26.

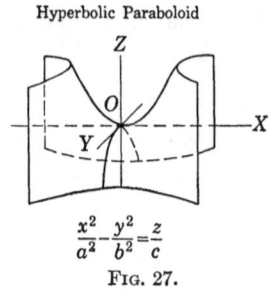

$$\frac{x^2}{a^2} - \frac{y^2}{b^2} = \frac{z}{c}$$

FIG. 27.

CURVES FOR REFERENCE

Circle

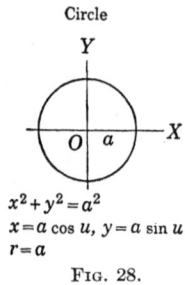

$$x^2 + y^2 = a^2$$
$$x = a \cos u, \; y = a \sin u$$
$$r = a$$

FIG. 28.

Ellipse

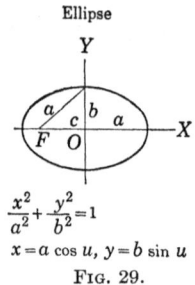

$$\frac{x^2}{a^2} + \frac{y^2}{b^2} = 1$$
$$x = a \cos u, \; y = b \sin u$$

FIG. 29.

Parabola

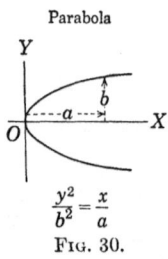

$$\frac{y^2}{b^2} = \frac{x}{a}$$

FIG. 30.

Equilateral Hyperbola

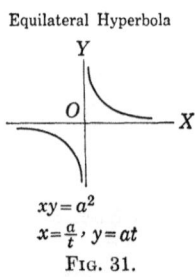

$$xy = a^2$$
$$x = \frac{a}{t}, \; y = at$$

FIG. 31.

Parabola

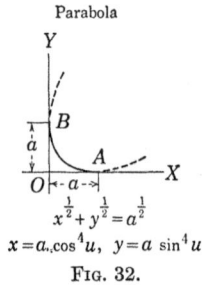

$$x^{\frac{1}{2}} + y^{\frac{1}{2}} = a^{\frac{1}{2}}$$
$$x = a \cos^4 u, \; y = a \sin^4 u$$

FIG. 32.

Hyperbola

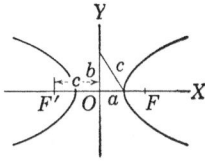

$$\frac{x^2}{a^2} - \frac{y^2}{b^2} = 1$$

$x = a \cosh u, \; y = b \sinh u$
$x = a \sec \varphi, \; y = a \tan \varphi$

Fig. 33.

Cubical Parabola

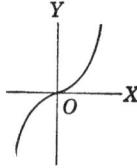

$a^2 y = x^3$

Fig. 34.

Semicubical Parabola

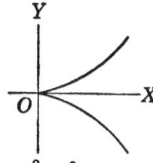

$ay^2 = x^3, \; a > 0$

Fig. 35.

Sine Curve

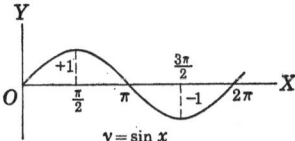

$y = \sin x$

Fig. 36.

Cosine Curve

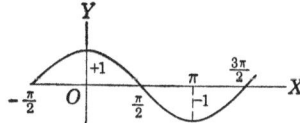

$y = \cos x$

Fig. 37.

Cosecant Curve

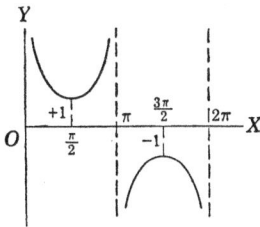

$y = \csc x$

Fig. 38.

Secant Curve

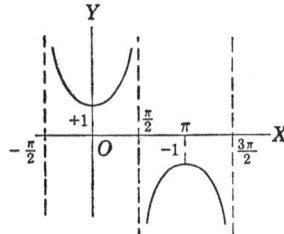

$y = \sec x$

Fig. 39.

Tangent Curve

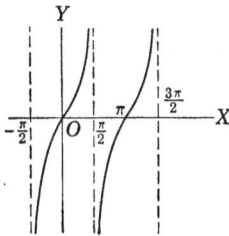

$y = \tan x$

Fig. 40.

Cotangent Curve

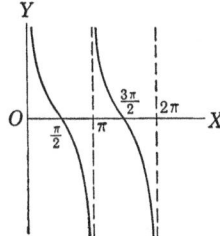

$y = \cot x$

Fig. 41.

Sine Wave

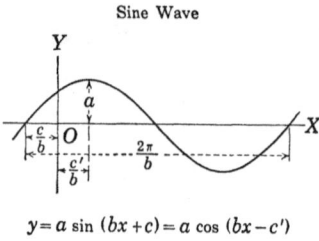

$$y = a \sin (bx + c) = a \cos (bx - c')$$
Fig. 42.

Oscillatory Wave of
Decreasing Amplitude

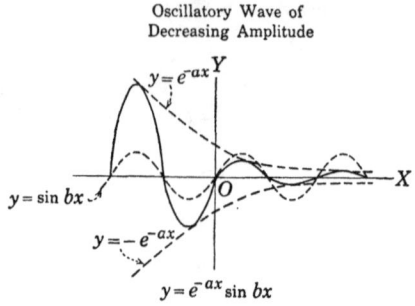

$$y = e^{-ax} \sin bx$$
Fig. 43.

Inverse Sine

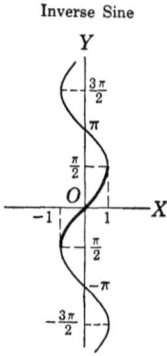

$$y = \sin^{-1} x = \text{arc sin } x$$
Fig. 44.

Inverse Cosine

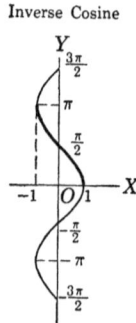

$$y = \cos^{-1} x = \text{arc cos } x$$
Fig. 45.

Inverse Cosecant

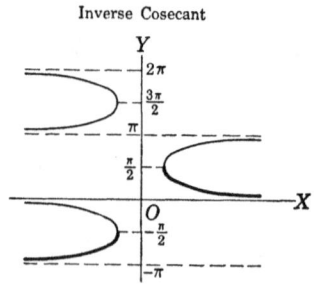

$$y = \csc^{-1} x = \text{arc csc } x$$
Fig. 46.

Inverse Secant

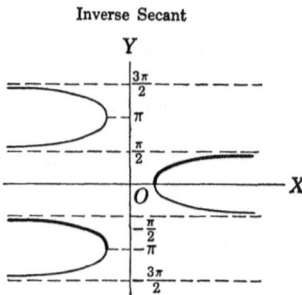

$$y = \sec^{-1} x = \text{arc sec } x$$
Fig. 47.

Inverse Tangent

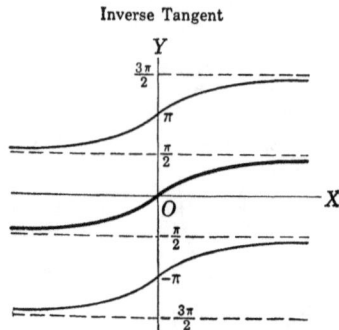

$$y = \tan^{-1} x = \text{arc tan } x$$
Fig. 48.

Inverse Cotangent

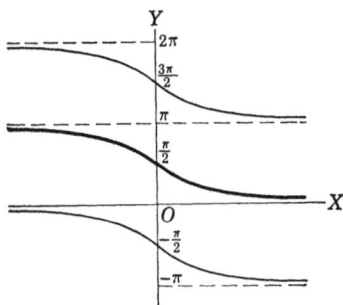

$y = \cot^{-1} x = $ arc cot x

FIG. 49.

Hyperbolic Sine

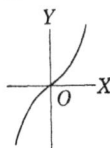

$y = \sinh x$

FIG. 50.

Hyperbolic Cosine

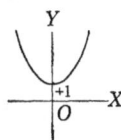

$y = \cosh x$

FIG. 51.

Hyperbolic Tangent

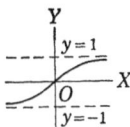

$y = \tanh x$

FIG. 52.

Hyperbolic Cosecant

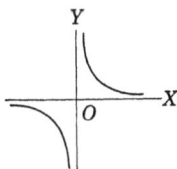

$y = \operatorname{csch} x$

FIG. 53.

Hyperbolic Secant

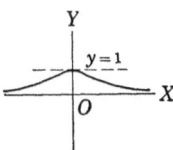

$y = \operatorname{sech} x$

FIG. 54.

Hyperbolic Cotangent

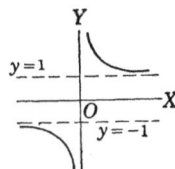

$y = \coth x$

FIG. 55.

Inverse Hyperbolic
Sine

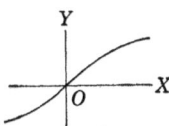

$y = \sinh^{-1} x$

FIG. 56.

Inverse Hyperbolic
Cosine

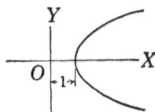

$y = \cosh^{-1} x$

FIG. 57.

Inverse Hyperbolic
Tangent

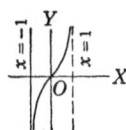

$y = \tanh^{-1} x$

FIG. 58.

Inverse Hyperbolic
Cosecant

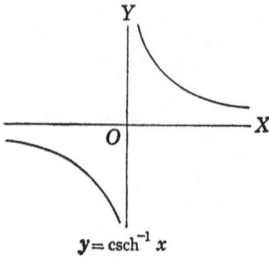

$y = \operatorname{csch}^{-1} x$

FIG. 59.

Inverse Hyperbolic
Secant

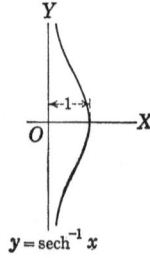

$y = \operatorname{sech}^{-1} x$

FIG. 60.

Inverse Hyperbolic
Cotangent

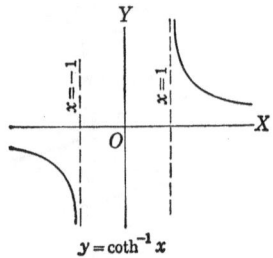

$y = \operatorname{coth}^{-1} x$

FIG. 61.

Gudermannian

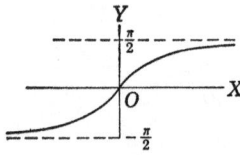

$y = \operatorname{gd} x$

FIG. 62.

Inverse
Gudermannian

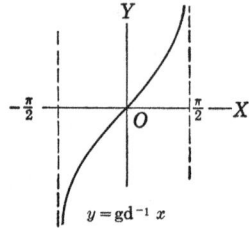

$y = \operatorname{gd}^{-1} x$

FIG. 63.

Limacon

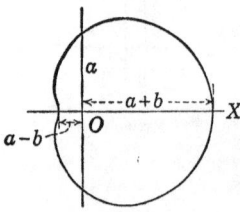

$r = a + b \cos \theta$
$a > b > 0$

FIG. 64.

Limacon

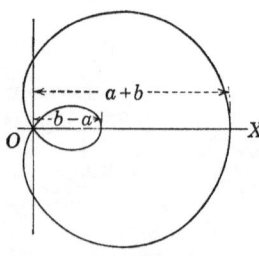

$r = a + b \cos \theta$
$b > a > 0$

FIG. 65.

Cardioid

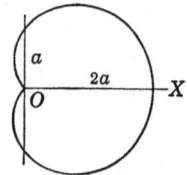

$r = a (1 + \cos \theta)$

FIG. 66.

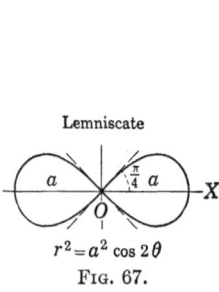

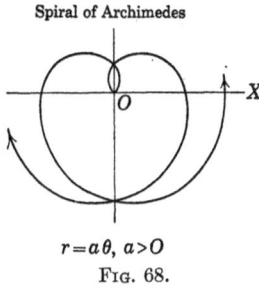

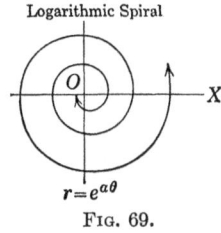

Spiral of Archimedes

Logarithmic Spiral

Lemniscate

$r^2 = a^2 \cos 2\theta$
Fig. 67.

$r = a\theta,\ a > 0$
Fig. 68.

$r = e^{a\theta}$
Fig. 69.

Circles

$r = 2a \cos \theta$
Fig. 70.

$r = a \cos \theta + b \sin \theta$
Fig. 71.

$r = 2a \sin \theta$
Fig. 72.

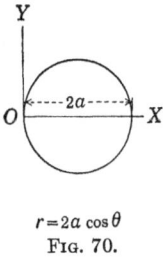

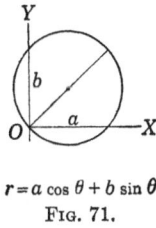

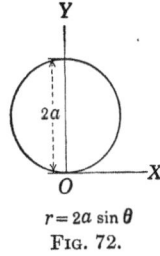

Three-leaved Roses

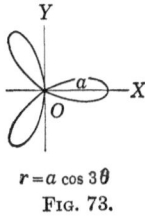

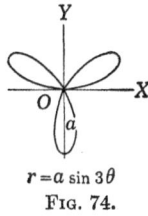

$r = a \cos 3\theta$
Fig. 73.

$r = a \sin 3\theta$
Fig. 74.

Four-leaved Roses

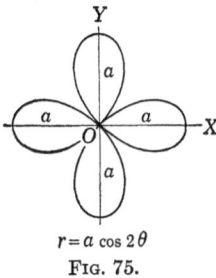

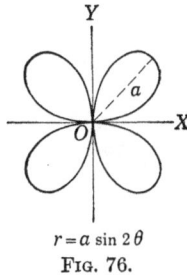

$r = a \cos 2\theta$
Fig. 75.

$r = a \sin 2\theta$
Fig. 76.

R o s e s

(*n* leaves if *n* is odd, 2*n* leaves if *n* is even)

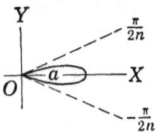

$r = a \cos n\theta$

FIG. 77.

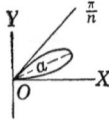

$r = a \sin n\theta$

FIG. 78.

Witch of Agnesi

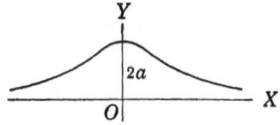

$y\,(x^2 + 4a^2) = 8a^3$

FIG. 79.

Cissoid of Diocles

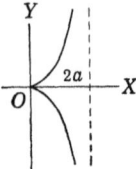

$y^2\,(2a - x) = x^3$
$x = 2a \sin^2 u$
$y = 2a \sin^2 u \tan u$

FIG. 80.

Folium of Descartes

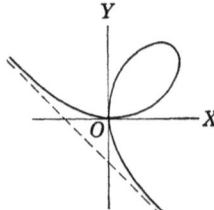

$x^3 + y^3 - 3axy = 0$
$x = \dfrac{3au}{1 + u^3}, \ y = \dfrac{3au^2}{1 + u^3}$

FIG. 81.

Strophoid

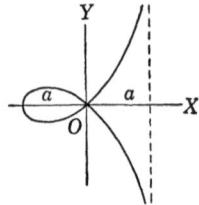

$y^2 = x^2 \dfrac{a + x}{a - x}$
$x = a \cos 2u$
$y = a \cos 2u \cot u$

FIG. 82.

Exponential Curve

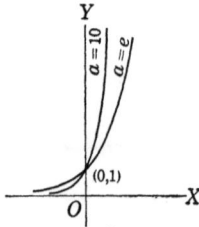

$y = a^x$

FIG. 83.

Probability Curve

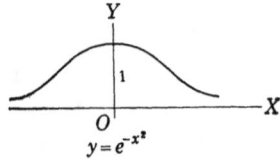

$y = e^{-x^2}$

FIG. 84.

Logarithmic Curve

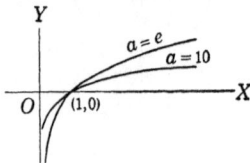

$y = \log_a x$

FIG. 85.

Cycloid

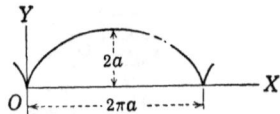

$x = a\,(\theta - \sin\theta), y = a\,(1 - \cos\theta)$

$x = a \text{ vers}^{-1} \dfrac{y}{a} - \sqrt{2ay - y^2}$

FIG. 86.

Hypocycloid

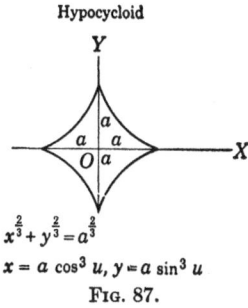

$$x^{\frac{2}{3}} + y^{\frac{2}{3}} = a^{\frac{2}{3}}$$

$x = a \cos^3 u, \; y = a \sin^3 u$

Fig. 87.

Catenary

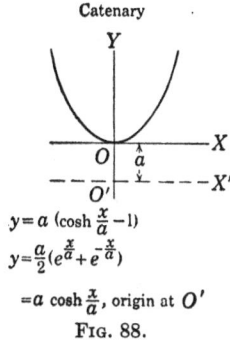

$y = a \left(\cosh \frac{x}{a} - 1\right)$

$y = \frac{a}{2}\left(e^{\frac{x}{a}} + e^{-\frac{x}{a}}\right)$

$= a \cosh \frac{x}{a}$, origin at O'

Fig. 88.

Tractrix

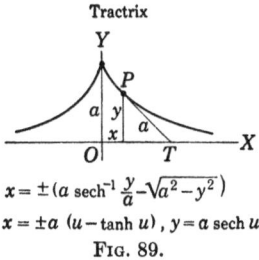

$x = \pm \left(a \operatorname{sech}^{-1} \frac{y}{a} - \sqrt{a^2 - y^2}\right)$

$x = \pm a \, (u - \tanh u), \; y = a \operatorname{sech} u$

Fig. 89.

Ellipse

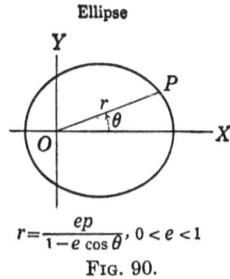

$r = \dfrac{ep}{1 - e \cos \theta}, \; 0 < e < 1$

Fig. 90.

Hyperbola

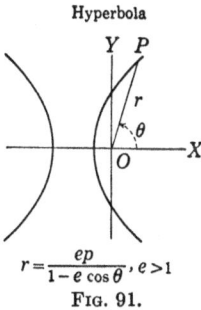

$r = \dfrac{ep}{1 - e \cos \theta}, \; e > 1$

Fig. 91.

Parabola

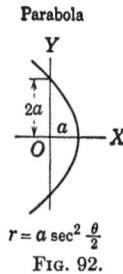

$r = a \sec^2 \frac{\theta}{2}$

Fig. 92.

Parabola

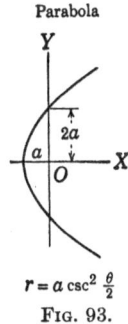

$r = a \csc^2 \frac{\theta}{2}$

Fig. 93.

DIFFERENTIAL CALCULUS

111. Derivative and Differential. Let $y = f(x)$. The *derivative* of y [or $f(x)$] with respect to x is denoted by $\dfrac{dy}{dx}$ and is defined to be

$$\frac{dy}{dx} = \lim_{\Delta x \to 0} \frac{\Delta y}{\Delta x} = \lim_{\Delta x \to 0} \frac{f(x + \Delta x) - f(x)}{\Delta x}.$$

This derivative is also denoted by y', $f'(x)$, or $D_x y$.

Higher derivatives of y with respect to x are defined and written as follows:

$$\frac{d^2y}{dx^2} = \frac{d}{dx}\left(\frac{dy}{dx}\right) = y'' = f''(x) = \frac{d}{dx}f'(x), \text{ (second derivative)};$$

$$\frac{d^ny}{dx^n} = \frac{d}{dx}\left(\frac{d^{n-1}y}{dx^{n-1}}\right) = y^{(n)} = f^{(n)}(x) = \frac{d}{dx}f^{(n-1)}(x), \text{ (nth derivative)}.$$

The value of $f^{(n)}(x)$ at $x = a$ is denoted by $f^{(n)}(a)$. The *rate of change* of $f(x)$ with respect to x at $x = a$ is $f'(a)$.

If the functional relation

$$y = f(x)$$

is represented by a curve (Fig. 94), the value of $\dfrac{dy}{dx}$ at any point

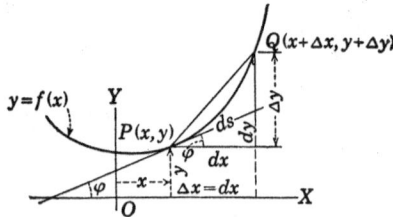

FIG. 94.

(x_0, y_0) on the curve is the *slope of the curve* (slope of the tangent to the curve) at that point.

The *differential* (dy) of y is

$$dy = y'\Delta x = y'\,dx = f'(x)\,dx.$$

The reciprocal of the derivative of y with respect to x is the derivative of x with respect to y:

$$\frac{dx}{dy} = 1 \div \frac{dy}{dx}.$$

If $y = f(u)$ and $u = g(x)$, then $\dfrac{dy}{dx} = \dfrac{dy}{du} \cdot \dfrac{du}{dx}$.

If $y = f(t)$ and $x = g(t)$, then $\dfrac{dy}{dx} = \dfrac{dy}{dt} \div \dfrac{dx}{dt} = \dfrac{f'(t)}{g'(t)}$.

112. Table of Differentials. In this table, u and v are functions of a variable x, or may themselves be independent variables; a and n denote constants; $e = 2.71828 \cdots$; and $\ln u = \log_e u$.

To obtain a table of derivatives, divide both members of each formula by du or by dx. Thus, formula 11 may be written

$$\frac{d}{du}(\sin u) = \cos u, \text{ or } \frac{d}{dx}(\sin u) = \cos u \frac{du}{dx}.$$

TABLE OF DIFFERENTIALS

1. $d\,a = 0.$

2. $d\,au = a\,du.$

3. $d\,(u + v) = du + dv.$

4. $d\,uv = u\,dv + v\,du.$

5. $d\dfrac{u}{v} = \dfrac{v\,du - u\,dv}{v^2}.$

6. $d\,u^n = nu^{n-1}\,du.$

7. $d\,e^u = e^u\,du.$

8. $d\,a^u = a^u \ln a\,du.$

9. $d\ln u = \dfrac{du}{u}.$

10. $d\log_a u = \dfrac{du}{u}\log_a e = \dfrac{du}{u \ln a}.$

11. $d\sin u = \cos u\,du.$

12. $d\cos u = -\sin u\,du.$

13. $d\tan u = \sec^2 u\,du.$

14. $d\cot u = -\csc^2 u\,du.$

15. $d\sec u = \sec u \tan u\,du.$

16. $d\csc u = -\csc u \cot u\,du.$

17. $d \operatorname{vers} u = \sin u\,du.$

18. $d\sin^{-1} u = \dfrac{du}{\sqrt{1 - u^2}}.$

19. $d\cos^{-1} u = -\dfrac{du}{\sqrt{1 - u^2}}.$

20. $d\tan^{-1} u = \dfrac{du}{1 + u^2}.$

21. $d\cot^{-1} u = -\dfrac{du}{1 + u^2}.$

22. $d\sec^{-1} u = \dfrac{du}{u\sqrt{u^2 - 1}}.$

23. $d\csc^{-1} u = -\dfrac{du}{u\sqrt{u^2 - 1}}.$

24. $d\sinh u = \cosh u\,du.$

25. $d\cosh u = \sinh u\,du.$

26. $d\tanh u = \operatorname{sech}^2 u\,du.$

27. $d\coth u = -\operatorname{csch}^2 u\,du.$

28. $d\operatorname{sech} u = -\operatorname{sech} u \tanh u\,du.$

29. $d\operatorname{csch} u = -\operatorname{csch} u \coth u\,du.$

30. $d\sinh^{-1} u = \dfrac{du}{\sqrt{u^2 + 1}}.$

31. $d\cosh^{-1} u = \dfrac{du*}{\sqrt{u^2 - 1}}.$

32. $d\tanh^{-1} u = \dfrac{du}{1 - u^2}.$

33. $d\coth^{-1} u = \dfrac{du}{1 - u^2}.$

34. $d\operatorname{sech}^{-1} u = -\dfrac{du\dagger}{u\sqrt{1 - u^2}}.$

35. $d \operatorname{gd} u = \operatorname{sech} u\,du.$

36. $d\operatorname{csch}^{-1} u = -\dfrac{du}{u\sqrt{1 + u^2}}.$

37. $d \operatorname{gd}^{-1} u = \sec u\,du.$

* $\cosh^{-1} u > 0$, see Fig. 57.

† $\operatorname{sech}^{-1} u > 0$, see Fig. 60.

38. $d \int_a^u f(u)du = f(u)du.$

39. $d \ln (u_1 u_2 \cdots u_n) = \dfrac{du_1}{u_1} + \dfrac{du_2}{u_2} + \cdots + \dfrac{du_n}{u_n}.$

40. $d(u_1 u_2 \cdots u_n) = (u_1 u_2 \cdots u_n)\left[\dfrac{du_1}{u_1} + \dfrac{du_2}{u_2} + \cdots + \dfrac{du_n}{u_n} \right].$

113. Tangent, Normal, Subtangent, and Subnormal, Rectangular Coordinates.

$$\tan \varphi = \frac{dy}{dx} = \text{slope of curve at } P,$$

AP = length of tangent (t) = $y \csc \varphi,$
PC = length of normal (n) = $y \sec \varphi,$

$$AB = \text{subtangent } (st) = y \cot \varphi = \frac{y}{y'},$$

$$BC = \text{subnormal } (sn) = y \tan \varphi = yy'.$$

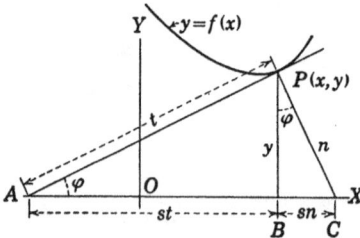

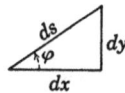

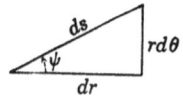

FIG. 95.   FIG. 96.   FIG. 97.

The *equation of the tangent line* to the curve $y = f(x)$ at the point (x_1, y_1) is

$$y - y_1 = f'(x_1)(x - x_1).$$

The *equation of the normal line* to the curve $y = f(x)$ at the point (x_1, y_1) is

$$f'(x_1)(y - y_1) = -(x - x_1).$$

114. Differential (ds) of Length of Arc. In rectangular coordinates (Fig. 96)

$$ds = \sqrt{(dx)^2 + (dy)^2} = \sqrt{1 + \left(\frac{dy}{dx}\right)^2}\, dx = \sqrt{1 + \left(\frac{dx}{dy}\right)^2}\, dy.$$

In polar coordinates (Fig. 97)

$$ds = \sqrt{(r\, d\theta)^2 + dr^2} = \sqrt{r^2 + \left(\frac{dr}{d\theta}\right)^2}\, d\theta = \sqrt{r^2\left(\frac{d\theta}{dr}\right)^2 + 1}\, dr.$$

If $x = f(t)$ and $y = g(t)$, then

$$ds = \sqrt{(dx)^2 + (dy)^2} = \sqrt{\left(\frac{dx}{dt}\right)^2 + \left(\frac{dy}{dt}\right)^2}\, dt.$$

For any curve $x = f(t)$, $y = g(t)$, $z = h(t)$

$$ds = \sqrt{(dx)^2 + (dy)^2 + (dz)^2} = \sqrt{\left(\frac{dx}{dt}\right)^2 + \left(\frac{dy}{dt}\right)^2 + \left(\frac{dz}{dt}\right)^2}\, dt.$$

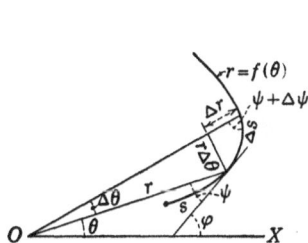
Fig. 98.

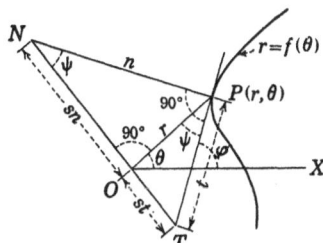
Fig. 99.

115. Tangent in Polar Coordinates.

$$\tan \psi = \frac{r\, d\theta}{dr} = \frac{r}{dr/d\theta}.$$

Slope of curve at $P = \tan \varphi = \tan (\psi + \theta)$.

116. Polar Tangent, Normal, Subtangent, and Subnormal.
PT = length of polar tangent $(t) = r \sec \psi$,
PN = length of polar normal $(n) = r \csc \psi$,

$$OT = \text{polar subtangent } (st) = r \tan \psi = r^2\frac{d\theta}{dr},$$

$$ON = \text{polar subnormal } (sn) = r \cot \psi = \frac{dr}{d\theta}.$$

117. Curvature and Radius of Curvature.   *Average curvature*

for arc $PQ = \dfrac{\Delta\varphi}{\Delta s}$.   The *curvature* (K) at P is the absolute value of

$\dfrac{d\varphi}{ds}$ at P (Fig. 100):

$$K = \left|\frac{d\varphi}{ds}\right| = \frac{|y''|}{(1 + y'^2)^{3/2}}.$$

The *radius of curvature* (*R*) at *P* is the reciprocal of *K* at *P*:

$$R = \frac{1}{K} = \left|\frac{ds}{d\varphi}\right| = \frac{(1 + y'^2)^{3/2}}{|y''|}.$$

If $x = g(y)$, then $K = \dfrac{|x''|}{(1 + x'^2)^{3/2}}.$

If $\begin{cases} x = f(t), \\ y = g(t), \end{cases}$ then $K = \dfrac{\left|\dfrac{dx}{dt}\dfrac{d^2y}{dt^2} - \dfrac{dy}{dt}\dfrac{d^2x}{dt^2}\right|}{[(dx/dt)^2 + (dy/dt)^2]^{3/2}}.$

If $r = f(\theta)$, then $K = \dfrac{\left|r^2 + 2\left(\dfrac{dr}{d\theta}\right)^2 - r\dfrac{d^2r}{d\theta^2}\right|}{[r^2 + (dr/d\theta)^2]^{3/2}}.$

The *center of curvature*, *C*, corresponding to any point *P* (Fig. 100) is the limiting position of the point of intersection of the normals at *P* and at a neighboring point *Q*, as *Q* is made to approach *P* along the curve. $CP = R$, and the coordinates (*u*, *v*) of the center of curvature are

(*a*) $\begin{cases} u = x - R \sin \varphi = x - \dfrac{dy}{d\varphi} = x - \dfrac{y'(1 + y'^2)}{y''}, \\ v = y + R \cos \varphi = y + \dfrac{dx}{d\varphi} = y + \dfrac{1 + y'^2}{y''}. \end{cases}$

The *evolute* of a curve is the locus of its centers of curvature. Equations (*a*) are parametric equations of the evolute of $y = f(x)$.

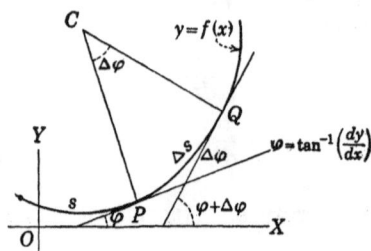

I. c. 100.

The normal to the curve at *P* is tangent to its evolute at *C*. Any one of the family of curves having a given curve as evolute is called an *involute* of that curve.

118. Partial Derivatives

(*a*) Let $u = f(x, y, \cdots)$. If all the variables except *x* are held fixed and $f(x, y, \ldots)$ is differentiated with respect to *x*, the *partial derivative* of $f(x, y, \ldots)$ with respect to *x* is obtained. It is denoted by

$$\frac{\partial u}{\partial x} \quad \text{or} \quad f_x(x, y, \ldots).$$

Similarly differentiation with respect to y, all the other variables being held fixed, gives

$$\frac{\partial u}{\partial y} \quad \text{or} \quad f_y(x, y, \ \cdots \ ), \text{ etc.}$$

Partial derivatives of higher orders are obtained in a similar manner and are denoted by

$$\frac{\partial^2 u}{\partial x^2} = \frac{\partial}{\partial x}\left(\frac{\partial u}{\partial x}\right) = f_{xx}, \qquad \frac{\partial^2 u}{\partial y^2} = \frac{\partial}{\partial y}\left(\frac{\partial u}{\partial y}\right) = f_{yy},$$

$$\frac{\partial^2 u}{\partial x \partial y} = \frac{\partial}{\partial x}\left(\frac{\partial u}{\partial y}\right) = f_{xy}.$$

If the derivatives concerned are continuous functions, the order of differentiation is immaterial: $f_{xy} = f_{yx}$.

(b) Let $u = f(x, y, \ \cdots \ )$ where
$x = g(r, s, \ \cdots \ ),\ y = h(r, s, \ \cdots \ ),\ \cdots \cdot$. Then

$$\frac{\partial u}{\partial r} = \frac{\partial u}{\partial x}\frac{\partial x}{\partial r} + \frac{\partial u}{\partial y}\frac{\partial y}{\partial r} + \cdots \ ,$$

$$\frac{\partial u}{\partial s} = \frac{\partial u}{\partial x}\frac{\partial x}{\partial s} + \frac{\partial u}{\partial y}\frac{\partial y}{\partial s} + \cdots \ .$$

119. Total Differential. (a) Let $u = f(x, \ y, \ \cdots \ )$. The total differential (du) of u is

$$du = \frac{\partial u}{\partial x}dx + \frac{\partial u}{\partial y}dy + \cdots \ .$$

(b) Let $u = f(x, \ y, \ \cdots \ )$, where $x = g(t),\ y = h(t),\ \cdots \cdot$. Then

$$\frac{du}{dt} = \frac{\partial u}{\partial x}\frac{dx}{dt} + \frac{\partial u}{\partial y}\frac{dy}{dt} + \cdots \ .$$

120. Differentiation of an Implicit Function. (a) Let y be defined implicitly as a function of x by

$$f(x, y) = 0.$$

Then

$$f_x(x, y)dx + f_y(x, y)dy = 0 \quad \text{and} \quad \frac{dy}{dx} = -\frac{f_x(x, y)}{f_y(x, y)}.$$

(b) Let u be defined implicitly as a function of x, y, z, . . . , by

$$f(u, x, y, z, \ldots) = 0.$$

Then

$$\frac{\partial f}{\partial x} + \frac{\partial f}{\partial u}\frac{\partial u}{\partial x} = 0 \quad \text{and} \quad \frac{\partial u}{\partial x} = -\frac{f_x(u, x, y, z, \cdots)}{f_u(u, x, y, z, \cdots)}.$$

121. Tangent Plane and Normal Line to a Surface. (a) If the equation of the surface is

$$z = f(x, y)$$

the equation of the plane tangent to the surface at the point (x_0, y_0, z_0) is

$$f_x(x_0, y_0)(x - x_0) + f_y(x_0, y_0)(y - y_0) - (z - z_0) = 0;$$

the equations of the line normal to the surface at this point are

$$\frac{x - x_0}{f_x(x_0, y_0)} = \frac{y - y_0}{f_y(x_0, y_0)} = \frac{z - z_0}{-1}.$$

(b) If the equation of the surface is

$$F(x, y, z) = 0,$$

the equation of the plane tangent to the surface at the point (x_0, y_0, z_0) is

$$F_x(x_0, y_0, z_0)(x - x_0) + F_y(x_0, y_0, z_0)(y - y_0)$$
$$+ F_z(x_0, y_0, z_0)(z - z_0) = 0;$$

the equations of the line normal to the surface at this point are

$$\frac{x - x_0}{F_x(x_0, y_0, z_0)} = \frac{y - y_0}{F_y(x_0, y_0, z_0)} = \frac{z - z_0}{F_z(x_0, y_0, z_0)}.$$

122. Tangent Line and Normal Plane to a Space Curve. (a) If the equations of the curve are

$$x = f(t), \qquad y = g(t), \qquad z = h(t),$$

the equations of the line tangent to the curve at the point (x_0, y_0, z_0) are

$$\frac{x - x_0}{f'(t_0)} = \frac{y - y_0}{g'(t_0)} = \frac{z - z_0}{h'(t_0)};$$

and the equation of the plane normal to the curve at this point is

$$f'(t_0)(x - x_0) + g'(t_0)(y - y_0) + h'(t_0)(z - z_0) = 0,$$

where $x_0 = f(t_0)$, $y_0 = g(t_0)$, $z_0 = h(t_0)$.

(b) If the equations of the curve are

$$y = f(x), \qquad z = g(x),$$

the equations of the line tangent to the curve at the point (x_0, y_0, z_0) are

$$y - y_0 = f'(x_0)(x - x_0), \qquad z - z_0 = g'(x_0)(x - x_0);$$

and the equation of the plane normal to the curve at this point is

$$x - x_0 + f'(x_0)(y - y_0) + g'(x_0)(z - z_0) = 0.$$

(c) If the equations of the curve are

$$f(x, y, z) = 0, \qquad g(x, y, z) = 0,$$

the equations of the line tangent to the curve at the point (x_0, y_0, z_0) are

$$\frac{x - x_0}{L(x_0, y_0, z_0)} = \frac{y - y_0}{M(x_0, y_0, z_0)} = \frac{z - z_0}{N(x_0, y_0, z_0)};$$

and the equation of the plane normal to the curve at this point is

$$L(x_0, y_0, z_0)(x - x_0) + M(x_0, y_0, z_0)(y - y_0) +$$
$$N(x_0, y_0, z_0)(z - z_0) = 0,$$

where

$$L = \begin{vmatrix} f_y & f_z \\ g_y & g_z \end{vmatrix}, \quad M = \begin{vmatrix} f_z & f_x \\ g_z & g_x \end{vmatrix}, \quad N = \begin{vmatrix} f_x & f_y \\ g_x & g_y \end{vmatrix}.$$

123. Law of the Mean.

$$f(b) - f(a) = (b - a)f'(x_0), \qquad a < x_0 < b.$$

124. Maclaurin's Series.

$$f(x) = f(0) + f'(0)x + \frac{f''(0)}{2!}x^2 + \frac{f'''(0)}{3!}x^3 + \cdots.$$

The remainder of this series after n terms is

$$R = \frac{f^{(n)}(x_0)}{n!}x^n, \qquad 0 < x_0 < x.$$

The necessary and sufficient condition that this series converge and represent the function $f(x)$ is that $\lim_{n \to \infty} R = 0$.

125. Taylor's Series for Functions of a Single Variable.

$$f(x) = f(a) + f'(a)(x - a) + \frac{f''(a)}{2!}(x - a)^2 + \cdots .$$

The remainder of this series after n terms is

$$R = \frac{f^{(n)}(x_0)}{n!}(x - a)^n, \qquad a < x_0 < x.$$

The necessary and sufficient condition that this series converge and represent the function $f(x)$ is that $\lim_{n \to \infty} R = 0$.

Taylor's series may also be written

$$f(a + h) = f(a) + f'(a)h + \frac{f''(a)}{2!}h^2 + \frac{f'''(a)}{3!}h^3 + \cdots .$$

126. Taylor's Series for Functions of Two Variables.

$$f(x, y) = f(a, b) + [f_x(a, b)(x - a) + f_y(a, b)(y - b)] \\ + \tfrac{1}{2}[f_{xx}(a, b)(x - a)^2 + 2f_{xy}(a, b)(x - a)(y - b) \\ + f_{yy}(a, b)(y - b)^2] + \cdots .$$

The nth term can be expressed symbolically as

$$\frac{1}{(n-1)!}\left[(x - a)\frac{\partial}{\partial x} + (y - b)\frac{\partial}{\partial y}\right]^{n-1}\left[f(x, y)\right]_{\substack{x = a \\ y = b}}$$

and the remainder after n terms as

$$R = \frac{1}{n!}\left[(x - a)\frac{\partial}{\partial x} + (y - b)\frac{\partial}{\partial y}\right]^{n}\left[f(x, y)\right]_{\substack{x = x_0, a < x_0 < x \cdot \\ y = y_0, b < y_0 < y}}$$

127. Maxima and Minima of Functions of One Variable.

Case I.   If $f'(a) = 0$ or ∞, the function $f(x)$ has a maximum or a minimum for $x = a$ according to whether the sign of $f'(x)$ changes from plus to minus or from minus to plus as x increases through $x = a$.

Case II.   If $f'(a) = f''(a) = \cdots = f^{(2n-1)}(a) = 0$,
$$f^{(2n)}(a) = k \neq 0,$$

the function $f(x)$ is

a maximum for $x = a$ if $k < 0$;
a minimum for $x = a$ if $k > 0$.

If the greatest and least values of $f(x)$ in an interval are desired, the maximum and minimum values of $f(x)$ within the interval must be compared with its values at the end points of the interval.

128. Points of Inflection. The curve $y = f(x)$ is

concave upward wherever $f''(x) > 0$;

concave downward wherever $f''(x) < 0$.

Case I.   If $f''(a) = 0$ and if $f''(x)$ changes sign as x varies continuously through $x = a$, the curve has a point of inflection at $x = a$.

Case II.   If $f''(a) = f'''(a) = \cdots = f^{(2n)}(a) = 0$,

$$f^{(2n+1)}(a) \neq 0,$$

the curve has a point of inflection at $x = a$.

129. Maxima and Minima of Functions of Two Variables. Let

$$z = f(x, y).$$

If for a pair of values $x = a, y = b$

$$\cdot \quad \frac{\partial z}{\partial x} = 0, \qquad \frac{\partial z}{\partial y} = 0 \qquad \text{and} \qquad \left(\frac{\partial^2 z}{\partial x \partial y}\right)^2 < \frac{\partial^2 z}{\partial x^2}\frac{\partial^2 z}{\partial y^2},$$

the function $f(x, y)$ has

a maximum for $x = a$, $\quad y = b \quad$ if $\quad \dfrac{\partial^2 z}{\partial x^2} < 0$;

a minimum for $x = a$, $\quad y = b \quad$ if $\quad \dfrac{\partial^2 z}{\partial x^2} > 0$.

130. Evaluation of Indeterminate Forms. Let $f(x)$ and $g(x)$ be two functions which either approach zero or become infinite as $x \to a$.

Case I.   If $\dfrac{f(a)}{g(a)} = \dfrac{0}{0}$ or $\dfrac{\infty}{\infty}$, use $\dfrac{f'(a)}{g'(a)}$ as the value of the fraction; if also $\dfrac{f'(a)}{g'(a)} = \dfrac{0}{0}$ or $\dfrac{\infty}{\infty}$, use $\dfrac{f''(a)}{g''(a)}$, and so on.

Case II.   If $f(a)\,g(a) = 0 \cdot \infty$, write $f(x)\,g(x) = \dfrac{f(x)}{\dfrac{1}{g(x)}}$ or $\dfrac{g(x)}{\dfrac{1}{f(x)}}$,

and proceed as in Case I.

Case III.   If $f(a) - g(a) = \infty - \infty$, reduce to Case I by combining the two terms.   This may be done by putting

$$f(x) - g(x) = \frac{\dfrac{1}{g(x)} - \dfrac{1}{f(x)}}{\dfrac{1}{f(x)\ g(x)}}.$$

Case IV.   If $f(a)^{g(a)} = 0^0$, ∞^0, or 1^∞, let $y = f(x)^{g(x)}$.   Then $\ln y = g(x) \ln f(x)$.   Evaluate $g(a) \ln f(a)$, using Case II.   If $g(a) \ln f(a) = m$, then use $f(a)^{g(a)} = e^m$.

The labor can frequently be lessened by writing the given expression as the product of (a factor which is determined for $x = a$) times (a factor which is indeterminate for $x = a$).   Sometimes it is easier to replace $f(x)$ and $g(x)$ by their expansions as series in powers of $(x - a)$.

INTEGRAL CALCULUS

131. Indefinite Integrals.   An *integral* of $f(x)$, denoted by $\int f(x)\ dx$, is any function $F(x)$ whose derivative is $f(x)$, or whose differential is $f(x)\ dx$.   In symbols

$$F(x) = \int f(x)\ dx \quad \text{if} \quad \frac{d\ F(x)}{dx} = f(x), \quad \text{or} \quad d\ F(x) = f(x)\ dx.$$

All the integrals of $f(x)$ are included in the expression

$$\int f(x)\ dx = F(x) + C,$$

where $F(x)$ is any particular integral of $f(x)$ and C is an arbitrary constant.

If u and v are functions of x and a is any constant, then

$$\int af(u)\ du = a \int f(u)\ du,$$

$$\int [f_1(u) + f_2(u) + \cdots + f_n(u)]\ du$$
$$= \int f_1(u)\ du + \int f_2(u)\ du + \cdots + \int f_n(u)\ du,$$
$$\int u\ dv = uv - \int v\ du.$$

132. Table of Fundamental Integrals.   The letter u denotes any function of an independent variable x, or may itself be an independent variable.   A constant of integration C should be added in each case.

TABLE OF FUNDAMENTAL INTEGRALS

1. $\int u^n \, du = \dfrac{u^{n+1}}{n+1}, n \neq -1.$ **2.** $\int \dfrac{du}{u} = \ln u.$

3. $\int e^u \, du = e^u.$ **4.** $\int a^u \, du = \dfrac{a^u}{\ln a}.$

5. $\int \sin u \, du = -\cos u.$ **6.** $\int \cos u \, du = \sin u.$

7. $\int \sec^2 u \, du = \tan u.$ **8.** $\int \csc^2 u \, du = -\cot u.$

9. $\int \sec u \tan u \, du = \sec u.$ **10.** $\int \csc u \cot u \, du = -\csc u.$

11. $\int \tan u \, du = \ln \sec u, \text{ or } -\ln \cos u.$

12. $\int \cot u \, du = \ln \sin u, \text{ or } -\ln \csc u.$

13. $\int \sec u \, du = \ln \tan \left(\dfrac{u}{2} + \dfrac{\pi}{4} \right) = \mathrm{gd}^{-1} u, \text{ or } \ln (\sec u + \tan u).$

14. $\int \csc u \, du = \ln \tan \dfrac{u}{2} = \mathrm{gd}^{-1} \left(u - \dfrac{\pi}{2} \right), \text{ or } \ln (\csc u - \cot u).$

15. $\int \sinh u \, du = \cosh u.$ **16.** $\int \cosh u \, du = \sinh u.$

17. $\int \mathrm{sech}^2 u \, du = \tanh u.$ **18.** $\int \mathrm{csch}^2 u \, du = -\coth u.$

19. $\int \mathrm{sech}\, u \tanh u \, du = -\mathrm{sech}\, u.$

20. $\int \mathrm{csch}\, u \coth u \, du = -\mathrm{csch}\, u.$

21. $\int \tanh u \, du = \ln \cosh u, \text{ or } -\ln \mathrm{sech}\, u.$

22. $\int \coth u \, du = \ln \sinh u, \text{ or } -\ln \mathrm{csch}\, u.$

23. $\int \mathrm{sech}\, u \, du = \mathrm{gd}\, u = \sin^{-1} \tanh u, \text{ or } 2 \tan^{-1} e^u.$

24. $\int \mathrm{csch}\, u \, du = \ln \tanh \dfrac{u}{2}.$

25. $\int \dfrac{du}{a^2 + u^2} = \dfrac{1}{a} \tan^{-1} \dfrac{u}{a}, \text{ or } -\dfrac{1}{a} \cot^{-1} \dfrac{u}{a}.$

TABLE OF FUNDAMENTAL INTEGRALS.—(Continued)

26. $\displaystyle\int \frac{du}{a^2 - u^2} = \frac{1}{a}\tanh^{-1}\frac{u}{a}$, or $\dfrac{1}{2a}\ln\dfrac{a+u}{a-u}$, if $u^2 < a^2$;

$\displaystyle\qquad = \frac{1}{a}\coth^{-1}\frac{u}{a}$, or $\dfrac{1}{2a}\ln\dfrac{u+a}{u-a}$, if $u^2 > a^2$.

27. $\displaystyle\int \frac{du}{\sqrt{a^2 - u^2}} = \sin^{-1}\frac{u}{a}$, or $-\cos^{-1}\frac{u}{a}$.

28. $\displaystyle\int \frac{du}{\sqrt{u^2 + a^2}} = \sinh^{-1}\frac{u}{a}$, or $\ln(u + \sqrt{u^2 + a^2})$.

29. $\displaystyle\int \frac{du}{\sqrt{u^2 - a^2}} = \cosh^{-1}\frac{u}{a}$, or $\ln(u + \sqrt{u^2 - a^2})$.

30. $\displaystyle\int \frac{du}{u\sqrt{u^2 - a^2}} = \frac{1}{a}\sec^{-1}\frac{u}{a}$, or $\frac{1}{a}\cos^{-1}\frac{a}{u}$.

31. $\displaystyle\int \frac{du}{u\sqrt{u^2 + a^2}} = -\frac{1}{a}\operatorname{csch}^{-1}\frac{u}{a}$, or $-\frac{1}{a}\sinh^{-1}\frac{a}{u}$.

32. $\displaystyle\int \frac{du}{u\sqrt{a^2 - u^2}} = -\frac{1}{a}\operatorname{sech}^{-1}\frac{u}{a}$, or $-\frac{1}{a}\cosh^{-1}\frac{a}{u}$.

133. Finding the Integral of a Function. To find the integral of a function $f(x)$, we must discover a function $F(x)$ whose derivative is recognized to be $f(x)$, or we must transform $f(x)$ into an expression whose integral is known. Some processes useful in making this transformation are discussed in the following paragraphs. See also the Table of Integrals, pages 90–123.

(a) *Integration by parts.* $\displaystyle\int u\,dv = uv - \int v\,du$.

Example. Find $I = \displaystyle\int x \sin x\,dx$.

Solution: Let $u = x$, $dv = \sin x\,dx$,
then $du = dx$, $v = -\cos x$.

$$I = -x\cos x + \int \cos x\,dx = -x\cos x + \sin x + C.$$

(b) *Integration by partial fractions.* If $f(x)$ and $g(x)$ are polynomials in x, then $\dfrac{f(x)}{g(x)}$ in a rational function of x. A rational function of x may be integrated by breaking it up into partial fractions (Art. 45) and integrating the result term by term. If the degree of the numerator is equal to or greater than that of the denominator, the fraction must be reduced to a mixed quantity in which the numerator of the fractional part is of lower degree than that of the denominator.

Example. Find $I = \int \dfrac{(x^4 - x^3 + 4x^2 - 6x + 4)dx}{x^3 + 4x}$.

Solution: $I = \int \left(x - 1 + \dfrac{4 - 2x}{x^3 + 4x} \right) dx = \int \left(x - 1 + \dfrac{1}{x} - \dfrac{x + 2}{x^2 + 4} \right) dx$

$$= \dfrac{x^2}{2} - x + \ln x - \ln \sqrt{x^2 + 4} - \tan^{-1} \dfrac{x}{2} + C.$$

(*c*) *Integration by substitution.* (1) An algebraic differential containing no radicals except fractional powers of $\dfrac{a + bx}{c + ex}$ can be reduced to a rational function of z by substituting

$$\dfrac{a + bx}{c + ex} = z^n, \qquad x = \dfrac{cz^n - a}{b - ez^n}, \qquad dx = \dfrac{n(bc - ae)z^{n-1}}{(b - ez^n)^2} dz,$$

where n is a common denominator of the fractional exponents of

$$\dfrac{a + bx}{c + ex}.$$

If $c = 1$, $e = 0$, then

$$a + bx = z^n, \qquad x = \dfrac{z^n - a}{b}, \qquad dx = \dfrac{nz^{n-1}}{b} dz.$$

Example. Find $I = \int \dfrac{dx}{(2 + 3x)^{\frac{1}{2}} + (2 + 3x)^{\frac{2}{3}}}$.

Solution: Let $2 + 3x = z^6$, then $dx = 2z^5\, dz$.

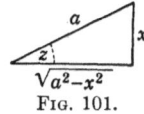
FIG. 101.

$I = \int \dfrac{2z^5\, dz}{z^3 + z^4} = \int \left(2z - 2 + \dfrac{2}{z + 1} \right) dz = z^2 - 2z + 2 \ln (z + 1)$

$= (2 + 3x)^{\frac{1}{3}} - 2(2 + 3x)^{\frac{1}{6}} + 2 \ln [(2 + 3x)^{\frac{1}{6}} + 1] + C.$

(2) Useful substitutions for differentials involving $\sqrt{a^2 - x^2}$ are

$$x = a \sin z, \qquad dx = a \cos z\, dz, \qquad \sqrt{a^2 - x^2} = a \cos z;$$
$$x = a \tanh z, \qquad dx = a \operatorname{sech}^2 z\, dz, \qquad \sqrt{a^2 - x^2} = a \operatorname{sech} z;$$
$$x = a \operatorname{sech} z, \qquad dx = -a \operatorname{sech} z \tanh z\, dz, \qquad \sqrt{a^2 - x^2} = a \tanh z.$$

Example. Find $I = \int \sqrt{a^2 - x^2}\, dx$.

Solution: Let $x = a \sin z$, then

$$I = a^2 \int \cos^2 z\, dz = \dfrac{a^2}{2} \int (\cos 2z + 1)dz = \dfrac{a^2}{2} (\sin z \cos z + z)$$

$$= \dfrac{x}{2} \sqrt{a^2 - x^2} + \dfrac{a^2}{2} \sin^{-1} \dfrac{x}{a} + C \text{ (see Fig. 101).}$$

(3) Useful substitutions for differentials involving $\sqrt{a^2 + x^2}$ are

$$x = a \sinh z, \quad dx = a \cosh z \, dz, \qquad \sqrt{a^2 + x^2} = a \cosh z;$$
$$x = a \tan z, \quad dx = a \sec^2 z \, dz, \qquad \sqrt{a^2 + x^2} = a \sec z;$$
$$x = a \operatorname{csch} z, \quad dx = -a \operatorname{csch} z \coth z \, dz, \quad \sqrt{a^2 + x^2} = a \coth z.$$

Example. Find $I = \int \sqrt{a^2 + x^2} \, dx.$

Solution: Let $x = a \sinh z$, then

$$I = a^2 \int \cosh^2 z \, dz = \frac{a^2}{2} \int (\cosh 2z + 1) dz = \frac{a^2}{2}(\sinh z \cosh z + z)$$

$$= \frac{x}{2} \sqrt{a^2 + x^2} + \frac{a^2}{2} \sinh^{-1} \frac{x}{a} + C.$$

(4) Useful substitutions for differentials involving $\sqrt{x^2 - a^2}$ are

$$x = a \cosh z, \quad dx = a \sinh z \, dz, \qquad \sqrt{x^2 - a^2} = a \sinh z;$$
$$x = a \coth z, \quad dx = -a \operatorname{csch}^2 z \, dz, \qquad \sqrt{x^2 - a^2} = a \operatorname{csch} z;$$
$$x = a \sec z, \quad dx = a \sec z \tan z \, dz, \qquad \sqrt{x^2 - a^2} = a \tan z.$$

Example. Find $I = \int \dfrac{dx}{x^3 \sqrt{x^2 - a^2}}.$

Solution: Let $x = a \sec z$, then

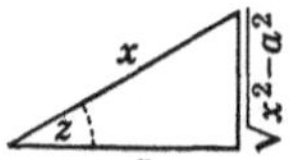

$$I = \frac{1}{a^3} \int \cos^2 z \, dz = \frac{1}{2a^3}(\sin z \cos z + z)$$

$$= \frac{\sqrt{x^2 - a^2}}{2a^2 x^2} + \frac{1}{2a^3} \sec^{-1} \frac{x}{a} + C \text{ (see Fig. 102).}$$

FIG. 102.

(5) If the integrand is a rational function of x and $\sqrt{ax^2 + bx + c}$, $f(x, \sqrt{ax^2 + bx + c})$, $a > 0$, the differential can be transformed into one which is rational in z by the substitution

$$x = \frac{z^2 - c}{2z\sqrt{a} + b}, \quad \text{whence} \quad dx = \frac{2z^2\sqrt{a} + 2bz + 2c\sqrt{a}}{(2z\sqrt{a} + b)^2} dz,$$

$$\sqrt{ax^2 + bx + c} = \frac{z^2\sqrt{a} + bz + c\sqrt{a}}{2z\sqrt{a} + b},$$

$$z = \sqrt{ax^2 + bx + c} + x\sqrt{a}.$$

The rational function obtained can be integrated by partial fractions.

The substitution $x = z - \dfrac{b}{2a}$, $dx = dz$, reduces the integrand

to $f\left(z - \dfrac{b}{2a}, \ \sqrt{a}\sqrt{z^2 + \dfrac{4ac - b^2}{4a^2}}\right)$ which can be rationalized by (3) or (4).

(6) If the integrand is a rational function of x and $\sqrt{-ax^2 + bx + c}$, $f(x, \ \sqrt{-ax^2 + bx + c})$, $a > 0$, the differential can be transformed into one which is rational in z by the substitution

$$x = \frac{\dfrac{b + D}{2} + \dfrac{b - D}{2a}z^2}{a + z^2}, \qquad dx = \frac{-2Dz\,dz}{(a + z^2)^2}, \qquad D = \sqrt{b^2 + 4ac}.$$

$$\sqrt{-ax^2 + bx + c} = \frac{Dz}{a + z^2}, \qquad z = \frac{\sqrt{-ax^2 + bx + c}}{x - \dfrac{b - D}{2a}}.$$

The rational function of z thus obtained can be integrated by partial fractions.

The substitution $x = z + \dfrac{b}{2a}$, $dx = dz$, reduces the integrand

to $f\left(z + \dfrac{b}{2a}, \ \sqrt{a}\sqrt{\dfrac{b^2 + 4ac}{4a^2} - z^2}\right)$ which can be rationalized by (2).

(7) If the integrand is a rational function of the trigonometric functions of x only, the differential can be transformed into one which is rational in z by the substitution

$$\tan \frac{x}{2} = z, \qquad \text{whence} \qquad dx = \frac{2dz}{1 + z^2},$$

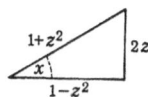

FIG. 103.

and the trigonometric functions of x in terms of z can be read from Fig. 103.

Example. Find $I = \displaystyle\int \frac{dx}{5 + 3\cos x}$.

Making the substitutions indicated above and simplifying, we have

$$I = \int \frac{dz}{4 + z^2} = \tfrac{1}{2}\tan^{-1}\frac{z}{2} = \tfrac{1}{2}\tan^{-1}\left(\tfrac{1}{2}\tan\frac{x}{2}\right) + C.$$

(8) If the integrand is a rational function of the hyperbolic functions of x only, the differential can be transformed into one which is rational in z, by the substitution

$$\tanh \frac{x}{2} = z, \qquad \text{whence} \qquad dx = \frac{2dz}{1 - z^2}, \text{ and}$$

$$\sinh x = \frac{2z}{1 - z^2}, \qquad \cosh x = \frac{1 + z^2}{1 - z^2}, \qquad \tanh x = \frac{2z}{1 + z^2}.$$

Example. Find $I = \displaystyle\int \frac{dx}{\sinh x + \tanh x}$. Making these substitutions, we

have $I = \displaystyle\int \frac{(1 + z^2)dz}{2z} = \frac{1}{2} \ln z + \frac{1}{4}z^2 = \frac{1}{2} \ln \tanh \frac{x}{2} + \frac{1}{4} \tanh^2 \frac{x}{2} + C.$

(d) *Integration by series.* If the integrand can be expressed as a convergent series in powers of x, and if the result of integrating this series term by term is also a convergent series, this last series may, within its interval of convergence, be used as the value of the integral (see Art. 7).

134. The Definite Integral. If $F(x) = \int f(x)dx$, the *definite integral* of $f(x)dx$ from $x = a$ to $x = b$ is denoted by $\int_a^b f(x)dx$, and has the value

$$\int_a^b f(x)dx = F(b) - F(a).$$

Also $\displaystyle\int_a^b f(x)dx = \lim_{n \to \infty} [f(x_1)\Delta x + f(x_2)\Delta x + \cdots + f(x_n)\Delta x],$

where $\Delta x = \dfrac{b - a}{n}$, and $x_1, x_2, \cdots, x_n$ are values of x, one in each of the n intervals into which $b - a$ is divided.

$$\int_a^b f(x)dx = -\int_b^a f(x)dx; \qquad \int_a^b f(x)dx = \int_a^c f(x)dx + \int_c^b f(x)dx.$$

$$\frac{d}{dx}\int_a^x f(x)dx = f(x); \qquad \frac{d}{d\alpha}\left[\int_a^b f(x, \alpha)dx\right] = \int_a^b \frac{\partial}{\partial\alpha}f(x, \alpha)dx.$$

$$\int_a^b \int_{f(x)}^{g(x)} \varphi(x, y)dy \; dx = \int_a^b \left[\int_{f(x)}^{g(x)} \varphi(x, y)dy\right]dx,$$

where x is held constant in the evaluation of the integral enclosed by brackets [ ].

135. Change of Limits. Let the variable of integration be changed by the substitution $x = g(y)$, where $y = \varphi(x)$. Then

$$\int_a^b f(x)dx = \int_{\varphi(a)}^{\varphi(b)} f[g(y)]g'(y)dy,$$

provided $g(y)$ and $\varphi(x)$ are single-valued functions of y and x, respectively.

Example. Find $I = \int_0^a x^2 \sqrt{a^2 - x^2}\, dx$. Let $x = a \sin y$, whence $y = \sin^{-1} \dfrac{x}{a}$, single valued for principal values of the inverse sine. When $x = 0$, $y = \sin^{-1} 0 = 0$; when $x = a$, $y = \sin^{-1} 1 = \dfrac{\pi}{2}$. Therefore

$$I = a^4 \int_0^{\frac{\pi}{2}} \sin^2 y \cos^2 y\, dy = \frac{\pi a^4}{16}.$$

136. Plane Area (A).*

(a) By single integration (see Figs. 104, 105, 106).

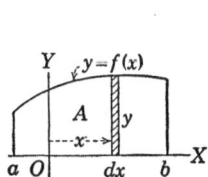
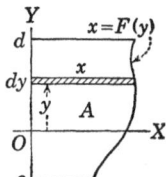
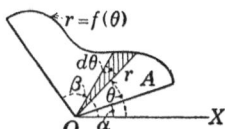

Fig. 104.   Fig. 105.   Fig. 106.

$$A = \int_a^b y\, dx \qquad A = \int_c^d x\, dy \qquad A = \tfrac{1}{2}\int_\alpha^\beta r^2\, d\theta$$
$$= \int_a^b f(x)\,dx. \qquad = \int_c^d F(y)\,dy. \qquad = \tfrac{1}{2}\int_\alpha^\beta f^2(\theta)\,d\theta.$$

(b) By double integration (see Figs. 107, 108, 109).

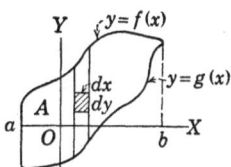
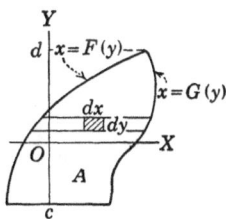
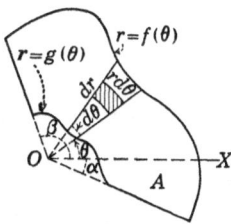

Fig. 107.   Fig. 108.   Fig. 109.

$$A = \int_a^b \int_{g(x)}^{f(x)} dy\, dx. \qquad A = \int_c^d \int_{F(y)}^{G(y)} dx\, dy. \qquad A = \int_\alpha^\beta \int_{g(\theta)}^{f(\theta)} r\, dr\, d\theta.$$

* If the quantity under the integral sign changes its algebraic sign between the limits of integration, break the interval of integration into sub-intervals in each of which this quantity does not change sign, and add the numerical values of the resulting integrals.

137. Length of Arc (s) of a Plane Curve.

$$s = \int_a^b \sqrt{1 + \left(\frac{dy}{dx}\right)^2}\, dx \qquad s = \int_\alpha^\beta \sqrt{r^2 + \left(\frac{dr}{d\theta}\right)^2}\, d\theta$$

$$= \int_c^d \sqrt{1 + \left(\frac{dx}{dy}\right)^2}\, dy. \qquad = \int_{r_1}^{r_2} \sqrt{1 + r^2\left(\frac{d\theta}{dr}\right)^2}\, dr.$$

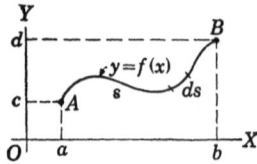

FIG. 110.

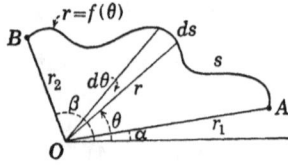

FIG. 111.

If

$$x = f(t), \qquad y = g(t),$$

the length of the curve from $t = t_1$ to $t = t_2$ is

$$s = \int_{t_1}^{t_2} \sqrt{f'^2(t) + g'^2(t)}\, dt.$$

138. Length of Arc of a Space Curve.   The length (s) of the curve

$$x = f(t), \qquad y = g(t), \qquad z = h(t)$$

from $t = t_1$ to $t = t_2$ is

$$s = \int_{t_1}^{t_2} \sqrt{f'^2(t) + g'^2(t) + h'^2(t)}\, dt.$$

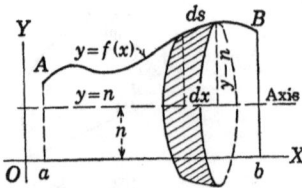

FIG. 112.

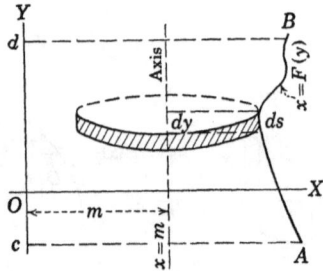

FIG. 113.

139. Area of Surface (S) of a Solid.   (a) Surface of revolution.

$$S = 2\pi \int_a^b (y - n)\sqrt{1 + \left(\frac{dy}{dx}\right)^2}\, dx \qquad \text{(Fig. 112)}$$

$$= 2\pi \int_{f(a)}^{f(b)} (y - n)\sqrt{1 + \left(\frac{dx}{dy}\right)^2}\, dy.$$

$$S = 2\pi \int_c^d (x - m)\sqrt{1 + \left(\frac{dx}{dy}\right)^2}\, dy \qquad \text{(Fig. 113)}$$

$$= 2\pi \int_{F(c)}^{F(d)} (x - m)\sqrt{1 + \left(\frac{dy}{dx}\right)^2}\, dx.$$

If the equation of the generating curve AB is $r = f(\theta)$, then

$$S = 2\pi \int_{\theta_B}^{\theta_A} (r \sin \theta - n)\sqrt{r^2 + \left(\frac{dr}{d\theta}\right)^2}\, d\theta. \qquad \text{(Fig. 112)}$$

$$S = 2\pi \int_{\theta_A}^{\theta_B} (r \cos \theta - m)\sqrt{r^2 + \left(\frac{dr}{d\theta}\right)^2}\, d\theta. \qquad \text{(Fig. 113)}$$

(b) *Any surface* $z = f(x, y)$.

$$S = \int\!\!\int_A \left[1 + \left(\frac{\partial z}{\partial x}\right)^2 + \left(\frac{\partial z}{\partial y}\right)^2\right]^{\frac{1}{2}} dy\, dx,$$

where the integration is to extend over the projection A of the surface on the xy-plane.

140. Volume (V) of a Solid.   (a) *Solid of revolution.*

(1) The volume generated by revolving the area bounded by the curve $y = f(x)$ and the lines $x = a$, $x = b$, $y = n$ about the line $y = n$ (see Fig. 112) is

$$V = \pi \int_a^b (y - n)^2 dx = \pi \int_a^b [f(x) - n]^2 dx.$$

(2) The volume generated by revolving the area bounded by the curve $x = F(y)$ and the lines $y = c$, $y = d$, $x = m$ about the line $x = m$ (see Fig. 113) is

$$V = \pi \int_c^d (x - m)^2 dy$$

$$= \pi \int_c^d [F(y) - m]^2 dy.$$

(b) *Solid with known parallel cross sections.* Let A_x (Fig. 114) be the area of the cross section of a solid made by a plane perpendicular to the x-axis.   Then

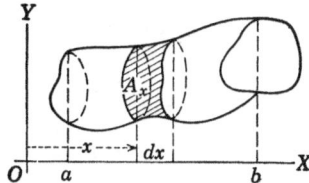

Fig. 114.

$$V = \int_a^b A_x\, dx.$$

This formula is often convenient when A_x can be readily expressed as a function of x.

(c) *Any solid.*   (1) By double integration.   If the volume lies under the surface $z = f(x, y)$, then

$$V = \int_A \int z \, dy \, dx = \int_A \int f(x, y) dy \, dx,$$

where A is the projection of the surface on the xy-plane.

(2) By triple integration (see Art. 99)

Rectangular coordinates $V = \int \int \int_V dz \, dy \, dx.$

Cylindrical coordinates $V = \int \int \int_V r' \, dr' \, d\theta \, dz.$

Spherical coordinates $V = \int \int \int_V r^2 \sin \phi \, d\theta \, d\phi \, dr.$

141. Mass (m) of a Body.*   If the density of a body is ρ, its mass is

$$m = \int dm,$$

where $dm = \rho \, ds, \, \rho \, dA, \, \rho \, dS,$ or $\rho \, dV.$

142. Statical Moment (M).   First Moment.*

(a) *Plane configuration.*

About OY $M_y = \int x \, dm = \int r \cos \theta \, dm.$

About OX $M_x = \int y \, dm = \int r \sin \theta \, dm.$

About O $M_o = \int \sqrt{x^2 + y^2} \, dm = \int r \, dm.$

(b) *Space configuration.*

With respect to the yz-plane $M_{yz} = \int x \, dm.$

With respect to the zx-plane $M_{zx} = \int y \, dm.$

With respect to the xy-plane $M_{xy} = \int z \, dm.$

143. Coordinates ($\bar{x}, \bar{y}, \bar{z}$) of Centroid or Center of Gravity.*

(a) *Plane configuration.*

$$\bar{x} = \frac{M_y}{m} = \frac{\int x \, dm}{m} = \frac{\int r \cos \theta \, dm}{m},$$

$$\bar{y} = \frac{M_x}{m} = \frac{\int y \, dm}{m} = \frac{\int r \sin \theta \, dm}{m},$$

(b) *Space configuration.*

$$\bar{x} = \frac{M_{yz}}{m} = \frac{\int x \, dm}{m}, \quad \bar{y} = \frac{M_{zx}}{m} = \frac{\int y \, dm}{m}, \quad \bar{z} = \frac{M_{xy}}{m} = \frac{\int z \, dm}{m}.$$

* The integration indicated by a single integral sign may involve a single, a double, or a triple integration, depending on the choice of the element. Limits of integration are to be supplied.

(c) *Composite body.*

$$\bar{x} = \frac{\bar{x}_1 m_1 + \bar{x}_2 m_2 + \cdots + \bar{x}_n m_n}{m_1 + \quad m_2 + \cdots + \quad m_n},$$

$$\bar{y} = \frac{\bar{y}_1 m_1 + \bar{y}_2 m_2 + \cdots + \bar{y}_n m_n}{m_1 + \quad m_2 + \cdots + \quad m_n},$$

$$\bar{z} = \frac{\bar{z}_1 m_1 + \bar{z}_2 m_2 + \cdots + \bar{z}_n m_n}{m_1 + \quad m_2 + \cdots + \quad m_n},$$

where $(\bar{x}_i, \bar{y}_i, \bar{z}_i)$ are the coordinates of the centroid of m_i.

144. Moment of Inertia (I). Second Moment.

(a) *Plane configuration.*

About OY $\quad I_y = \int x^2 \, dm = \int r^2 \cos^2 \theta \, dm.$

About OX $\quad I_x = \int y^2 \, dm = \int r^2 \sin^2 \theta \, dm.$

About O $\quad\quad I_o = I_x + I_y = \int (x^2 + y^2) \, dm = \int r^2 \, dm.$

(b) *Space configuration.*

With respect to the yz-plane $\quad I_{yz} = \int x^2 \, dm.$

With respect to the zx-plane $\quad I_{zx} = \int y^2 \, dm.$

With respect to the xy-plane $\quad I_{xy} = \int z^2 \, dm.$

About OX $\quad\quad\quad\quad\quad\quad I_x = I_{zx} + I_{xy} = \int (y^2 + z^2) dm.$

About OY $\quad\quad\quad\quad\quad\quad I_y = I_{yz} + I_{xy} = \int (x^2 + z^2) dm.$

About OZ $\quad\quad\quad\quad\quad\quad I_z = I_{yz} + I_{zx} = \int (x^2 + y^2) dm.$

(c) *Theorem of parallel axes.* If L is any line in space and L_g is a line parallel to L, passing through the centroid of the body, then

$$I_L = I_{L_g} + d^2 m,$$

where d is the distance between the two parallel lines L and L_g.

145. Radius of Gyration (k). The length k defined by the equation

$$k^2 m = I$$

is the *radius of gyration* of the quantity m with respect to the line or plane of reference.

(a) *Composite body.*

$$k^2 = \frac{k_1^2 m_1 + k_2^2 m_2 + \cdots + k_n^2 m_n}{m_1 + \quad m_2 + \cdots + \quad m_n},$$

where k_i is the radius of gyration of m_i with respect to the line or plane of reference.

(b) *Theorem of parallel axes.* If L is any line in space and L_g is a line parallel to L, passing through the centroid of m, then

$$k_L^2 = k_{L_g}^2 + d^2,$$

where d is the distance between the parallel lines L and L_g.

146. Product of Inertia (P). (a) *Plane configuration.*
With respect to the x- and y-axes $P_{xy} = \int xy\, dm.$
  (b) *Space configuration.*
With respect to the xz- and yz-planes $P_{xy} = \int xy\, dm.$
With respect to the xz- and xy-planes $P_{yz} = \int yz\, dm.$
With respect to the yz- and xy-planes $P_{zz} = \int zx\, dm.$
  (c) *Theorem of parallel axes for a plane area (A).*

$$P_{xy} = \bar{P}_{xy} + \bar{x}\bar{y}A,$$

where P_{xy} is the product of inertia of A with respect to any pair of rectangular axes in the plane of A and $\bar{P}_{xy}$ is the product of inertia with respect to a pair of parallel axes through the centroid ($\bar{x}$, $\bar{y}$) of the area A.

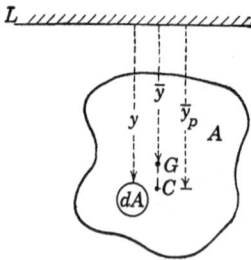

FIG. 115.

147. Pressure. Moment of Pressure. Center of Pressure. (a) Let the plane area A be vertically submerged in a liquid of constant specific weight w, intersecting the surface in the line L.

Let $\bar{y}$ = distance from L to G, the centroid of A,
  $\bar{y}_p$ = distance from L to C, the center of pressure on A,
  k_L^2 = squared radius of gyration of A about L,
  k_g^2 = squared radius of gyration of A about a line through G parallel to L.

Pressure against A $\displaystyle P = \int_A wy\, dA = w\bar{y}A.$

Moment of P about L $\displaystyle (M_p)_L = \int_A wy^2\, dA = wk_L^2A.$

Depth of center of pressure $\displaystyle \bar{y}_p = \frac{\int y^2\, dA}{\int y\, dA} = \frac{k_L^2}{\bar{y}} = \bar{y} + \frac{k_g^2}{\bar{y}}.$

(b) Let the plane of A make an angle α with the horizontal.

Pressure against A $\displaystyle P = \int_A wy \sin \alpha\, dA = w\bar{y}A \sin \alpha.$

Moment of P about L $\displaystyle (M_p)_L = \int_A wy^2 \sin \alpha\, dA = wk_L^2 A \sin \alpha.$

Distance from L to

center of pressure $\bar{y}_p = \dfrac{\int y^2 \, dA}{\int y \, dA} = \dfrac{k_L^2}{\bar{y}} = \bar{y} + \dfrac{k_{\bar{g}}^2}{\bar{y}}.$

Since $\bar{y} \sin \alpha$ is the depth of the centroid of the submerged area, the total pressure is the product of the specific weight of the liquid, the submerged area, and the depth of the centroid of this area.

148. Work (U) Done by a Force (F).   Let the force move a particle from position $s = s_0$ to $s = s_1$, where s is length of path measured for a fixed point on the path.

(a) Force constant and in the direction of motion of the particle.

$$U = F(s_1 - s_0).$$

(b) Force constant in magnitude and making a constant angle θ with the path of the particle

$$U = F(s_1 - s_0) \cos \theta.$$

(c) Force variable in magnitude, or in direction, or in both magnitude and direction

$$U = \int_{s_0}^{s_1} F \cos \theta \, ds,$$

where θ is the angle the line of action of the force makes with the path of the particle.

149. Approximate Integration.   (a) *Simpson's Rule.*   Let the interval of integration from $x = a$ to $x = b$ be divided into an even number of equal intervals.   Let $h = \dfrac{b - a}{n}$ denote the length of each interval and denote by $x_1, x_2, \ldots, x_{n-1}$ the points of subdivision of the interval of integration.   Simpson's rule for the approximate evaluation of a definite integral is

$$\int_a^b f(x)dx = \frac{h}{3}[f(a) + 4f(x_1) + 2f(x_2) + \cdots$$

$$+ 2f(x_{n-2}) + 4f(x_{n-1}) + f(b)] = \frac{h}{3}\sum cf(x),$$

where $c = 1, 4, 2, 4, \cdots, 2, 4, 1$.

(b) *Trapezoidal Rule.*   For any number n of intervals the trapezoidal rule is

$$\int_a^b f(x) \, dx = h\left[\frac{f(a) + f(b)}{2} + f(x_1) + f(x_2) + \cdots + f(x_{n-1})\right].$$

150. Mean Value of a Function. The mean value of a function $f(x)$ for the interval $x = a$ to $x = b$ is

$$\frac{1}{b - a} \int_a^b f(x)dx.$$

DIFFERENTIAL EQUATIONS

151. Ordinary Differential Equations of the First Order.

(a) *Variables separated:* $f(x)dx + f(y)dy = 0.$

Solution: $\int f(x)dx + \int f(y)dy = C.$

(b) *Variables separable:* $f(x)G(y)dx + F(x)g(y)dy = 0.$

Solution: $\int \frac{f(x)}{F(x)}dx + \int \frac{g(y)}{G(y)}dy = C.$

(c) *Exact:*

$$f(x, y)dx + g(x, y)dy = 0, \quad \text{where} \quad f_y(x, y) = g_x(x, y) = P.$$

Solution: $\int_{(y)} f(x, y)dx + \int \left[g(x, y) - \int_{(y)} P\, dx \right] dy = C, \quad \text{or}$

$\int_{(x)} g(x, y)dy + \int \left[f(x, y) - \int_{(x)} P\, dy \right] dx = C$, where (y) means that y is held constant during that integration.

(d) *Homogeneous:* $f(x, y)dx + g(x, y)dy = 0,$ where
$f(tx, ty) = t^n f(x, y)$ and $g(tx, ty) = t^n g(x, y).$

Solution: The substitution $y = vx$ (or $x = vy$) reduces this equation to one in which the variables are separable, type (b).

If the equation is written in the form $\dfrac{dy}{dx} = F\left(\dfrac{y}{x}\right)$, the solution is

$$\ln x = \int \frac{dv}{F(v) - v} + C, \quad \text{and} \quad v = \frac{y}{x}.$$

Or, the equation becomes exact (c) when multiplied through by

$$\frac{1}{xf(x, y) + yg(x, y)}.$$

(e) *Linear:* $\dfrac{dy}{dx} + P(x)y = Q(x).$

Solution: The equation becomes exact (c) when multiplied through by the integrating factor $e^{\int P\, dx}$. The solution is

$$y = e^{-\int P\, dx} \left[\int e^{\int P\, dx} Q(x)dx + C \right].$$

(f) *Bernoulli's Equation:* $\dfrac{dy}{dx} + P(x)y = y^nQ(x)$,

Solution: The substitution $y^{1-n} = z$ gives

$$\frac{dz}{dx} + (1 - n)P(x)z = (1 - n)Q(x),$$

which is linear in x and z, type (e).

(g) *Clairaut's Equation:* $y = x\dfrac{dy}{dx} + f\left(\dfrac{dy}{dx}\right).$

General solution: $y = Cx + f(C)$.

Singular solution: Eliminate p between the two equations

$$y = xp + f(p) \qquad \text{and} \qquad x + \frac{d}{dp}f(p) = 0.$$

(h) *Integrating factor:* Any multiplier of a differential equation [as $e^{\int P\,dx}$ in (e)] which makes the equation exact is an *integrating factor.* An equation of the form $f(x, y)dx + g(x, y)dy = 0$ has an unlimited number of integrating factors. This factor may be found by inspection in many cases.

152. Ordinary Differential Equations of the Second Order. Special cases.

(a) $$\frac{d^2y}{dx^2} = f(x).$$

Solution:

$$y = \int\left[\int f(x)dx\right]dx + C_1x + C_2 = x\int f(x)dx - \int xf(x)dx + c_1x + c_2.$$

(b) $$\frac{d^2y}{dx^2} = f(y).$$

Solution: $x = \displaystyle\int \frac{dy}{\sqrt{C_1 + 2\int f(y)dy}} + C_2.$

(c) $$\frac{d^2y}{dx^2} = f\left(\frac{dy}{dx}\right).$$

Solution: Parametric equations of the solution are

$$x = \int \frac{dp}{f(p)} + C_1, \qquad y = \int \frac{p\,dp}{f(p)} + C_2.$$

(d) $f\left(x, \dfrac{dy}{dx}, \dfrac{d^2y}{dx^2}\right) = 0$ (not containing y explicitly).

Solution: Let $\dfrac{dy}{dx} = p, \dfrac{d^2y}{dx^2} = \dfrac{dp}{dx}$, reducing the equation to one of the first order in p and x. Then use the methods of Art. 151.

(e) $f\left(y, \dfrac{dy}{dx}, \dfrac{d^2y}{dx^2}\right) = 0$ (not containing x explicitly).

Solution: Reduce to an equation of the first order by putting $\dfrac{dy}{dx} = p, \dfrac{d^2y}{dx^2} = p\dfrac{dp}{dy}$, then use the methods of Art. 151.

(f) $$\dfrac{d^2y}{dx^2} + f(x)\dfrac{dy}{dx} + g(x)y = h(x).$$

Solution: If $y = F(x)$ is a particular solution of the reduced equation

$$\dfrac{d^2y}{dx^2} + f(x)\dfrac{dy}{dx} + g(x)y = 0,$$

the substitution $y = vF(x)$ transforms the original equation into one which does not contain v explicitly. Then use the methods of (d).

153. Second-order Linear Equation with Constant Coefficients. If $x = x_p$ is any particular solution of the equation

$$\dfrac{d^2x}{dt^2} + 2m\dfrac{dx}{dt} \pm k^2x = f(t), \tag{1}$$

and $x = x_c$ is the general solution of the *reduced equation*

$$\dfrac{d^2x}{dt^2} + 2m\dfrac{dx}{dt} \pm k^2x = 0, \tag{2}$$

then $x = x_c + x_p$ is the general solution of equation (1). The nature of the solution depends on the nature of the roots, r_1 and r_2, of the auxiliary algebraic equation

$$r^2 + 2mr \pm k^2 = 0. \tag{3}$$

Case I. $\dfrac{d^2x}{dt^2} + k^2x = f(t)$; $m = 0, r_1 = kj, r_2 = -kj$.

Value of $f(t)$	Value of x
Zero (Simple harmonic motion)	$x_c = C_1 \sin kt + C_2 \cos kt$ $\quad = A \sin(kt + \theta_1)$ $\quad = A \cos(kt + \theta_2)$
$a + bt$	$x_c + \dfrac{a + bt}{k^2}$
$a \sin nt + b \cos nt,\ n \neq k$	$x_c + \dfrac{a \sin nt + b \cos nt}{k^2 - n^2}$
$a \sin kt + b \cos kt$	$x_c + \dfrac{t}{2k}(b \sin kt - a \cos kt)$
ae^{bt}	$x_c + \dfrac{ae^{bt}}{b^2 + k^2}$

Case II. $\dfrac{d^2x}{dt^2} - k^2 x = f(t);\ m = 0,\ r_1 = k,\ r_2 = -k.$

Value of $f(t)$	Value of x
Zero	$x_c = C_1 e^{kt} + C_2 e^{-kt}$ $\quad = A \sinh kt + B \cosh kt$
$a + bt$	$x_c - \dfrac{a + bt}{k^2}$
$a \sinh nt + b \cosh nt,\ n \neq k$	$x_c + \dfrac{a \sinh nt + b \cosh nt}{n^2 - k^2}$
$a \sinh kt + b \cosh kt$	$x_c + \dfrac{t}{2k}(b \sinh kt + a \cosh kt)$
$ae^{bt},\ b \neq k$	$x_c + \dfrac{ae^{bt}}{b^2 - k^2}$
ae^{kt}	$x_c + \dfrac{t}{2k}ae^{kt}$

Case III. Roots of auxiliary equation (3) real and equal.

$$\frac{d^2x}{dt^2} + 2m\frac{dx}{dt} + m^2 x = f(t);\ m \neq 0,\ r_1 = r_2 = -m.$$

Value of $f(t)$	Value of x
Zero	$x_c = e^{-mt}(C_1 + C_2 t)$
$a + bt$	$x_c + \dfrac{a}{m^2} + \dfrac{b}{m^2}\left(t - \dfrac{2}{m}\right)$
$ae^{bt},\ b \neq -m$	$x_c + \dfrac{ae^{bt}}{(b + m)^2}$
ae^{-mt}	$x_c + \tfrac{1}{2}at^2 e^{-mt}.$

Case IV.   Roots of auxiliary equation (3) real and distinct.

(a) $$\frac{d^2x}{dt^2} + 2m\frac{dx}{dt} + k^2x = f(t), \qquad m^2 > k^2.$$

Let $p = \sqrt{m^2 - k^2}$;   then $r_1 = -m + p$, $r_2 = -m - p$.

Value of $f(t)$	Value of x
Zero	$x_c = C_1e^{r_1t} + C_2e^{r_2t}$
	$\quad = e^{-mt}(A \sinh pt + B \cosh pt)$
$a + bt$	$x_c + \dfrac{a}{k^2} + \dfrac{b}{k^2}\left(t - \dfrac{2m}{k^2}\right)$
ae^{bt}, $b \neq r_1$, $b \neq r_2$	$x_c + \dfrac{ae^{bt}}{b^2 + 2bm + k^2}$
ae^{r_1t}	$x_c + \dfrac{ate^{r_1t}}{2(r_1 + m)}$
$e^{ct}(a \sinh nt + b \cosh nt)$, not both $c = -m$ and $n = p$	$x_c + \dfrac{e^{ct}}{R^2 - S^2}[(Ra - Sb) \sinh nt$ $\qquad + (Rb - Sa) \cosh nt],$ $R = n^2 + k^2 + c^2 + 2cm,$ $\qquad\qquad S = 2n(m + c)$
$e^{-mt}(a \sinh pt + b \cosh pt)$	$x_c + \dfrac{te^{-mt}}{2p}(b \sinh pt + a \cosh pt)$

(b) $$\frac{d^2x}{dt^2} + 2m\frac{dx}{dt} - k^2x = f(t).$$

Replace k^2 by $-k^2$ in formulas of (a) above.

Case V.   Roots of auxiliary equation (3) conjugate imaginaries.

$$\frac{d^2x}{dt^2} + 2m\frac{dx}{dt} + k^2x = f(t), \qquad m^2 < k^2.$$

Let $q = \sqrt{k^2 - m^2}$;   then $r_1 = -m + qj$, $r_2 = -m - qj$.

Value of $f(t)$	Value of x
Zero (damped vibration)	$x_c = e^{-mt}(C_1 \sin qt + C_2 \cos qt)$
	$\quad = Ae^{-mt} \sin (qt + \theta_1)$
	$\quad = Ae^{-mt} \cos (qt + \theta_2)$
$a + bt$	$x_c + \dfrac{a}{k^2} + \dfrac{b}{k^2}\left(t - \dfrac{2m}{k^2}\right)$
ae^{bt}	$x_c + \dfrac{ae^{bt}}{b^2 + 2mb + k^2}$

Value of $f(t)$ | Value of x

$e^{ct}(a \sin nt + b \cos nt)$, $x_c + \dfrac{e^{ct}}{R^2 + S^2}[(Ra + Sb) \sin nt$

not both $c = -m$ and $n = q$ $+ (Rb - Sa) \cos nt]$,

$$R = k^2 - n^2 + c^2 + 2mc,$$
$$S = 2n(m + c)$$

$e^{-mt}(a \sin qt + b \cos qt)$ $x_c + \dfrac{te^{-mt}}{2q}(b \sin qt - a \cos qt)$.

The value of the particular solution x_p may be found as follows:

Case III, $x_p = e^{-mt}[t \int f(t)e^{mt}\, dt - \int tf(t)e^{mt}\, dt]$;

Case IV, $x_p = \dfrac{1}{r_1 - r_2}\left[e^{r_1 t} \int f(t)e^{-r_1 t}\, dt - e^{r_2 t} \int f(t)e^{-r_2 t}\, dt \right]$;

Case V, $x_p = \dfrac{e^{-mt}}{q}\left[\sin qt \int f(t)e^{mt} \cos qt\, dt \right.$

$$\left. - \cos qt \int f(t)e^{mt} \sin qt\, dt \right];$$

Case I = Case V with $m = 0$; Case II = Case IV with $m = 0$, $r_1 = -r_2$. See also Art. 154(b).

If $f(t) = f_1(t) + f_2(t)$, and if x_{p_1} and x_{p_2} are particular solutions corresponding to $f_1(t)$ and $f_2(t)$ respectively, then $x_{p_1} + x_{p_2}$ is a particular solution corresponding to $f(t)$.

154. Linear Differential Equation with Constant Coefficients. If $x = x_p$ is any particular solution of the equation

$$\frac{d^n x}{dt^n} + a_1\frac{d^{n-1}x}{dt^{n-1}} + a_2\frac{d^{n-2}x}{dt^{n-2}} + \cdots + a_n x = f(t), \qquad (1)$$

and $x = x_c$ is the general solution of the reduced equation

$$\frac{d^n x}{dt^n} + a_1\frac{d^{n-1}x}{dt^{n-1}} + a_2\frac{d^{n-2}x}{dt^{n-2}} + \cdots + a_n x = 0, \qquad (2)$$

then $x = x_c + x_p$ is the general solution of equation (1). The expression x_c is called the complementary function of (1).

The nature of the solution depends on the nature of the roots of the auxiliary algebraic equation

$$r^n + a_1 r^{n-1} + a_2 r^{n-2} + \cdots + a_n = 0. \qquad (3)$$

(a) *The homogeneous equation* (2). The complementary function x_c is the sum of n terms obtained as follows.

For each distinct real root $r = s$ of (3) there is a term Ce^{st}.

If s occurs twice among the roots of (3), the corresponding terms of the solution are $e^{st}(C_1 + C_2 t)$; if s occurs three times, the corresponding terms are $e^{st}(C_1 + C_2 t + C_3 t^2)$, and so on.

For each pair of complex roots of (3), $a + bj$ and $a - bj$, the terms in the solution are

$$e^{at}(C_1 \sin bt + C_2 \cos bt).$$

If the same pair of complex roots occurs twice, the corresponding terms in the solution are

$$e^{at}[(C_1 + C_2 t) \sin bt + (C_3 + C_4 t) \cos bt],$$

and so on.

In building up the solution in any particular case, the n coefficients C must be indicated by different symbols, as C_1, C_2, C_3, . . . , C_n.

(b) *The nonhomogeneous equation* (1). First find the complementary function x_c. If successive differentiation of the terms of $f(t)$ (neglecting any numerical coefficients) will yield only a finite number of terms of distinct functional form, the particular solution x_p may be found by the method of undetermined coefficients as follows.

If no one of these terms (the terms of $f(t)$ and their derivatives) occurs in x_c, set x_p equal to the sum of these terms each multiplied by an undetermined coefficient.

If any of these terms occurs also in x_c, arrange the terms in groups, each group being made up of one term of $f(t)$ and its derivatives. If a term of any group occurs in x_c, multiply each term of the group by the lowest integral power of t that will make every term of the group different from any term in x_c, and use these instead of the original terms in making up the particular integral x_p.

Substitute this assumed value of x_p in equation (1) and determine the coefficients by equating coefficients of like terms in the two members of the resulting equation.

The solution of (1) is $x = x_c + x_p$.

Methods for finding the particular solution in terms of integrals (see page 213) or by the method of "variable parameters" may be found in treatises on differential equations.

155. Simultaneous Linear Differential Equations with Constant Coefficients.

$$a_0\frac{d^n x}{dt^n} + \cdots + a_n x + b_0\frac{d^m y}{dt^m} + \cdots + b_m y = f(t),$$

$$p_0\frac{d^r x}{dt^r} + \cdots + p_r x + q_0\frac{d^s y}{dt^s} + \cdots + q_s y = g(t).$$

Replacing $\frac{d^i}{dt^i}$ by D^i these equations become

$$(a_0 D^n + a_1 D^{n-1} + \cdots + a_n)x + (b_0 D^m + b_1 D^{m-1} + \cdots$$
$$+ b_m)y = f(t),$$
$$(p_0 D^r + p_1 D^{r-1} + \cdots + p_r)x + (q_0 D^s + q_1 D^{s-1} + \cdots$$
$$+ q_s)y = g(t).$$

Treating these two equations as two simultaneous algebraic equations in x and y, and eliminating one of these variables a linear differential equation in the other variable of the type just treated is obtained. After the first variable is found, the other should be determined (if possible) without further integration in order not to introduce extraneous constants of integration. If additional constants are introduced, it may be necessary to substitute the values of x and y in the original equations to determine the relations among the constants.

156. Linear Partial Differential Equation of the First Order.

$$P(x, y, z)\frac{\partial z}{\partial x} + Q(x, y, z)\frac{\partial z}{\partial y} = R(x, y, z).$$

Solution: Find two independent integrals of

$$\frac{dx}{P} = \frac{dy}{Q} = \frac{dz}{R}\left[= \frac{a\,dx + b\,dy + c\,dz}{aP + bQ + cR}\right].$$

Let them be $u(x, y, z) = c_1$, $v(x, y, z) = c_2$. Then

$$\varphi(u, v) = 0,$$

where φ is an arbitrary function, is a solution of the partial differential equation.

VECTORS

157. Scalars and Vectors. A *scalar* is any real number. A *scalar quantity* is any quantity that is completely determined to scale by a single real number.

A *vector* is a directed line segment (arrow) anywhere in space. Any quantity that is completely specified by a vector is a *vector quantity*.

If the line of the vector is fixed in space, it is a *localized* vector; if its magnitude and direction are fixed, but it is allowed to assume any of its parallel positions, it is a *free* vector. Two vectors are *equal* if they are parallel, have the same sense, and are of the same length. Vector $AB = -$vector BA.

If s is a scalar and V is a vector, then the product sV is a vector which is parallel to V and $|s|$-times as long; its sense is the same as, or opposite to that of V according as s is positive or negative.

158. Addition of Vectors. If the initial point of a vector V is placed at the terminal point of a vector U, the vector drawn from the initial point of U to the terminal point of V is the vector sum of U and V.

FIG. 116.

Since the diagonal of the parallelogram having vectors U and V as sides is the vector sum of U and V, vectors are said to be added by the parallelogram law.

To subtract a vector U from a vector V, add to V the negative of U. If the initial points of U and V coincide, $V - U$ is the vector from the tip of U to the tip of V.

159. A Vector in Terms of Its Components. If a, b, c are the projections of a vector V on the x-, y-, z-axes, respectively, and i, j, k are vectors of unit length having the respective directions of these axes, then

$$V = ai + bj + ck.$$

The magnitude of V is $|V| = \sqrt{a^2 + b^2 + c^2}$, and direction numbers of V are a, b, c.

If the vector lies in the xy-plane

$$|V| = \sqrt{a^2 + b^2}, \qquad \theta = \tan^{-1}\frac{b}{a} = \sin^{-1}\frac{b}{\sqrt{a^2 + b^2}}.$$

160. Vector Sum (V) of n Vectors. If $V_i = a_i i + b_i j + c_i k$,

$$V = V_1 + V_2 + \cdots + V_n = (a_1 + a_2 + \cdots + a_n)i$$
$$+ (b_1 + b_2 + \cdots + b_n)j + (c_1 + c_2 + \cdots + c_n)k.$$

161. Product of a Vector (V) by a Scalar (s).

$$sV = (sa)i + (sb)j + (sc)k.$$
$$Vs = sV, \qquad (s_1 + s_2)V = s_1V + s_2V.$$

162. The Scalar (or Dot) Product of Two Vectors: $V_1 \cdot V_2$.
$V_1 \cdot V_2 = |V_1||V_2| \cos \varphi$, where φ is the angle between V_1 and V_2.

Example. If the point of application of a force F moves along a vector distance x, the work done by F during this displacement is $F \cdot x$.

$$V_1 \cdot V_2 = a_1a_2 + b_1b_2 + c_1c_2 = V_2 \cdot V_1.$$
$$(V_1 + V_2) \cdot V_3 = V_1 \cdot V_3 + V_2 \cdot V_3 = V_3 \cdot (V_1 + V_2).$$
$$i \cdot i = j \cdot j = k \cdot k = 1; \qquad i \cdot j = j \cdot k = k \cdot i = 0.$$

163. The Vector (or Cross) Product of Two Vectors: $V_1 \times V_2$.
$V_1 \times V_2 = 1|V_1||V_2| \sin \varphi$, where φ is the angle from V_1 to V_2 and 1 is a unit vector perpendicular to the plane of V_1 and V_2 and so directed that a right-handed screw advancing along 1 would turn V_1 toward V_2.

Example. If a force F has a moment arm x about a point O, the moment of F about O is $F \times x$.

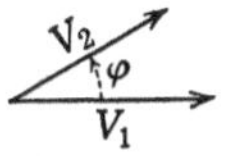

$$V_1 \times V_2 = -V_2 \times V_1 = \begin{vmatrix} i & j & k \\ a_1 & b_1 & c_1 \\ a_2 & b_2 & c_2 \end{vmatrix}$$

Fig. 117.

$$(V_1 + V_2) \times V_3 = V_1 \times V_3 + V_2 \times V_3 = -V_3 \times (V_1 + V_2).$$
$$V_1 \times (V_2 \times V_3) = V_2(V_1 \cdot V_3) - V_3(V_1 \cdot V_2).$$
$$i \times i = j \times j = k \times k = 0, \quad i \times j = k, \quad j \times k = i, \quad k \times i = j.$$
$$V_1 \cdot (V_2 \times V_3) = (V_1 \times V_2) \cdot V_3 = V_2 \cdot (V_3 \times V_1) = \begin{vmatrix} a_1 & a_2 & a_3 \\ b_1 & b_2 & b_3 \\ c_1 & c_2 & c_3 \end{vmatrix}$$
$$(V_1 \times V_2) \cdot (V_3 \times V_4) = (V_1 \cdot V_3)(V_2 \cdot V_4) - (V_1 \cdot V_4)(V_2 \cdot V_3).$$

164. Differentiation of Vectors. If, in the vector $V = ai + bj + ck$, a, b, and c are functions of a *scalar* variable t, then

$$\frac{dV}{dt} = \frac{da}{dt}i + \frac{db}{dt}j + \frac{dc}{dt}k.$$
$$\frac{d}{dt}(V_1 + V_2 + \cdots + V_n) = \frac{dV_1}{dt} + \frac{dV_2}{dt} + \cdots + \frac{dV_n}{dt}.$$
$$\frac{d}{dt}(SV) = S\frac{dV}{dt} + V\frac{dS}{dt}$$

$$\frac{d}{dt}(V_1 \cdot V_2) = \frac{dV_1}{dt} \cdot V_2 + V_1 \cdot \frac{dV_2}{dt}.$$

$$V \cdot \frac{dV}{dt} = |V|\frac{d|V|}{dt}.$$

$$\frac{d}{dt}(V_1 \times V_2) = \frac{dV_1}{dt} \times V_2 + V_1 \times \frac{dV_2}{dt}.$$

$$\nabla = i\frac{\partial}{\partial x} + j\frac{\partial}{\partial y} + k\frac{\partial}{\partial z}, \text{ an operator.}$$

$$\text{grad } S \equiv \nabla S \equiv \frac{\partial S}{\partial x}i + \frac{\partial S}{\partial y}j + \frac{\partial S}{\partial z}k,$$

where S is a scalar.

$$\text{div } V \equiv \nabla \cdot V \equiv \frac{\partial a}{\partial x} + \frac{\partial b}{\partial y} + \frac{\partial c}{\partial z} \text{ (divergence of } V\text{).}$$

$$\text{curl } V \equiv \text{rot } V \equiv \begin{vmatrix} i & j & k \\ \frac{\partial}{\partial x} & \frac{\partial}{\partial y} & \frac{\partial}{\partial z} \\ a & b & c \end{vmatrix} = \nabla \times V.$$

$$\text{div grad } S \equiv \nabla^2 S \equiv \frac{\partial^2 S}{\partial x^2} + \frac{\partial^2 S}{\partial y^2} + \frac{\partial^2 S}{\partial z^2}.$$

$$\nabla^2 V = i\nabla^2 a + j\nabla^2 b + k\nabla^2 c.$$

$$\text{curl}^2 V = \nabla \text{ div } V - \nabla^2 V.$$

$$\text{curl grad } S = 0, \qquad \text{div curl } V = 0.$$

165. Vector Equations for the Motion of a Rigid Body. Consider the body as an aggregate of particles of masses m_1, m_2, m_3, ... Let

r_i = vector from origin of coordinates to particle of mass m_i;

$\bar{r}$ = vector from origin to the centroid of the body;

F_i = the external vector force applied to m_i;

$v_i = \dfrac{dr_i}{dt}$ = the vector velocity of m_i.

The moment of momentum (σ) of the body (system of particles) is defined to be

$$\sigma = \Sigma m_i r_i \times v_i.$$

The equations of motion for the body are

$$\sum m_i \frac{d^2 r_i}{dt^2} = \sum m_i \frac{d^2 \bar{r}}{dt^2} = \sum F_i;$$

$$\frac{d\sigma}{dt} = \sum m_i r_i \times \frac{dv_i}{dt} = \sum r_i \times F_i.$$

MECHANICS

STATICS

166. Notation. (F, θ) at (x, y) will be used to designate a force in the xy-plane of magnitude F with line of action through

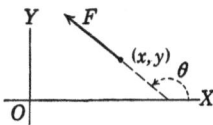

Fig. 118.

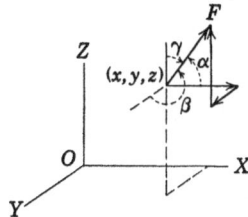

Fig. 119.

the point (x, y) directed at the angle θ with the positive direction of the x-axis.

$(F; \alpha, \beta, \gamma)$ at (x, y, z) will be used to designate a force of magnitude F with line of action through the point (x, y, z) and with direction angles α, β, γ.

167. Components of a Force. If the force F is the vector sum of forces $F_1, F_2, \ldots , F_n$, these forces are components of F.

The rectangular components parallel to the x- and y-axes of a force (F, θ) are, respectively,

$$X = F \cos \theta, \qquad Y = F \sin \theta.$$

The rectangular components parallel to the x-, y-, and z-axes of a force $(F; \alpha, \beta, \gamma)$ are, respectively,

$$X = F \cos \alpha, \qquad Y = F \cos \beta, \qquad Z = F \cos \gamma.$$

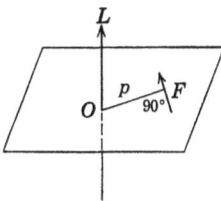

Fig. 120.

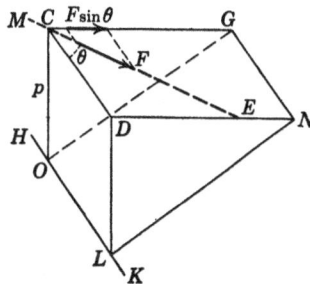

Fig. 121.

168. Moment of a Force. The moment M of a force F about a line L which is perpendicular to a plane containing the line of action of the force is the product of the force by the length of the common perpendicular p between the line and the line of action of the force (Fig. 120):

$$M_0 = M_L = Fp.$$

The moment M of a force F about any line L is the moment about L of its rectangular component in a plane perpendicular to L (Fig. 121):

$$M_L = pF \sin \theta,$$

where p is the common perpendicular between L and line of action of F.

The moment of a force about any line is equal to the sum of the moments of its components about the line. If the force F has components X, Y, Z parallel to the coordinate axes, then

$$M_x = Yz - Zy,$$
$$M_y = Zx - Xz,$$
$$M_z = Xy - Yx.$$

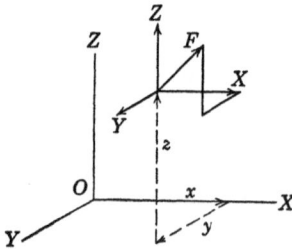

FIG. 122.

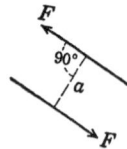

FIG. 123.

169. Couple. Two parallel forces having equal magnitudes, but differing in sense and in line of action, form a couple. The moment of a couple is the product of one of the forces of the couple by the arm of the couple (the perpendicular distance between the lines of action of the two forces).

The sum of the moments of the forces forming the couple, about any line perpendicular to the plane of the couple, is equal to the moment of the couple (Fa).

170. Coplanar Forces. Let X and Y denote the sum of the components, parallel to the x- and y-axes, of a system of coplanar forces $F_1, F_2, \ldots , F_n$, and let M_0 denote the sum of the moments of these forces about the origin of coordinates. Let the force (F_i, θ_i) at (x_i, y_i) have components X_i, Y_i. Then (Arts. 167 and 168)

$$X = X_1 + X_2 + \cdots + X_n = \sum_{i=1}^{n} F_i \cos \theta_i,$$

$$Y = Y_1 + Y_2 + \cdots + Y_n = \sum_{i=1}^{n} F_i \sin \theta_i,$$

$$M_0 = \sum_{i=1}^{n} (Y_i x_i - X_i y_i) = \sum_{i=1}^{n} (x_i F_i \sin \theta_i - y_i F_i \cos \theta_i).$$

(a) *The resultant* of a system of coplanar forces is a single force having components X and Y, acting at an arbitrarily chosen point in the plane of the forces, and a couple whose moment is equal to the sum of the moments of the given forces about that point.

If X and Y are not both zero, the resultant is a single force (F, θ) of magnitude

$$F = \sqrt{X^2 + Y^2} = X \sec \theta = Y \csc \theta, \qquad \left(\tan \theta = \frac{Y}{X} \right)$$

with line of action

$$Yx - Xy = M_0.$$

If the forces $F_1, F_2, \ldots, F_n$ are concurrent, the line of action of their resultant passes through the point common to the lines of action of the forces.

If X and Y are both zero, the resultant of the system of forces is a couple of moment $M = M_0$.

(b) *The conditions of equilibrium* for a system of coplanar forces are

$$X = 0, \qquad Y = 0, \qquad M = 0,$$

where M denotes the sum of the moments of the given forces about any point in the plane of the forces.

Alternative equations of equilibrium are

$$X = 0, \qquad M_A = 0, \qquad M_B = 0,$$

where A and B are two points in the plane of the forces, not both on the same perpendicular to the direction of X; or

$$M_A = 0, \qquad M_B = 0, \qquad M_C = 0,$$

where A, B, C are three points in the plane of the forces, not all on the same straight line.

If the forces are *concurrent*, the equations of equilibrium are

$$X = 0, \qquad Y = 0.$$

Force triangle. If three concurrent forces are in equilibrium, their vector sum is zero, and the magnitudes and directions of the forces are represented by the sides of the force triangle

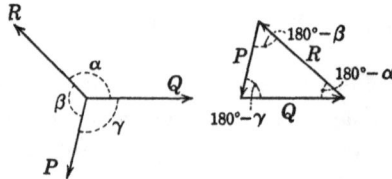

FIG. 124.

obtained in their vector addition. The following relations are satisfied by the three forces P, Q, R:

$$\frac{P}{\sin \alpha} = \frac{Q}{\sin \beta} = \frac{R}{\sin \gamma}.$$

171. Forces in Space of Three Dimensions.   Let $(F_1; \alpha_1, \beta_1, \gamma_1)$ at (x_1, y_1, z_1), $(F_2; \alpha_2, \beta_2, \gamma_2)$ at (x_2, y_2, z_2), . . . , $(F_n; \alpha_n, \beta_n, \gamma_n)$ at (x_n, y_n, z_n) be any system of forces. The sum of the components, parallel to the coordinate axes, of the forces of this system are

$$X = \sum X_i = \sum_{i=1}^{n} F_i \cos \alpha_i, \qquad Y = \sum Y_i = \sum_{i=1}^{n} F_i \cos \beta_i,$$

$$Z = \sum Z_i = \sum_{i=1}^{n} F_i \cos \gamma_i;$$

and the sum of the moments, about the coordinate axes, of the forces of this system are

$$M_x = \sum_{i=1}^{n} (Y_i z_i - Z_i y_i), \qquad M_y = \sum_{i=1}^{n} (Z_i x_i - X_i z_i),$$

$$M_z = \sum_{i=1}^{n} (X_i y_i - Y_i x_i).$$

(a) *The resultant* of this system of forces is a single force $(F; \alpha, \beta, \gamma)$ acting through an arbitrarily chosen point of space, and a couple $(M; \lambda, \mu, \nu)$, where

$$F = \sqrt{X^2 + Y^2 + Z^2},$$

$$\cos \alpha = \frac{X}{F}, \qquad \cos \beta = \frac{Y}{F}, \qquad \cos \gamma = \frac{Z}{F};$$

and, if the origin of coordinates is on the line of action of F,

$$M = \sqrt{M_x^2 + M_y^2 + M_z^2},$$

$$\cos \lambda = \frac{M_x}{M}, \qquad \cos \mu = \frac{M_y}{M}, \qquad \cos \nu = \frac{M_z}{M}.$$

If the forces are concurrent, the resultant is

$$F = \sqrt{X^2 + Y^2 + Z^2},$$

$$\cos \alpha = \frac{X}{F}, \qquad \cos \beta = \frac{Y}{F}, \qquad \cos \gamma = \frac{Z}{F},$$

with line of action through the point common to the lines of action of the forces.

The wrench. If the line of action of F is taken along the line

$$\frac{M_x - Yz + Zy}{X} = \frac{M_y - Zx + Xz}{Y} = \frac{M_z - Xy + Yx}{Z},$$

the couple $(M'; \alpha, \beta, \gamma)$ of the wrench has its axis parallel to the line of action of F, and

$$M' = M \cos \varphi,$$

where $\cos \varphi = \cos \alpha \cos \lambda + \cos \beta \cos \mu + \cos \gamma \cos \nu$.

(b) *The conditions of equilibrium* for this system of forces are

$$X = 0, \qquad Y = 0, \qquad Z = 0; \qquad M_x = 0, \qquad M_y = 0, \qquad M_z = 0$$

If the forces are concurrent, the equations of equilibrium are

$$X = 0, \qquad Y = 0, \qquad Z = 0.$$

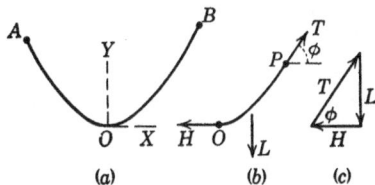

Fɪɢ. 125.

172. Cables. In Fig. 125(a) is shown a flexible, inextensible cable suspended from two points A and B, O being the point where the tangent line to the cable is horizontal; in (b) is shown a free-body diagram of a piece OP of the cable; in (c) is shown a force triangle for the forces holding OP in equilibrium, L being the total load (assumed vertical) carried by OP. Then, for the axes shown,

$$\frac{dy}{dx} = \tan \varphi = \frac{L}{H}.$$

(a) *Parabolic cable.* Let the cable carry a uniform load of w pounds per horizontal foot. Then $L = wx$ and

$$y = \frac{wx^2}{2H}.$$

(b) *Catenary cable.* Let the cable carry a load of w pounds per foot of cable. Then

$$L = ws,$$

where s is the length of OP, and

$$y = \frac{H}{w}\left(\cosh \frac{wx}{H} - 1\right), \qquad s = \frac{H}{w} \sinh \frac{wx}{H}, \qquad L = H \sinh \frac{wx}{H},$$

$$T = H \cosh \frac{wx}{H}, \qquad T = H + wy, \qquad \left(y + \frac{H}{w}\right)^2 - s^2 = \left(\frac{H}{w}\right)^2.$$

See also Fig. 88.

173. Friction. If a body W is at rest on an inclined plane under the action of gravity W and the reaction R of the plane on the body, the reaction R is equal to the weight W of the block and makes the angle φ with the normal to the plane.

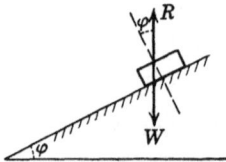

FIG. 126.

The component of R along the plane is the force of friction F and the component of R normal to the plane is the normal reaction N of the plane on the body:

$$F = R \sin \varphi, \qquad N = R \cos \varphi.$$

The angle which R makes with the normal to the plane when motion is impending is the angle of friction, φ_m. The tangent of φ_m is the coefficient of static friction f_s for the two substances in contact:

$$f_s = \tan \varphi_m.$$

If a body W is at rest in contact with a surface S, and P is the resultant of all the forces acting on the body except its weight

W and the reaction R of the surface on the body, then P, W, and R are three concurrent forces in equilibrium and

$$\frac{R}{\cos (\theta + \alpha)} = \frac{P}{\sin (\theta + \varphi)} = \frac{W}{\cos (\alpha - \varphi)}.$$

The force of friction is

$$F = R \sin \varphi$$

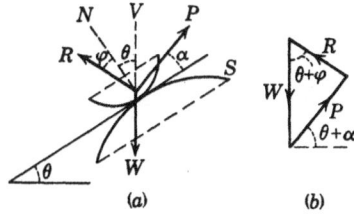

Fig. 127.

which is equal to $f_s N = f_s(R \cos \varphi)$ when motion is impending.

After motion has begun, the angle φ assumes a value which is usually smaller than its value for impending motion. The tangent of this angle is the coefficient of kinetic friction.

Belt friction. Let a belt or cord be in contact with a rough circular cylinder along a right section of its surface. If the pull on one end is T_2 and on the other end T_1, $T_2 > T_1$, then

$$T_2 = T_1 e^{f\alpha}$$

where f is the coefficient of friction, and α (radians) is the angle of contact.

MOTION OF A POINT

Notation

(a) *Rectilinear motion.* At time t

s = distance from a fixed point,

$v = \dfrac{ds}{dt}$ = linear velocity,

$a = \dfrac{dv}{dt}$ = linear acceleration.

(b) *Circular motion.* At time t

θ = angular distance from a fixed line,

$\omega = \dfrac{d\theta}{dt}$ = angular velocity,

$\alpha = \dfrac{d\omega}{dt}$ = angular acceleration.

174. Equations of Motion.   Rectilinear and Circular Motion.

Rectilinear motion	Circular motion
$a = \dfrac{dv}{dt} = v\dfrac{dv}{ds} = \dfrac{d^2s}{dt^2}.$	$\alpha = \dfrac{d\omega}{dt} = \omega\dfrac{d\omega}{d\theta} = \dfrac{d^2\theta}{dt^2}.$
If a is constant, and	If α is constant, and
$v = v_0$, $s = 0$ when $t = 0$	$\omega = \omega_0$, $\theta = 0$ when $t = 0$

Rectilinear motion   Circular motion

$v = v_0 + at,$ $\omega = \omega_0 + \alpha t,$

$s = v_0 t + \frac{1}{2}at^2,$ $\theta = \omega_0 t + \frac{1}{2}\alpha t^2,$

$v^2 = v_0^2 + 2as,$ $\omega^2 = \omega_0^2 + 2\alpha\theta,$

$s = \dfrac{v_0 + v}{2}t.$ $\theta = \dfrac{\omega_0 + \omega}{2}t.$

If $a = f(s)$, and $v = v_0$, $s = s_0$ when $t = 0$, then

$$v^2 = v_0^2 + 2\int_{s_0}^{s} a\,ds, \qquad t = \int_{s_0}^{s} \frac{ds}{\left[v_0^2 + 2\int_{s_0}^{s} a\,ds\right]^{\frac{1}{2}}}.$$

If $a = f(t)$, and $v = v_0$, $s = s_0$ when $t = 0$, then

$$v = v_0 + \int_0^t a\,dt, \qquad s = \int_0^t \left(v_0 + \int_0^t a\,dt\right)dt.$$

If $a = f(v)$, and $v = v_0$, $s = s_0$ when $t = 0$, then

$$t = \int_{v_0}^{v} \frac{dv}{a}, \qquad s = s_0 + \int_{v_0}^{v} \frac{v\,dv}{a}.$$

Similar equations are obtained for circular motion by replacing a, v, s by α, ω, θ, respectively.

175. Simple Harmonic Motion. Any rectilinear motion for which $a = -k^2 s$ is called simple harmonic motion. If $v = v_0$ and $s = s_0$ when $t = 0$, then

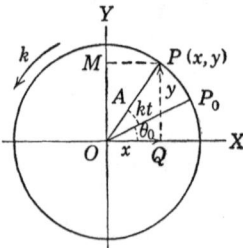

FIG. 128.

$$s = s_0 \cos kt + \frac{v_0}{k} \sin kt,$$

$$v = -k\,s_0 \sin kt + v_0 \cos kt$$

$$= \pm k\sqrt{s_0^2 + \frac{v_0^2}{k^2} - s^2}.$$

The *period* of the motion is $T = \dfrac{2\pi}{k}$, its *amplitude* is $\left(s_0^2 + \dfrac{v_0^2}{k^2}\right)^{\frac{1}{2}}$.

As a point P moves with constant speed on the circumference of a circle, its projection Q on any fixed line in the plane of the circle moves with simple harmonic motion. Thus, if P is at P_0 (Fig. 128) when $t = 0$, and if OP is rotating at k radians per second, then at time t

$$x = OQ = A \cos (kt + \theta_0), \qquad a_x = \frac{d^2x}{dt^2} = -k^2x;$$

$$y = OM = A \sin (kt + \theta_0), \qquad a_y = \frac{d^2y}{dt^2} = -k^2y.$$

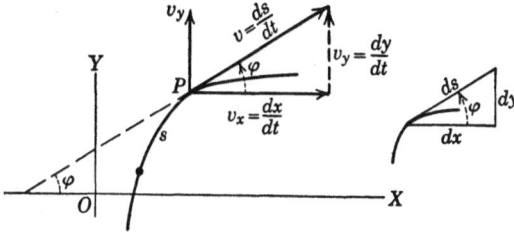

Fig. 129.

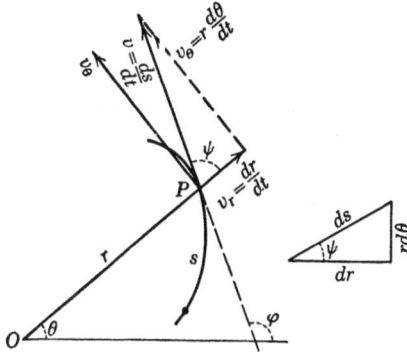

Fig. 130.

176. Components of Velocity and Acceleration.

Rectangular coordinates   Polar coordinates

$$v_x = v \cos \varphi = \frac{dx}{dt}, \qquad\qquad v_r = v \cos \psi = \frac{dr}{dt},$$

$$v_y = v \sin \varphi = \frac{dy}{dt}; \qquad\qquad v_\theta = v \sin \psi = r \frac{d\theta}{dt};$$

$$a_x = \frac{dv_x}{dt} = \frac{d^2x}{dt^2}, \qquad\qquad a_r = \frac{d^2r}{dt^2} - r\left(\frac{d\theta}{dt}\right)^2,$$

$$a_y = \frac{dv_y}{dt} = \frac{d^2y}{dt^2}; \qquad\qquad a_\theta = \frac{1}{r}\frac{d}{dt}\left(r^2 \frac{d\theta}{dt}\right).$$

$$a_t = \frac{dv}{dt} = \frac{d^2s}{dt^2} \text{ (along tangent line)},$$

$$a_n = \frac{v^2}{R} \text{ (along normal toward center of curvature)},$$

where $R\left(=\dfrac{ds}{d\varphi}\right)$ is the radius of curvature at the point.

For motion in a circle of radius R

$$v = R\omega, \qquad a_t = R\alpha, \qquad a_n = R\omega^2.$$

For any curve in space

$$v_x = \frac{dx}{dt}, \qquad v_y = \frac{dy}{dt}, \qquad v_z = \frac{dz}{dt};$$

$$a_x = \frac{d^2x}{dt^2}, \qquad a_y = \frac{d^2y}{dt^2}, \qquad a_z = \frac{d^2z}{dt^2}.$$

177. Relative Velocities and Accelerations in Plane Motion.

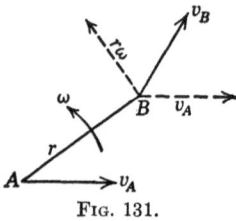

FIG. 131.

A body is said to have plane motion when each point in it moves in a fixed plane.

Let A and B be any two points of a rigid body in plane motion both moving in the same fixed plane.

The velocity v_B of B is equal to the vector sum of the velocity v_A of A and the velocity $v_{B/A}$ of B relative to A. This relation is expressed by the vector equation

$$v_B = v_A + v_{B/A} = v_A + r\omega.$$

The acceleration a_B of B is equal to the vector sum of the acceleration a_A of A and the acceleration $a_{B/A}$ of B relative to A. This relation is expressed by the vector equation

$$a_B = a_A + a_{B/A} = a_A + a_t + a_n$$
$$= a_A + r\alpha + r\omega^2.$$

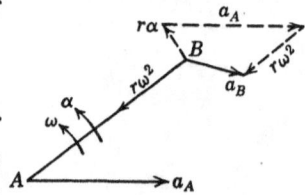

FIG. 132.

178. Instantaneous Center. (*a*)

Lines drawn through any two points in a plane of motion, perpendicular respectively to the directions of motion of the two points at any instant, intersect at the instantaneous center (of zero velocity). The line through this point perpendicular to the plane of motion is the instantaneous axis of rotation. Any point in the body r units from this axis has an instantaneous velocity $r\omega$, perpendicular to the direction of r.

(b) If a_A and a_B are the accelerations of any two points in a plane of motion, the point whose coordinates, referred to a_A as polar axis, are

$$r = \frac{a_A}{\sqrt{\alpha^2 + \omega^4}}, \qquad \theta = \tan^{-1}\frac{\alpha}{\omega^2},$$

is the instantaneous center of zero acceleration.

KINETICS OF A PARTICLE
179. Newton's Laws of Motion.

I. Every body continues in its state of rest or of uniform motion in a straight line, except in so far as it may be compelled by force to change that state.

II. Change of motion (time rate of change of momentum) is proportional to the force applied and takes place in the direction of the straight line in which the force acts.

III. To every action there is always an equal and opposite reaction; or the mutual actions of any two bodies are always equal and oppositely directed.

180. Equations of Motion of a Particle.

The sum F of the components in any direction of all the forces acting on a particle of mass m is equal to m times the component a of the linear acceleration of the particle in that direction:

$$F = ma = \frac{W}{g}a,$$

where W is the weight of the particle and g is the value of the acceleration of gravity in the particular locality.

If X, Y, and Z are the sums of the components, parallel to the x-, y-, and z-axes, respectively, of all the forces acting on the particle, then

$$X = ma_x = \frac{W}{g}a_x, \qquad Y = ma_y = \frac{W}{g}a_y, \qquad Z = ma_z = \frac{W}{g}a_z,$$

where

$$a_x = \frac{dv_x}{dt} = \frac{v_x\,dv_x}{dx} = \frac{d^2x}{dt^2},$$

$$a_y = \frac{dv_y}{dt} = \frac{v_y\,dv_y}{dy} = \frac{d^2y}{dt^2},$$

$$a_z = \frac{dv_z}{dt} = \frac{v_z\,dv_z}{dz} = \frac{d^2z}{dt^2}.$$

181. Elastic Springs. (*a*) *Free vibration.* Let a weight of W pounds be suspended by an elastic spring, stretching it e feet. Then a force F will stretch it s feet, where $F = \dfrac{W}{e}s$.

If the body is given a vertical motion, it will oscillate in simple harmonic motion (frictional resistances being neglected). Denote by x (positive downward) the displacement of the body from its equilibrium position at any time. Then

$$s = e + x, \qquad a = -\frac{g}{e}x.$$

If $v = v_0$ and $x = x_0$ when $t = 0$, then

$$x = x_0 \cos \sqrt{\frac{g}{e}}\, t + v_0 \sqrt{\frac{e}{g}} \sin \sqrt{\frac{g}{e}}\, t.$$

The period of the motion is $T = 2\pi \sqrt{\dfrac{e}{g}}$,

its amplitude is $\left(x_0^2 + \dfrac{v_0^2 e}{g} \right)^{\frac{1}{2}}$.

(*b*) *Damped vibration.* If the motion in (*a*) is subjected to a resistance cv, where v is the velocity of the body in feet per second, then

$$a = -\frac{g}{e}x - \frac{cg}{W}v, \qquad \text{or} \qquad \frac{d^2x}{dt^2} + \frac{cg}{W}\frac{dx}{dt} + \frac{g}{e}x = 0.$$

For the solution of this equation, see Art. 153.

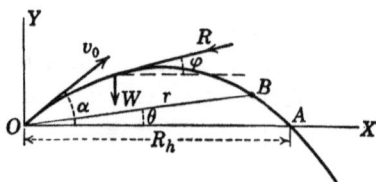

FIG. 133.

182. Motion of a Projectile. It is assumed that the projectile moves as if it were a particle. Let the xy-plane be the plane of motion and let the body be projected from O with an initial velocity v_0 at an angle α with the horizontal (Fig. 133).

(*a*) *Air resistance neglected.* The equations of motion are

$$a_x = 0, \qquad a_y = -g,$$

whence

$$x = v_0 t \cos \alpha,$$
$$y = v_0 t \sin \alpha - \tfrac{1}{2}gt^2.$$

The rectangular equation for the path is

$$y = x \tan \alpha - \frac{gx^2}{2v_0^2 \cos^2 \alpha}.$$

The time t_A from O to A is

$$t_A = \frac{2v_0 \sin \alpha}{g}.$$

The horizontal range R_h is

$$R_h = \frac{v_0^2 \sin 2\alpha}{g}.$$

The time t_B from O to B is

$$t_B = \frac{2v_0 \sin (\alpha - \theta)}{g \cos \theta}.$$

The range r on the plane inclined θ to the horizontal and passing through the point of projection is

$$r = \frac{2v_0^2 \cos \alpha \sin (\alpha - \theta)}{g \cos^2 \theta}.$$

The greatest height h is reached when $t = \dfrac{v_0 \sin \alpha}{g}$ and is

$$h = \frac{v_0^2 \sin^2 \alpha}{2g}.$$

(b) *Resistance R acting along the tangent to the path.* In this event

$$X = -R \cos \varphi, \qquad Y = -R \sin \varphi - W,$$

where φ is the inclination of the tangent line at any point of the path. The equations of motion now reduce to

$$a_x = -\frac{gR \cos \varphi}{W}, \qquad a_y = -\frac{gR \sin \varphi}{W} - g.$$

If $R = kv$ where k is a constant and v is the velocity of the projectile in its path, then

$$x = \frac{Wv_0 \cos \alpha}{gk}(1 - e^{-\frac{gk}{W}t}),$$

$$y = \frac{W}{gk}\left(v_0 \sin \alpha + \frac{W}{k}\right)(1 - e^{-\frac{gk}{W}t}) - \frac{W}{k}t.$$

183. Impulse and Momentum. The impulse Q of a force F acting for an interval of time from $t = t_0$ to $t = t_1$ is defined by the equation

$$Q = \int_{t_0}^{t_1} F \, dt.$$

The momentum of a particle of mass m moving with velocity v is defined by the equation

$$\text{Momentum} = mv.$$

Let F be the sum of the components in a given direction of all the forces acting on a particle of mass m. If v_0 is the velocity of the particle in the given direction at time t_0 and v_1 its velocity at time t_1, then

$$Q = \int_{t_0}^{t_1} F \, dt = m(v_1 - v_0) = \frac{W}{g}(v_1 - v_0).$$

184. Impact. Conservation of Momentum. Let a body of mass m_1 having velocity u_1 collide with a body of mass m_2 having velocity u_2 so as to produce direct central impact. If v_1 is the velocity of m_1 and v_2 the velocity of m_2 after the impact, then

$$m_1 v_1 + m_2 v_2 = m_1 u_1 + m_2 u_2,$$
$$v_1 - v_2 = -e(u_1 - u_2),$$

where e is the coefficient of restitution.

The first of the two equations expresses the principle of conservation of momentum.

If h is the height of rebound of a sphere dropped from a height H on a horizontal surface, then $e = \sqrt{\dfrac{h}{H}}$ for the materials of the two bodies. If the bodies are inelastic, $e = 0$; if perfectly elastic, $e = 1$.

The loss of kinetic energy due to the impact is

$$\frac{m_1 m_2}{m_1 + m_2}(1 - e^2)\frac{(u_1 - u_2)^2}{2}.$$

185. Work and Kinetic Energy. The work U done by a force in moving a particle along a curve from a position $s = s_0$ to a position $s = s_1$ is defined by the equation:

$$U = \int_{s_0}^{s_1} F_t \, ds,$$

where F_t is the tangential component of the force.

The kinetic energy of a particle of mass m moving with velocity v is defined by the equation:

$$\text{Kinetic energy} = \tfrac{1}{2}mv^2 = \frac{W}{2g}v^2.$$

Let F be the sum of the components in the direction of motion of all the forces acting on a particle of mass m. If v_0 is the velocity of the particle in its path at a position $s = s_0$ and v_1 its velocity at a position $s = s_1$, then

$$\int_{s_0}^{s_1} F \, ds = \frac{m}{2}(v_1^2 - v_0^2) = \frac{W}{2g}(v_1^2 - v_0^2).$$

This equation express the work-energy relation.

186. Power. The work U done by a force F which acts in the direction of motion and moves a particle from position s_0 to position s is

$$U = \int_{s_0}^{s} F \, ds;$$

the rate of doing work is

$$\frac{dU}{dt} = \frac{dU}{ds}\frac{ds}{dt} = Fv.$$

If F is expressed in pounds and v in feet per second, the horse-power developed is

$$\text{Horsepower} = \frac{Fv}{550}.$$

DYNAMICS OF A RIGID BODY

187. Effective Forces. If any particle of a body were considered apart from the rest of the body, the single force which would cause the particle to move as it does move is the effective force for the particle. If a particle of mass m has acceleration a, the effective force for the particle is ma. The resultant of the effective forces for all the particles of a body is the effective-force system for the body.

D'Alembert's principle states that the effective-force system for a body is equivalent to the system of external forces acting on the body.

The reversed effective-force system, or the inertia force system, for a body is the system obtained by reversing the sense of each force of the system of effective forces.

The system of external forces and the system of reversed effective forces are in equilibrium.

188. Motion of the Centroid.   The motion of the centroid of a body is the same as the motion of a particle having the same mass as the body and acted on by forces identical in magnitude, direction, and sense with the external forces acting on the body.

If X, Y, and Z are the sums of the components of the external forces parallel to the coordinate axes, the equations of motion of the centroid are:

$$X = m\bar{a}_x = \frac{W}{g}\bar{a}_x, \qquad Y = m\bar{a}_y = \frac{W}{g}\bar{a}_y, \qquad Z = m\bar{a}_z = \frac{W}{g}\bar{a}_z,$$

where $\bar{a}_x$, $\bar{a}_y$, and $\bar{a}_z$ are the components of the acceleration of the centroid parallel to these axes.

If F_t, F_n, and F_3 are the components of the external forces in the directions, respectively, of axes parallel to the tangent to the curve in which the centroid is moving, parallel to the normal line through the center of curvature, and in a direction perpendicular to the plane determined by these two, then

$$F_t = m\bar{a}_t = \frac{W}{g}\frac{d\bar{v}}{dt}, \qquad F_n = m\bar{a}_n = \frac{W}{g}\frac{\bar{v}^2}{R}, \qquad F_3 = \frac{W}{g}\bar{a}_3.$$

If the path is a plane curve, then $F_3 = 0$.

189. Translation of a Rigid Body.   Let a body of mass m have a motion of translation only.   If a is the acceleration of the body the effective force F for the body is

$$F = ma$$

with line of action through the centroid of the body.

The equations of the motion can be obtained by applying the conditions of equilibrium to the system of forces consisting of the external forces and the reversed effective force.

190. Rotation of a Rigid Body about a Fixed Axis.   The moment M about a fixed axis of rotation L of the external forces acting on a rigid body is equal to the product of the angular acceleration α of the body and the moment of inertia I of the body about the axis of rotation:

$$M_L = I_L\alpha.$$

191. Equations of Plane Motion. Let the xy-plane be the plane of motion for a rigid body having plane motion. Let X and Y denote the sum of the components, parallel to the x- and y-axes, of the external forces acting on the body, and let M_g denote the sum of the moments of these forces about an axis through the centroid perpendicular to the plane of motion. The equations of this motion are

$$X = m\bar{a}_x = \frac{W}{g}\bar{a}_x, \qquad Y = m\bar{a}_y = \frac{W}{g}\bar{a}_y, \qquad M_g = I_g\alpha = \frac{W}{g}k_g^2\alpha,$$

where $\bar{a}_x$ and $\bar{a}_y$ are the components of the acceleration of the centroid parallel to the x- and y-axes.

192. Work Done in Rotation. Let a rigid body of mass m rotate about a fixed axis. If M is the moment about this axis of the external forces acting on the body, the work done by the forces in rotating the body from a position $\theta = \theta_0$ to a position $\theta = \theta_1$ is

$$\text{Work done in rotation} = \int_{\theta_0}^{\theta_1} M\, d\theta.$$

If I is the moment of inertia of the body about the axis of rotation, the kinetic energy of rotation of the body at the instant the angular velocity is ω is

$$\text{Kinetic energy of rotation} = \tfrac{1}{2}I\omega^2.$$

The work done in a rotation is equal to the increase in the kinetic energy of rotation:

$$\int_{\theta_0}^{\theta_1} M\, d\theta = \tfrac{1}{2}I(\omega_1^2 - \omega_0^2) = \frac{W}{2g}k^2(\omega_1^2 - \omega_0^2),$$

where $\omega = \omega_0$ at $\theta = \theta_0$ and $\omega = \omega_1$ at $\theta = \theta_1$.

193. Kinetic Energy of a Rigid Body Having Plane Motion. The kinetic energy (K. E.) of a rigid body of mass m having plane motion is equal to the kinetic energy due to the motion of its centroid, considered as a particle of mass m, plus the kinetic energy of rotation of the body about an axis through the centroid, perpendicular to the plane of motion:

$$\text{K.E.} = \tfrac{1}{2}m\bar{v}^2 + \tfrac{1}{2}I_g\omega^2 = \frac{W}{2g}\bar{v}^2 + \frac{W}{2g}k_g^2\omega^2.$$

194. Work-energy Relation in Plane Motion. The work done by the forces acting on the body during the plane motion is equal to the increase in the kinetic energy of the body during this motion. In the notation of the preceding articles

$$\int_{\bar{x}_0}^{\bar{x}_1} X\,dx + \int_{\bar{y}_0}^{\bar{y}_1} Y\,dy + \int_{\theta_0}^{\theta_1} M_g\,d\theta = \frac{W}{2g}(\bar{v}_1^2 - \bar{v}_0^2) + \frac{W}{2g}k_g^2(\omega_1^2 - \omega_0^2).$$

195. Impulse and Momentum in Plane Motion. Any system of external forces acting on a rigid body can be replaced by a force F acting through the centroid, and a couple M.

The linear impulse of F during the time t is $\int_0^t F\,dt$; the angular impulse of M during the time t is $\int_0^t M\,dt$.

The linear impulse of a force F during the time t has the same measure as the resulting increase in linear momentum:

$$\int_0^t F\,dt = m(\bar{v} - \bar{v}_0) = \frac{W}{g}(\bar{v} - \bar{v}_0).$$

The angular impulse has the same measure as the resulting increase in angular momentum:

$$\int_0^t M_g\,dt = I_g(\omega - \omega_0) = \frac{W}{g}k_g^2(\omega - \omega_0).$$

196. Power Transmitted by a Shaft. The work done per minute by a shaft subjected to a mean twisting moment of T pound-feet, and making N revolutions per minute, is $2\pi NT$ foot-pounds.

The horsepower (H.P.) transmitted by this shaft is

$$\text{H.P.} = \frac{2\pi NT}{33,000}.$$

From Art. 202,

$$T = \frac{f_sI_0}{R}.$$

197. Equations of Free Motion. Let a rigid body of mass m be acted on by a system of forces whose resultant (Art. 171) is a force (X, Y, Z) and a couple (M_x, M_y, M_z). Then

$$X = m\bar{a}_x = \frac{W}{g}\frac{d^2\bar{x}}{dt^2}, \qquad Y = m\bar{a}_y = \frac{W}{g}\frac{d^2\bar{y}}{dt^2}, \qquad Z = m\bar{a}_z = \frac{W}{g}\frac{d^2\bar{z}}{dt^2},$$

where $(\bar{x}, \bar{y}, \bar{z})$ are the coordinates of the centroid of the body at any time; and

$$M_x = \sum\left(z\frac{d^2y}{dt^2} - y\frac{d^2z}{dt^2}\right)\Delta m = \frac{d}{dt}\sum\left(z\frac{dy}{dt} - y\frac{dz}{dt}\right)\Delta m,$$

$$M_y = \sum\left(x\frac{d^2z}{dt^2} - z\frac{d^2x}{dt^2}\right)\Delta m = \frac{d}{dt}\sum\left(x\frac{dz}{dt} - z\frac{dx}{dt}\right)\Delta m,$$

$$M_z = \sum\left(y\frac{d^2x}{dt^2} - x\frac{d^2y}{dt^2}\right)\Delta m = \frac{d}{dt}\sum\left(y\frac{dx}{dt} - x\frac{dy}{dt}\right)\Delta m.$$

Since $X = m\bar{a}_x = \dfrac{d(m\bar{v}_x)}{dt}$, the time rate of change of the linear momentum of a rigid body in any direction is equal to the sum of the components of all the external forces in that direction.

From any of the last three equations, the time rate of change of angular momentum of a rigid body about any axis is equal to the sum of the moments of all the external forces about that axis.

If H_x, H_y, H_z are the components of the angular momentum and $\omega_x, \omega_y, \omega_z$ are the angular velocities of the body about the x-, y-, z-axes, respectively, then

$$H_x = \sum\left(z\frac{dy}{dt} - y\frac{dz}{dt}\right)\Delta m = I_x\omega_x - P_{xy}\omega_y - P_{xz}\omega_z,$$

$$H_y = \sum\left(x\frac{dz}{dt} - z\frac{dx}{dt}\right)\Delta m = -P_{xy}\omega_x + I_y\omega_y - P_{yz}\omega_z,$$

$$H_z = \sum\left(y\frac{dx}{dt} - x\frac{dy}{dt}\right)\Delta m = -P_{xz}\omega_x - P_{yz}\omega_y + I_z\omega_z.$$

If the rotation viewed by an observer on the axis looking in the negative direction of the axis is counterclockwise, the component angular velocity is positive.

SOME PROPERTIES OF PLANE AREAS

Any Area (A)

Moment of Inertia (I)

$$I = k^2 A$$
$$I_0 = I_x + I_y$$
$$I_{BB} = I_{AA} + Ad^2$$
$$I_{OC} = I_x \cos^2\theta + I_y \sin^2\theta - 2P_{xy}\sin\theta\cos\theta$$

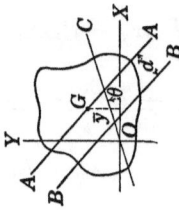

Squared Radius of Gyration (k²)

$$k_0^2 = k_x^2 + k_y^2$$
$$k_{BB}^2 = k_{AA}^2 + d^2$$
$$k_{OC}^2 = \frac{I_{OC}}{A}$$

Figure	Area (A) / Product of inertia (P)	Centroid (G)	Moment of inertia (I)	Squared radius of gyration (k²)
Square	$A = a^2$ $P_{xy} = \dfrac{a^4}{4}$	$\bar{x} = \dfrac{a}{2}$ $\bar{y} = \dfrac{a}{2}$	$I_{AA} = I_{BB} = \dfrac{a^4}{12}$ $I_x = I_y = \dfrac{a^4}{3}$ $I_0 = I_x + I_y$	$k_{AA}^2 = k_{BB}^2 = \dfrac{a^2}{12}$ $k_x^2 = k_y^2 = \dfrac{a^2}{3}$ $k_0^2 = \dfrac{2}{3}a^2$
Hollow square	$A = a^2 - b^2$ $P_{xy} = \dfrac{a^2(a^2 - b^2)}{4}$	$\bar{x} = \dfrac{a}{2}$ $\bar{y} = \dfrac{a}{2}$	$I_{AA} = I_{BB} = \dfrac{a^4 - b^4}{12}$ $I_x = I_y = I_{AA} + A\dfrac{a^2}{4}$ $I_0 = I_x + I_y$	$k_{AA}^2 = k_{BB}^2 = \dfrac{a^2 + b^2}{12}$ $k_x^2 = k_y^2 = \dfrac{4a^2 + b^2}{12}$ $k_0^2 = \dfrac{1}{6}(4a^2 + b^2)$

Rectangle	$A = bh$ $P_{xy} = \dfrac{b^2 h^2}{4}$	$\bar{x} = \dfrac{b}{2}$ $\bar{y} = \dfrac{h}{2}$	$I_{AA} = \dfrac{bh^3}{12}$ $I_{BB} = \dfrac{hb^3}{12}$ $I_x = \dfrac{bh^3}{3}$ $I_y = \dfrac{hb^3}{3}$ $I_0 = \dfrac{bh(b^2+h^2)}{3}$	$k^2_{AA} = \dfrac{h^2}{12}$ $k^2_{BB} = \dfrac{b^2}{12}$ $k^2_x = \dfrac{h^2}{3}$ $k^2_y = \dfrac{b^2}{3}$ $k^2_0 = \dfrac{b^2+h^2}{3}$
Hollow rectangle	$A = bh - rs$ $P_{xy} = \dfrac{bh(bh-rs)}{4}$	$\bar{x} = \dfrac{b}{2}$ $\bar{y} = \dfrac{h}{2}$	$I_{AA} = \dfrac{bh^3 - rs^3}{12}$ $I_{BB} = \dfrac{hb^3 - sr^3}{12}$ $I_x = I_{AA} + A\dfrac{h^2}{4}$ $I_y = I_{BB} + A\dfrac{b^2}{4}$	$k^2_{AA} = \dfrac{bh^3 - rs^3}{12(bh - rs)}$ $k^2_{BB} = \dfrac{hb^3 - sr^3}{12(bh - rs)}$ $k^2_x = k^2_{AA} + \dfrac{h^2}{4}$ $k^2_y = k^2_{BB} + \dfrac{b^2}{4}$

SOME PROPERTIES OF PLANE AREAS.—(Continued)

Figure	Area (A) Product of inertia (P)	Centroid (G)	Moment of inertia (I)	Squared radius of gyration (k^2)
Triangle	$A = \dfrac{bh}{2}$	$\bar{y} = \dfrac{h}{3}$	$I_{AA} = \dfrac{bh^3}{36}$ $I_{CC} = \dfrac{bh^3}{12}$ $I_{DD} = \dfrac{bh^3}{4}$	$k_{AA}^2 = \dfrac{h^2}{18}$ $k_{CC}^2 = \dfrac{h^2}{6}$ $k_{DD}^2 = \dfrac{h^2}{2}$
Trapezoid	$A = \dfrac{h(a+b)}{2}$	$\bar{y} = \dfrac{h}{3}\dfrac{2a+b}{a+b}$	$I_{AA} = \dfrac{h^3(a^2+4ab+b^2)}{36(a+b)}$ $I_{CC} = \dfrac{h^3(3a+b)}{12}$	$k_{AA}^2 = \dfrac{h^2(a^2+4ab+b^2)}{18(a+b)^2}$ $k_{CC}^2 = \dfrac{h^2(3a+b)}{6(a+b)}$
Circle	$A = \pi a^2$ $P_{xy} = \pi a^4$	$\bar{x} = a$ $\bar{y} = a$	$I_{AA} = \dfrac{\pi a^4}{4}$ $I_x = I_y = \dfrac{5\pi a^4}{4}$ $I_G = \dfrac{\pi a^4}{2}$	$k_{AA}^2 = \dfrac{a^2}{4}$ $k_x^2 = k_y^2 = \dfrac{5a^2}{4}$ $k_G^2 = \dfrac{a^2}{2}$

	A and P_{xy}	Centroid	I	k^2
Hollow circle 	$A = \pi(a^2 - b^2)$ $P_{xy} = \pi a^2(a^2 - b^2)$	$\bar{x} = a$ $\bar{y} = a$	$I_{AA} = \dfrac{\pi(a^4 - b^4)}{4}$ $I_x = I_y = I_{AA} + Aa^2$ $I_G = \dfrac{\pi(a^4 - b^4)}{2}$	$k_{AA}^2 = \dfrac{a^2 + b^2}{4}$ $k_x^2 = k_y^2 = k_{AA}^2 + a^2$ $k_G^2 = \dfrac{a^2 + b^2}{2}$
Semicircle 	$A = \dfrac{\pi a^2}{2}$ $P_{xy} = \dfrac{2a^4}{3}$	$\bar{x} = a$ $\bar{y} = \dfrac{4a}{3\pi}$	$I_{AA} = \dfrac{a^4(9\pi^2 - 64)}{72\pi}$ $= 0.1098a^4$ $I_x = \dfrac{\pi a^4}{8}$ $I_y = \dfrac{5\pi a^4}{8}$	$k_{AA}^2 = \dfrac{a^2(9\pi^2 - 64)}{36\pi^2}$ $= 0.0699a^2$ $k_x^2 = \dfrac{a^2}{4}$ $k_y^2 = \dfrac{5a^2}{4}$
Hollow semicircle 	$A = \dfrac{\pi}{2}(R^2 - r^2)$ $P_{xy} = 0$	$\bar{y} = \dfrac{4}{3\pi}\dfrac{R^3 - r^3}{R^2 - r^2}$ $\bar{x} = 0$	$I_x = I_y = \dfrac{\pi}{8}(R^4 - r^4)$ $I_{AA} = I_x - \bar{y}^2 A$ $I_0 = \dfrac{\pi}{4}(R^4 - r^4)$	$k_x^2 = k_y^2 = \dfrac{R^2 + r^2}{4}$ $k_{AA}^2 = k_x^2 - \bar{y}^2$ $k_0^2 = \dfrac{R^2 + r^2}{2}$

SOME PROPERTIES OF PLANE AREAS.—(Continued)

Figure	Area (A) Product of inertia (P)	Centroid (G)	Moment of inertia (I)	Squared radius of gyration (k^2)
Circular sector	$A = a^2\theta$ $P_{xy} = 0$	$\bar{x} = \dfrac{2a}{3}\dfrac{\sin\theta}{\theta}$ $\bar{y} = 0$	$I_x = \dfrac{a^4}{4}(\theta - \sin\theta\cos\theta)$ $I_y = \dfrac{a^4}{4}(\theta + \sin\theta\cos\theta)$ $I_0 = I_x + I_y = \dfrac{a^4\theta}{2}$	$k_x^2 = \dfrac{a^2}{4}\dfrac{\theta - \sin\theta\cos\theta}{\theta}$ $k_y^2 = \dfrac{a^2}{4}\dfrac{\theta + \sin\theta\cos\theta}{\theta}$ $k_0^2 = k_x^2 + k_y^2 = \dfrac{a^2}{2}$
Circular segment	$A = a^2(\theta - \sin\theta\cos\theta)$ $P_{xy} = 0$	$\bar{x} = \dfrac{2a}{3}\dfrac{\sin^3\theta}{\theta - \sin\theta\cos\theta}$ $\bar{y} = 0$	$I_x = \dfrac{Aa^2}{4}\times$ $\left(1 - \dfrac{2\sin^3\theta\cos\theta}{3(\theta - \sin\theta\cos\theta)}\right)$ $I_y = \dfrac{Aa^2}{4}\times$ $\left(1 + \dfrac{2\sin^3\theta\cos\theta}{\theta - \sin\theta\cos\theta}\right)$	$k_x^2 = \dfrac{a^2}{4}\times$ $\left(1 - \dfrac{2\sin^3\theta\cos\theta}{3(\theta - \sin\theta\cos\theta)}\right)$ $k_y^2 = \dfrac{a^2}{4}\times$ $\left(1 + \dfrac{2\sin^3\theta\cos\theta}{\theta - \sin\theta\cos\theta}\right)$

Ellipse	$A = \pi ab$ $P_{xy} = \pi a^2 b^2$	$\bar{x} = a$ $\bar{y} = b$	$I_{AA} = \dfrac{\pi ab^3}{4}$ $I_{BB} = \dfrac{\pi ba^3}{4}$ $I_G = \pi ab\dfrac{a^2 + b^2}{4}$	$k^2_{AA} = \dfrac{b^2}{4}$ $k^2_{BB} = \dfrac{a^2}{4}$ $k^2_G = \dfrac{a^2 + b^2}{4}$
Hollow ellipse	$A = \pi(ab - cd)$ $P_{xy} = \pi ab(ab - cd)$	$\bar{x} = a$ $\bar{y} = b$	$I_{AA} = \dfrac{\pi}{4}(ab^3 - cd^3)$ $I_{BB} = \dfrac{\pi}{4}(ba^3 - dc^3)$ $I_G = I_{AA} + I_{BB}$	$k^2_{AA} = \dfrac{ab^3 - cd^3}{4(ab - cd)}$ $k^2_{BB} = \dfrac{ba^3 - dc^3}{4(ab - cd)}$ $k^2_G = k^2_{AA} + k^2_{BB}$
Semiellipse	$A = \dfrac{\pi ab}{2}$ $P_{xy} = 0$	$\bar{x} = 0$ $\bar{y} = \dfrac{4b}{3\pi}$	$I_x = \dfrac{\pi ab^3}{8}$ $I_y = \dfrac{\pi ba^3}{8}$ $I_{AA} = I_x - \bar{y}^2 A$	$k^2_x = \dfrac{b^2}{4}$ $k^2_y = \dfrac{a^2}{4}$ $k^2_{AA} = \dfrac{b^2}{4} - \bar{y}^2$

SOME PROPERTIES OF PLANE AREAS.—*(Continued)*

Figure	Area (A) Product of inertia (P)	Centroid (G)	Moment of inertia (I)	Squared radius of gyration (k^2)
Hollow semiellipse	$A = \dfrac{\pi(ab - cd)}{2}$ $P_{xy} = 0$	$\bar{x} = 0$ $\bar{y} = \dfrac{4}{3\pi}\dfrac{ab^2 - cd^2}{ab - cd}$	$I_x = \dfrac{\pi}{8}(ab^3 - cd^3)$ $I_y = \dfrac{\pi}{8}(ba^3 - dc^3)$ $I_{AA} = I_x - A\bar{y}^2$	$k_x^2 = \dfrac{ab^3 - cd^3}{4(ab - cd)}$ $k_y^2 = \dfrac{ba^3 - dc^3}{4(ab - cd)}$ $k_{AA}^2 = k_x^2 - \bar{y}^2$
Parabolic segment	$A = \dfrac{4ab}{3}$ $P_{xy} = 0$	$\bar{x} = \dfrac{3}{5}a$ $\bar{y} = 0$	$I_x = \dfrac{4}{15}ab^3$ $I_y = \dfrac{4}{7}ba^3$ $I_{BB} = \dfrac{16}{175}ba^3$	$k_x^2 = \dfrac{b^2}{5}$ $k_y^2 = \dfrac{3a^2}{7}$ $k_{BB}^2 = \dfrac{12a^2}{175}$
Parabolic half segment	$A = \dfrac{2ab}{3}$ $P_{xy} = \dfrac{a^2b^2}{6}$	$\bar{x} = \dfrac{3}{5}a$ $\bar{y} = \dfrac{3}{8}b$	$I_x = \dfrac{2}{15}ab^3$ $I_y = \dfrac{2}{7}ba^3$	$k_x^2 = \dfrac{b^2}{5}$ $k_y^2 = \dfrac{3a^2}{7}$

TABLE OF SOME PROPERTIES OF BODIES
OF CONSTANT DENSITY

SOME PROPERTIES OF BODIES OF CONSTANT DENSITY

Any Solid of Mass m, density ρ — Moment of Inertia (I) — Squared Radius of Gyration (k^2)

$$I_{BB} = I_{AA} + md^2$$

$$I_{OC} = I_x \cos^2 \alpha + I_y \cos^2 \beta + I_z \cos^2 \gamma - 2P_{zy} \cos \alpha \cos \beta - 2P_{yz} \cos \beta \cos \gamma - 2P_{zz} \cos \alpha \cos \gamma$$

$$k^2_{BB} = k^2_{AA} + d^2$$

$$k^2_{OC} = \frac{I_{OC}}{m}$$

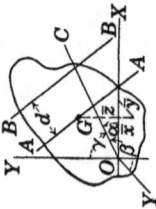

Solid	Mass (m) Product of inertia (P)	Centroid (G)	Moment of inertia (I)	Squared radius of gyration (k^2)
Straight rod	$m = a\rho$	$\bar{x} = \dfrac{a}{2}$	$I_{BB} = \dfrac{ma^2}{12}$ $\quad$ $I_{CC} = \dfrac{ma^2}{3}$	$k^2_{BB} = \dfrac{a^2}{12}$ $\quad$ $k^2_{CC} = \dfrac{a^2}{3}$
Broken rod	$m = 2a\rho$	$\bar{x} = \dfrac{a \sin \theta}{2}$ $\quad$ $\bar{y} = 0$	$I_x = \dfrac{ma^2 \cos^2 \theta}{3}$ $\quad$ $I_y = \dfrac{ma^2 \sin^2 \theta}{3}$ $\quad$ $I_0 = \dfrac{ma^2}{3}$ $\quad$ $I_G = ma^2 \left(\dfrac{1}{3} - \dfrac{\sin^2 \theta}{4} \right)$	$k^2_x = \dfrac{a^2 \cos^2 \theta}{3}$ $\quad$ $k^2_y = \dfrac{a^2 \sin^2 \theta}{3}$ $\quad$ $k^2_0 = \dfrac{a^2}{3}$ $\quad$ $k^2_G = a^2 \left(\dfrac{1}{3} - \dfrac{\sin^2 \theta}{4} \right)$

Circular rod		$m = 2a\theta\rho$	$\bar{x} = \dfrac{a \sin \theta}{\theta}$ $\bar{y} = 0$	$I_x = \dfrac{ma^2(\theta - \sin\theta\cos\theta)}{2\theta}$ $I_y = \dfrac{ma^2(\theta + \sin\theta\cos\theta)}{2\theta}$ $I_0 = ma^2$	$k_x^2 = \dfrac{a^2(\theta - \sin\theta\cos\theta)}{2\theta}$ $k_y^2 = \dfrac{a^2(\theta + \sin\theta\cos\theta)}{2\theta}$ $k_0^2 = a^2$
Cube		$m = a^3\rho$ $P_{xy} = P_{yz} = P_{zx} = \dfrac{ma^2}{4}$	$\bar{x} = \dfrac{a}{2}$ $\bar{y} = \dfrac{a}{2}$ $\bar{z} = \dfrac{a}{2}$	$I_{AA} = \dfrac{ma^2}{6}$ $I_z = \dfrac{2ma^2}{3}$ $I_{BB} = \dfrac{5ma^2}{12}$	$k_{AA}^2 = \dfrac{a^2}{6}$ $k_z^2 = \dfrac{2a^2}{3}$ $k_{BB}^2 = \dfrac{5a^2}{12}$
Hoop		$m = 2\pi a\rho$	$\bar{x} = a$ $\bar{y} = a$	$I_G = ma^2$ $I_{\text{element}} = 2ma^2$	$k_G^2 = a^2$ $k_{\text{element}}^2 = 2a^2$

SOME PROPERTIES OF BODIES OF CONSTANT DENSITY.—(Continued)

Solid	Mass (m) Product of inertia (P)	Centroid (G)	Moment of inertia (I)	Squared radius of gyration (k^2)
Rectangular prism	$m = abc\rho$ $$P_{xy} = \frac{mab}{4}$$ $$P_{yz} = \frac{mbc}{4}$$ $$P_{zx} = \frac{mca}{4}$$	$\bar{x} = \dfrac{a}{2}$ $\bar{y} = \dfrac{b}{2}$ $\bar{z} = \dfrac{c}{2}$	$I_{AA} = m\dfrac{a^2 + b^2}{12}$ $I_z = m\dfrac{a^2 + b^2}{3}$ $I_{BB} = m\dfrac{b^2 + 4c^2}{12}$	$k^2_{AA} = \dfrac{a^2 + b^2}{12}$ $k^2_z = \dfrac{a^2 + b^2}{3}$ $k^2_{BB} = \dfrac{b^2 + 4c^2}{12}$
Right circular cylinder	$m = \pi a^2 h\rho$ $P_{xy} = 0$ $P_{yz} = 0$ $P_{zx} = 0$	$\bar{x} = 0$ $\bar{y} = 0$ $\bar{z} = \dfrac{h}{2}$	$I_x = I_y$ $= m\dfrac{3a^2 + 4h^2}{12}$ $I_z = \dfrac{ma^2}{2}$ $I_{BB} = m\dfrac{3a^2 + h^2}{12}$	$k^2_x = k^2_y$ $= \dfrac{3a^2 + 4h^2}{12}$ $k^2_z = \dfrac{a^2}{2}$ $k^2_{BB} = \dfrac{3a^2 + h^2}{12}$

Hollow right circular cylinder 	$m = \pi h(R^2 - r^2)\rho$ $P_{xy} = 0$ $P_{yz} = 0$ $P_{zz} = 0$	$\bar{x} = 0$ $\bar{y} = 0$ $\bar{z} = \dfrac{h}{2}$	$I_x = I_y$ $\quad = m\left(\dfrac{R^2 + r^2}{4} + \dfrac{h^2}{3}\right)$ $I_z = m\dfrac{R^2 + r^2}{2}$ $I_{BB} = m\left(\dfrac{R^2 + r^2}{4} + \dfrac{h^2}{12}\right)$ $I_{\text{element}} = m\dfrac{3R^2 + r^2}{2}$	$k_x^2 = k_y^2$ $\quad = \dfrac{R^2 + r^2}{4} + \dfrac{h^2}{3}$ $k_z^2 = \dfrac{R^2 + r^2}{2}$ $k_{BB}^2 = \dfrac{R^2 + r^2}{4} + \dfrac{h^2}{12}$ $k_{\text{element}}^2 = \dfrac{3R^2 + r^2}{2}$
Cylindrical shell 	$m = 2\pi r h\rho$ $P_{xy} = 0$ $P_{yz} = 0$ $P_{zx} = 0$	$\bar{x} = 0$ $\bar{y} = 0$ $\bar{z} = \dfrac{h}{2}$	$I_x = I_y = m\left(\dfrac{r^2}{2} + \dfrac{h^2}{3}\right)$ $I_z = mr^2$ $I_{BB} = m\left(\dfrac{r^2}{2} + \dfrac{h^2}{12}\right)$ $I_{\text{element}} = 2mr^2$	$k_x^2 = k_y^2 = \dfrac{r^2}{2} + \dfrac{h^2}{3}$ $k_z^2 = r^2$ $k_{BB}^2 = \dfrac{r^2}{2} + \dfrac{h^2}{12}$ $k_{\text{element}}^2 = 2r^2$

SOME PROPERTIES OF BODIES OF CONSTANT DENSITY.—(Continued)

Solid	Mass (m) Product of inertia (P)	Centroid (G)	Moment of inertia (I)	Squared radius of gyration (k^2)
Right rectangular pyramid	$m = \dfrac{abh\rho}{3}$ $P_{xy} = 0$ $P_{yz} = 0$ $P_{zx} = 0$	$\bar{x} = 0$ $\bar{y} = 0$ $\bar{z} = \dfrac{h}{4}$	$I_x = m\left(\dfrac{b^2}{20} + \dfrac{h^2}{10}\right)$ $I_y = m\left(\dfrac{a^2}{20} + \dfrac{h^2}{10}\right)$ $I_z = m\left(\dfrac{a^2+b^2}{20}\right)$ $I_{BB} = m\left(\dfrac{b^2}{20} + \dfrac{3h^2}{80}\right)$	$k_x^2 = \dfrac{b^2 + 2h^2}{20}$ $k_y^2 = \dfrac{a^2 + 2h^2}{20}$ $k_z^2 = \dfrac{a^2 + b^2}{20}$ $k_{BB}^2 = \dfrac{4b^2 + 3h^2}{80}$
Right circular cone	$m = \dfrac{\pi r^2 h\rho}{3}$ $P_{xy} = 0$ $P_{yz} = 0$ $P_{zz} = 0$	$\bar{x} = 0$ $\bar{y} = 0$ $\bar{z} = \dfrac{h}{4}$	$I_z = I_y = m\left(\dfrac{3r^2}{20} + \dfrac{h^2}{10}\right)$ $I_z = \dfrac{3mr^2}{10}$ $I_{BB} = m\left(\dfrac{3r^2}{20} + \dfrac{3h^2}{80}\right)$ $I_{CC} = m\left(\dfrac{3r^2}{20} + \dfrac{3h^2}{5}\right)$	$k_z^2 = k_y^2 = \dfrac{3r^2 + 2h^2}{20}$ $k_z^2 = \dfrac{3r^2}{10}$ $k_{BB}^2 = \dfrac{12r^2 + 3h^2}{80}$ $k_{CC}^2 = \dfrac{3r^2 + 12h^2}{20}$

Body	Mass and products of inertia	Centroid	Moments of inertia	Radii of gyration
Frustum of cone 	$m = \dfrac{\pi h \rho}{3}(R^2 + Rr + r^2)$ $P_{xy} = 0$ $P_{yz} = 0$ $P_{zx} = 0$	$\bar{x} = 0$ $\bar{y} = 0$ $\bar{z} = \dfrac{h(R^2 + 2Rr + 3r^2)}{4(R^2 + Rr + r^2)}$	$I_z = \dfrac{3m}{10}\dfrac{R^5 - r^5}{R^3 - r^3}$	$k_z^2 = \dfrac{3}{10}\dfrac{R^5 - r^5}{R^3 - r^3}$
Sphere 	$m = \dfrac{4\pi r^3 \rho}{3}$ $P_{xy} = 0$ $P_{yz} = 0$ $P_{zx} = 0$	$\bar{x} = 0$ $\bar{y} = 0$ $\bar{z} = 0$	$I_x = I_y = I_z = \dfrac{2mr^2}{5}$	$k_x^2 = k_y^2 = k_z^2 = \dfrac{2r^2}{5}$
Hollow sphere 	$m = \dfrac{4\pi}{3}(R^3 - r^3)\rho$ $P_{xy} = 0$ $P_{yz} = 0$ $P_{zx} = 0$	$\bar{x} = 0$ $\bar{y} = 0$ $\bar{z} = 0$	$I_x = I_y = I_z =$ $\dfrac{2m}{5}\dfrac{R^5 - r^5}{R^3 - r^3}$	$k_x^2 = k_y^2 = k_z^2 =$ $\dfrac{2}{5}\dfrac{R^5 - r^5}{R^3 - r^3}$

SOME PROPERTIES OF BODIES OF CONSTANT DENSITY.—(*Continued*)

Solid	Mass (m) Product of inertia (P)	Centroid (G)	Moment of inertia (I)	Squared radius of gyration (k^2)
Spherical shell	$m = 4\pi r^2 \rho$ $P_{xy} = 0$ $P_{yz} = 0$ $P_{zx} = 0$	$\bar{x} = 0$ $\bar{y} = 0$ $\bar{z} = 0$	$I_x = I_y = I_z = \dfrac{2mr^2}{3}$	$k_x^2 = k_y^2 = k_z^2 = \dfrac{2r^2}{3}$
Hemisphere	$m = \dfrac{2\pi r^3 \rho}{3}$ $P_{xy} = 0$ $P_{yz} = 0$ $P_{zx} = 0$	$\bar{x} = 0$ $\bar{y} = 0$ $\bar{z} = \dfrac{3r}{8}$	$I_x = I_y = I_z = \dfrac{2mr^2}{5}$	$k_x^2 = k_y^2 = k_z^2 = \dfrac{2r^2}{5}$

Body	Figure	Mass & Products	Centroid	Moments of Inertia	Radii of Gyration
Ellipsoid		$m = \pi abc\rho$ $P_{xy} = 0$ $P_{yz} = 0$ $P_{zx} = 0$	$\bar{x} = 0$ $\bar{y} = 0$ $\bar{z} = 0$	$I_x = m\dfrac{b^2 + c^2}{5}$ $I_y = m\dfrac{a^2 + c^2}{5}$ $I_z = m\dfrac{a^2 + b^2}{5}$	$k_x^2 = \dfrac{b^2 + c^2}{5}$ $k_y^2 = \dfrac{a^2 + c^2}{5}$ $k_z^2 = \dfrac{a^2 + b^2}{5}$
Paraboloid of revolution		$m = \dfrac{\pi r^2 h \rho}{2}$ $P_{xy} = 0$ $P_{yz} = 0$ $P_{zz} = 0$	$\bar{x} = \dfrac{2h}{3}$ $\bar{y} = 0$ $\bar{z} = 0$	$I_x = \dfrac{mr^2}{3}$ $I_y = I_z = m\left(\dfrac{r^2}{6} + \dfrac{h^2}{2}\right)$ $I_{BB} = m\left(\dfrac{r^2}{6} + \dfrac{h^2}{18}\right)$	$k_x^2 = \dfrac{r^2}{3}$ $k_y^2 = k_z^2 = \dfrac{r^2 + 3h^2}{6}$ $k_{BB}^2 = \dfrac{3r^2 + h^2}{18}$
Elliptic paraboloid		$m = \dfrac{\pi abc\rho}{2}$ $P_{xy} = 0$ $P_{yz} = 0$ $P_{zx} = 0$	$\bar{x} = \dfrac{2a}{3}$ $\bar{y} = 0$ $\bar{z} = 0$	$I_x = m\left(\dfrac{b^2 + c^2}{6}\right)$ $I_z = m\left(\dfrac{b^2}{6} + \dfrac{a^2}{2}\right)$ $I_{BB} = m\left(\dfrac{b^2}{6} + \dfrac{a^2}{18}\right)$	$k_x^2 = \dfrac{b^2 + c^2}{6}$ $k_z^2 = \dfrac{b^2 + 3a^2}{6}$ $k_{BB}^2 = \dfrac{3b^2 + a^2}{18}$

PROPERTIES OF MATERIALS

198. Stress. The intensity of stress f produced by a force F uniformly distributed over a plane section of area A is

$$f = \frac{F}{A}.$$

Let A be the area of a right section of a bar (Fig. 134) which is subjected to an axial force F.

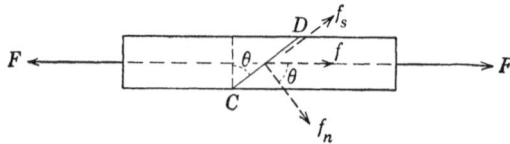

Fig. 134.

On an oblique section CD making an angle θ with a right section A the intensity of stress f in the direction of F is

$$f = \frac{F}{A} \cos \theta;$$

the intensity of shearing stress f_s along CD is

$$f_s = \frac{F}{A} \cos \theta \sin \theta = \frac{F}{2A} \sin 2\theta;$$

the intensity of stress f_n normal to CD is

$$f_n = \frac{F}{A} \cos^2 \theta.$$

199. Strain. The strain e produced in a body of length L by a longitudinal force which causes a change in length of ΔL is

$$e = \frac{\Delta L}{L}.$$

200. Modulus of Elasticity. Within the elastic limit, the ratio of the stress f to the strain e is a constant E:

$$\frac{f}{e} = E.$$

This is known as Hooke's law. The constant E is called the modulus of elasticity.

201. Some Physical Constants.

Material	Weight, pounds per cubic foot	Ultimate strength, pounds per square inch			Elastic limit, pounds per square inch	Mod. elasticity, pounds per square inch
		Tension	Compression	Shear		
Structural steel........	490	60,000	60,000	50,000	30,000	30×10^6
Cast iron..............	450	30,000	90,000	20,000	6,000	15×10^6
Wrought iron..........	480	50,000	40,000	40,000	25,000	28×10^6
Yellow pine (dry)......	40	9,000	7,000	1,500	3,000	15×10^5
Oak (dry).............	48	10,000	6,000	4,000	3,000	15×10^5

202. Torsion. A circular shaft of outer radius R is twisted by a couple of magnitude T so that any straight line parallel to the axis of the shaft is deformed into a curve meeting the elements of the cylindrical surface at a constant angle φ. The radius of a right section distant l from a fixed section is turned through an angle θ. Denote by f_s the greatest intensity of shearing stress set up, by I_0 the polar moment of inertia of a right section about its center and by G a constant, the modulus of rigidity for the material of the shaft. Then

$$\frac{T}{I_0} = \frac{f_s}{R} = \frac{G\theta}{l} = \frac{G\varphi}{R}.$$

The shearing strain is φ.

BEAMS

203. Definitions. A beam is a horizontal bar in equilibrium under the action of a system of couples and vertical forces (loads and supporting forces) all acting in the same plane, called the *plane of bending*. When the beam is bent, the concave side is in compression, the convex side in tension. The plane of zero stress in tension and compression is the *neutral plane*. The intersection of the neutral plane with the plane of bending is the *elastic curve* of the beam. Any plane section of the beam perpendicular to the elastic curve cuts the neutral plane in a horizontal line called the *neutral axis of the section*.

204. Shearing Force. The shearing force S at any section of a beam is the sum of the external forces acting on the beam on

one side of the section.   The shearing force at a section is positive
when the external forces tend to move upward the part of
the beam to the left of the section and tend to move downward
the part to the right of the section.

The shearing force S_x at a distance x from the left end of the
beam is

$$S_x = \Sigma R_i - \Sigma W_i, \quad \text{or} \quad S_x = \Sigma W_j - \Sigma R_j,$$

where R_i and W_i are the supporting forces and loads, respectively,
to the left of the section, and R_j and W_j are the supporting forces
and loads to the right of the section.

205. Bending Moment.   The bending moment M at any
section of a beam is the sum of the moments of the external
couples and forces acting on the beam on one side of the section,
about a horizontal line in the section.   A moment that tends to
cause compression in the upper fibers of the beam is positive.

If the beam carries loads W_i acting at $x = x_i$ from the left end,
and there are reactions R_i at $x = a_i$ from the left end, the bending
moment (M_x) at a section x from the left end is

$$M_x = M_0 + \Sigma R_i(x - a_i) - \Sigma W_i(x - x_i),$$

where M_0 is the bending moment at $x = 0$ and the summations
are for loads and reactions to the left of the section.

Relations between shearing force and bending moment are

$$\frac{dM_x}{dx} = S_x, \quad M_b - M_a = \int_a^b S_x \, dx.$$

To the left of the section

$$M_x = M_0 + \text{(area above } x\text{-axis)} - \text{(area below } x\text{-axis)}.$$

To the right of the section

$$M_x = M_l + \text{(area below } x\text{-axis)} - \text{(area above } x\text{-axis)}.$$

206. Bending.   The equations of simple bending are

$$\frac{f}{y} = \frac{M}{I} = \frac{E}{R},$$

where f is the intensity of stress (tensile or compressive) at a
distance y from the neutral axis of the section;
M is the bending moment at the section;

I is the moment of inertia of the area of the section about its neutral axis;

E is the modulus of elasticity for the material of the beam;

and R is the radius of curvature of the elastic curve at the point where it intersects the section.

The neutral axis is the horizontal line in the section, passing through its centroid.

207. Section Modulus. Let y_t denote the distance from the neutral axis to the fibers in greatest tensile stress and y_c the corresponding distance for compressive stress. Then I/y_t is the section modulus for tension, I/y_c is the section modulus for compression. If the beam is symmetrical about its neutral plane, the section modulus for the beam is I/c, where c is the half-depth of the beam.

208. Equation of the Elastic Curve. The differential equation of the elastic curve of a beam is

$$EI\frac{d^2y}{dx^2} = M.$$

In this equation E is the modulus of elasticity, I the moment of inertia of the cross section of the beam about its neutral axis, and M the bending moment at any section due to the given loads.

The slope of the elastic curve is $\dfrac{dy}{dx} = \displaystyle\int \frac{M}{EI}dx + C_1$; the equation of the elastic curve is $y = \displaystyle\int\left[\int \frac{M}{EI}dx\right]dx + C_1x + C_2,$ where C_1 and C_2 are to be determined from known relations among x, $\dfrac{dy}{dx}$, and y.

For coordinate axes used in the tables that follow, the *deflection* of the loaded beam at $x = a$ is the negative of y when $x = a$.

The magnitude of the deflection at any point may also be found from the formula

$$d = \int_0^l \frac{Mm}{EI}dx,$$

where m is the bending moment at any section due to a unit load at the point for which the deflection is to be computed, and l is the total span of the beam. Note that since m has different forms on opposite sides of the unit load, and M may have different forms in different portions of the beam, d is the sum of two or more integrals.

209. Theorems of Three Moments. If a continuous beam rests on three or more supports on the same level, and carries a uniformly distributed load over each span, then

$$M_1 l_1 + 2M_2(l_1 + l_2) + M_3 l_2 = -\tfrac{1}{4}(w_1 l_1^3 + w_2 l_2^3),$$

where M_1, M_2, M_3 are the bending moments over any three consecutive supports, l_1 and l_2 are the spans and w_1 and w_2, respectively, are the loads per unit length on these spans.

If, instead of a distributed load, the beam carries a concentrated load w_1 in the first of the two spans at the distance $k_1 l_1$ from the first of the three supports, and a concentrated load w_2 in the second span at the distance $k_2 l_2$ from the second of the three supports, then

$$M_1 l_1 + 2M_2(l_1 + l_2) + M_3 l_2 = -w_1 l_1^2(k_1 - k_1^3)$$
$$- w_2 l_2^2(2k_2 - 3k_2^2 + k_2^3).$$

BEAMS OF UNIFORM CROSS SECTION, LOADED TRANSVERSELY

Beam, shearing-force and bending-moment curves	Supporting forces (P and Q)	Shearing force (S) Bending moment (M)	Maximum deflection (d)
	$P = \dfrac{W}{2}$ $Q = \dfrac{W}{2}$	$S_x = \dfrac{W}{2}, x < \dfrac{l}{2}$ $M_x = \dfrac{Wx}{2}, x \lessgtr \dfrac{l}{2}$ $M_{\frac{l}{2}} = \dfrac{Wl}{4}$	$d_{\frac{l}{2}} = \dfrac{Wl^3}{48EI}$
	$P = W$	$S_x = W$ $M_x = Wx - Wl$ $M_0 = -Wl$	$d_l = \dfrac{Wl^3}{3EI}$
	$P = \dfrac{Wb}{l}$ $Q = \dfrac{Wa}{l}$	$S_x = \dfrac{Wb}{l}, x < a$ $M_x = \dfrac{Wb}{l}x, 0 \lessgtr x \lessgtr a$ $M_x = \dfrac{Wa}{l}(l - x),$ $\qquad a \lessgtr x \lessgtr l$ $M_a = \dfrac{Wab}{l}$	$d =$ $\dfrac{Wb}{3EIl}\left[\dfrac{a(l+b)}{3}\right]^{3/2}$ at $x = \sqrt{\dfrac{a(l+b)}{3}}$

BEAMS OF UNIFORM CROSS SECTION, LOADED TRANSVERSELY.—(*Continued*)

Beam, shearing-force and bending-moment curves	Sup-porting forces (P and Q)	Shearing force (S) Bending moment (M)	Maximum deflection (d)
	$P = \frac{11}{16}W$ $Q = \frac{5}{16}W$	$S_x = \frac{11W}{16},\ x < \frac{l}{2}$ $M_x = \frac{W}{16}(-3l + 11x),$ $0 \lessgtr x \lessgtr \frac{l}{2}$ $M_x = \frac{5W}{16}(l - x),$ $\frac{l}{2} \lessgtr x \lessgtr l$ $M_0 = -\frac{3Wl}{16}$	$d = \frac{Wl^3}{48\sqrt{5}EI}$ at $x = \left(1 - \frac{1}{\sqrt{5}}\right)l$ $= 0.5528l$
	$P = \frac{W}{2}$ $Q = \frac{W}{2}$	$S_x = \frac{W}{2},\ x < \frac{l}{2}$ $M_x = \frac{Wx}{2} - \frac{Wl}{8},$ $0 \lessgtr x \lessgtr \frac{l}{2}$ $M_x = -\frac{Wx}{2} + \frac{3Wl}{8},$ $\frac{l}{2} \lessgtr x \lessgtr l$ $M_0 = -\frac{Wl}{8}$ $M_{\frac{l}{2}} = \frac{Wl}{8}$	$d_{\frac{l}{2}} = \frac{Wl^3}{192EI}$
	$P = W$ $Q = W$	$S_x = W,\ x < a$ $M_x = Wx,\ 0 \lessgtr x \lessgtr a$ $M_x = Wa,$ $a \lessgtr x \lessgtr l - a$ $M_x = W(l - x),$ $l - a \lessgtr x \lessgtr l$	$d_{\frac{l}{2}} =$ $\frac{Wa(3l^2 - 4a^2)}{24EI}$

BEAMS OF UNIFORM CROSS SECTION, LOADED TRANSVERSELY.—*(Continued)*

Beam, shearing-force and bending-moment curves	Supporting forces (P and Q)	Shearing force (S) Bending moment (M)	Maximum deflection (d)
	$P = W$ $Q = W$	$S_x = -W,$ $\qquad -c \gtreqless x < 0$ $S_x = 0,$ $\qquad 0 < x < (l - 2c)$ $M_x = -W(c + x),$ $\qquad -c \gtreqless x \gtreqless 0$ $M_x = -Wc,$ $\qquad 0 \gtreqless x \gtreqless l - 2c$ $M_x = W(x - l + c),$ $\qquad l - 2c \gtreqless x \gtreqless l - c$ $M_{\max.} = Wc$	$d_{\text{end}} =$ $\dfrac{Wc^2(3l - 4c)}{6EI}$ $d_{\text{middle}} =$ $\dfrac{Wc(l - 2c)^2}{8EI}$
	$P = \dfrac{wl}{2}$ $Q = \dfrac{wl}{2}$	$S_x = \dfrac{wl}{2} - wx$ $M_x = \dfrac{wx}{2}(l - x)$ $M_{\frac{l}{2}} = \dfrac{wl^2}{8} = \dfrac{Wl}{8}$	$d_{\frac{l}{2}} = \dfrac{5wl^4}{384EI}$ $\qquad = \dfrac{5Wl^3}{384EI},$ $W = wl$
	$P = wl$	$S_x = w(l - x)$ $M_x = -\dfrac{w}{2}(l - x)^2$ $M_0 = -\dfrac{wl^2}{2} = -\dfrac{Wl}{2}$	$d_l = \dfrac{wl^4}{8EI}$ $\qquad = \dfrac{Wl^3}{8EI},$ $W = wl$

BEAMS OF UNIFORM CROSS SECTION, LOADED TRANSVERSELY.—*(Continued)*

Beam, shearing: force and bending-moment curves	Supporting forces (P and Q)	Shearing forces (S) Bending moment (M)	Maximum deflection (d)
	$P = \frac{5}{8}wl$ $Q = \frac{3}{8}wl$	$S_x = \frac{5}{8}wl - wx$ $M_x = -\frac{w}{8}(4x^2 - 5lx + l^2)$ $M_0 = -\frac{wl^2}{8}$ $M_{\frac{5}{8}l} = \frac{9wl^2}{128}$	$d = \frac{0.00541wl^4}{EI}$ $= \frac{wl^4}{185EI}$ $= \frac{Wl^3}{185EI}$ at $x = 0.5785l$
	$P = \frac{wl}{2}$ $Q = \frac{wl}{2}$	$S_x = \frac{w}{2}(l - 2x)$ $M_x = -\frac{w}{12}(6x^2 - 6lx + l^2)$ $M_{\frac{l}{2}} = \frac{wl^2}{24}$ $M_0 = -\frac{wl^2}{12}$	$d_{\frac{l}{2}} = \frac{wl^4}{384EI}$ $= \frac{Wl^3}{384EI},$ $W = wl$
	$P = \frac{W}{3}$ $Q = \frac{2W}{3}$	$S_x = \frac{W}{3l^2}(l^2 - 3x^2)$ $M_x = \frac{Wx}{3l^2}(l^2 - x^2)$ $M_{\frac{l}{\sqrt{3}}} = \frac{2Wl}{9\sqrt{3}}$ $= 0.128Wl$	$d = \frac{0.01304Wl^3}{EI}$ at $x = 0.519l$

INDEX

263